The Mapping of Ohio

Thomas H. Smith

The Mapping of Ohio

The delineation of the State of Ohio
through the use of manuscript maps, printed maps,
and plats, sketches and plans from original
map makers with a narrative which describes each
map from contemporary sources.

Kent State University Press

ISBN: 0-87338-054-1
Library of Congress Catalog Card Number: 75-99081

Manufactured in the United States of America
Designed by Harold Stevens

Library of Congress Cataloging in Publication Data

Smith, Thomas H 1936—
The mapping of Ohio.

Includes bibliography and index.
1. Ohio—Maps. 2. Ohio—Historical geography—Maps.
3. Cartography—Ohio. I. Title.
G1395.S61976 526'.09771 75-99081
ISBN 0-87338-054-1 (cloth)

TO JO AND BRAD

Table of Contents

Introduction

The idea of putting together a collection of maps of Ohio came to me while I was trying to fulfill a classroom need. After teaching Ohio history for several years, I became aware that many students lacked a fundamental knowledge of the state's topography. One could easily describe, with all the flair and drama of an artful storyteller, the hazards and perils faced on the Ohio frontier by such adventurers as Christopher Gist, Samuel Brady, and Simon Kenton; the problems and frustrations felt by those who planned and built the Ohio-Erie Canal; the flow of immigration into Ohio; and the relationship between the economic development of Ohio and that of the nation as a whole. But if one could not locate the Forks of the Ohio, Pickaway Plains, Lower Sandusky, the Newark Summit or the Hanging Rock Iron District, the material under discussion would lack perspective. The easiest way to make up for this shortcoming would be to illustrate for the students the location of these historical sites, and it seemed to me that a collection of historic maps of the state would best serve the purpose.

Once the maps, sketches and plans, along with descriptive materials, were assembled for this collection, it soon became evident that the project had carried beyond its original intent. As I analyzed each map or sketch, I realized that the history of a particular area was revealed by its cartography. Set in lines, angles and curves was a two-dimensional story that graphically explained the extension of exploration, the density of Indian populations, the conquest of a territory, the growth of political subdivisions, the spread of urbanization, and the growth of internal developments.

Historic maps are part of the nation's cultural heritage. Often, they were made by well-known cartographers or draftsmen who studied their subject either first hand or through the records of eye witnesses. Their work, an act of expression, becomes a form of art: the maps, in many instances accompanied by a descriptive narrative, travelog or report, are part of the country's literature. Certainly, Charlevoix's history of North America, Mores's geography, and John Melish's journal of his travel throughout the United States cannot be excluded from the realm of literature. These works were printed by famous publishing houses, including Shepard Kollock of Elizabethtown, and H. S. Tanner, H. C. Carey, and Thomas, Cowperthwait and Company of Philadelphia, which were all part of the development of the cultural heritage of the United States.

Collecting historic Ohio cartography was a fascinating and exciting research experience. It did present numerous problems, however, and material was not always abundant. Unlike other fields of historical research, the collection of historical cartography was not encouraged (with a few exceptions, of course) until about twenty-five years ago. Before that time, few private, public, or university libraries—or archives for that matter—were much concerned with maps; they were either acquired as part of larger manuscript collections, or salvaged and isolated from their original sources. To the librarian, maps were burdensome: they demanded special storage facilities, and there was no way to catalog them properly because there were no adequate bibliographies, finding aids or publishers' lists.[1]

The Map Division of the Library of Congress took the lead in encouraging the systematic collection of maps by libraries across the country. Just prior to World War II, Clara Egli LeGear, curator for the Map Division of the Library of Congress, tried to stimulate interest in map collecting and suggested to professional librarians the method of cataloging employed by the Library of Congress.[2] Articles began to appear in *Library Journal,* explaining professional techniques for map acquisition, storage and use.[3] In 1954 the Special Libraries Association conducted a survey of map collections in North America and reported that there were 527 private and public map depositories in the United States and Canada.[4] Walter W. Ristow, who currently is Chief of the Map Division of the Library of Congress, concluded that the data from this survey were "reassuring evidence that cartographic publications are now accepted as legitimate and useful reference aids."[5] Despite these advances, however, as late as 1965 a survey conducted by the Map Division of the Library of Congress to identify the numerous county atlases in each state revealed that many state historical libraries had lagged behind in their map and atlas procurement and that many did not possess in their holdings an atlas for each of their own counties.[6]

Historic cartography of Ohio is not difficult to locate. The libraries of the state's historical societies have map collections, though none of the state's university libraries can boast of any sizeable historic map collections.

To be sure, most of the universities possess large cartographic collections located either in their libraries or geography departments, but their holdings are usually limited to modern or twentieth-century technical cartography (geological, hydrographical, aeronautical, or topographical) and are not addressed to the collection of maps of historical value.

The significant historical map collections in Ohio having statewide interest are located at the Western Reserve Historical Society in Cleveland and at the Ohio Historical Society in Columbus. Thanks to the efforts of Charles C. Baldwin, secretary and later president of the Western Reserve Historical Society during the latter part of the nineteenth century, who collected manuscript and printed maps over many years,[7] the Western Reserve Historical Society possesses the largest map collection in the state. Although concerned with the cartography of Cleveland and the Western Reserve, this collection is rich in material that has statewide as well as national interest. The Ohio Historical Society has an excellent collection that is well cataloged, available, and useable. Its maps are primarily limited to Ohio, but the variety of its holdings makes it the best collection in the state. Both of these historical societies have substantial county atlas holdings, although the Ohio Historical Society does not possess a complete collection of the state's eighty-eight political subdivisions. Other historical society libraries in the state have small map collections, which usually contain a limited number of items of statewide interest, while their major emphasis is limited to the cartography of local interest.

Outside the state, the largest collection of general interest to the historical cartography of Ohio is, of course, found in the Map Division of the Library of Congress,[8] which has innumerable manuscript and printed maps and atlases. More specialized collections are in the Map Division of the National Archives, which possesses maps that deal with government land surveys, federal military activity and the diagrams and plans for internal improvements.[9] A number of Ohio maps are in the large collection at the William L. Clements Library at the University of Michigan in Ann Arbor.[10] Of course, historical Ohio maps are not limited only to these collections outside the state, but the collections found in these libraries are of such quantity and variety that sufficient examples are available to offer the researcher a satisfying representation of Ohio's historic cartography.

In this study, it was determined that seven broad categories should be used: early maps of North America; maps or diagrams of the state's elaborate earthworks; plans, sketches and diagrams of forts and military engagements; works produced by the surveyors in Ohio; maps of the state; urban maps; and plans and maps that depicted the state's internal improvements.

Specific criteria were adopted as guidelines to determine a map's usefulness to this study. First, the map itself had to illustrate a wide accumulation of knowledge. Second, it had to include a cartographic norm or establish a base that was accepted and copied by other cartographers. Third, it had to be unique and warrant special consideration apart from the patterns of the cartography considered as normal for a particular category or time period. Fourth, it had to be the first of its kind drawn or one that represented a particular style of cartography. Fifth, it had to be part of a series of historic cartography which illustrated a specific pattern of growth or development within the state. Each map did not necessarily have to satisfy all the criteria, but had to meet at least one.

Hundreds of maps were reviewed, and perhaps some that were eliminated could have been used. The selection process, however, rested solely with the author, who believes that this collection represents the story of the mapping of Ohio. There is, of course, always the possibility that an as yet undiscovered map might turn up that is better than the one I have used, but this is an unavoidable risk for anyone compiling a selection such as this.

Once a map was selected for this study, certain questions about it became critical. For the manuscript maps and the early maps of North America, it was necessary to determine the purpose of their construction, the sources used by the author in their construction, and the historical circumstances of the cartographer himself. These maps were analyzed, and the sources used by the cartographer were consulted to describe each map. In some cases, the manuscript maps were isolated from material that could have identified them, and the search for particular information about them was difficult. Most of the printed maps were easier to identify and describe, but some of them, too, had been separated from their original sources and were hard to trace. The surveyed plans of the earthworks and mounds in Ohio were selected because they represent an important aspect of the state's history that is many times forgotten, and they illustrate prehistoric sites which, in the majority of instances, are no longer extant. Also, the collection of the sketches and plans of forts and military posts in Ohio help to shed light on a topic that has fired the imagination of generations of Ohioans.

Putting this collection of maps together required the assistance of many people, and I owe a great deal to their helpfulness and patience. Foremost, I want to thank my good friend, Mr. Conrad Weitzel, Research Librarian at the Ohio Historical Library, for without his assistance this project would have been much more difficult. Further, I want to acknowledge the help of Dr. Raymond Baby, Chief Archaeologist of the Ohio Historical Society; Dr. Richard Knopf, Professor of History, Ashtabula Campus, Kent State University; Mrs. Lee Majors, former Librarian at the Cincinnati Historical Library; Mr. Patrick J. Mullin of the Dawes Memorial Library, Marietta College; Mrs. Mary A. Bradley, Librarian, New Harmony Workingman's Institute Library, New Harmony, Indiana; Mrs. Virginia Hawley, General Reference Librarian, Western Reserve Historical Library; Mr. S. M. "Ben" Heer, Deputy Supervisor of School and Ministerial Lands, Record Center, Land Office, State Auditor's Office; Mr. Thomas G. DeClaire, Mr. Andy Modelski and Mr. Frank Tharin of the Map Division of the Library of Congress; Mr. A. P. Muntz, Chief, Cartographic Branch, National Archives; Mr. Elmer O. Parker, Assistant Director of the Old Military Records Division, National Archives; and Mr. A. W. Murdoch, Director, Archives Branch, Department of Public Records and Archives, Ontario. Lastly, I want to acknowledge the assistance of my wife, Joanne, who read the manuscript and offered helpful suggestions.

1. Walter W. Ristow, "Maps," *The Library of Congress Quarterly Journal of Current Acquisitions,* XI (August 1953), 239.

2. Clara Egli LeGear, "Divisions of Maps at the Library of Congress," *Education,* LX (December 1939), 220–24.

3. G. D. Hubbard, "So the Map Collection Can Grow: Arrangement and Care," *Library Journal,* LXXI (January 15, 1946), 91–93; L. A. Brown, "Those Rare Maps Are Not Curios," *Library Journal,* LXXV (March 15, 1950), 459–61; A. B. Carlson, "Old Maps Recorded History," *Library Journal,* LXXV (March 15, 1950), 461–63.

4. Special Libraries Association, *Map Collection in the United States and Canada* (New York, 1954). For Ohio, see pp. 108–12.

5. Walter W. Ristow, "Maps," *The Library of Congress Quarterly Journal of Acquisitions,* XI (August 1954), 239.

6. Walter W. Ristow and Catherine I. Bahn, "Maps," *The Quarterly Journal of the Library of Congress,* XXII (July 1965), 219–21.

7. Charles C. Baldwin, "Early Maps of Ohio and the West," *Tract No. 25* of the Western Reserve Historical Society, April 1875.

8. Philip Lee Phillips, *A List of Maps of America in the Library of Congress* (Washington, D. C., 1901), with Ohio's listings on pp. 626–36; Philip Lee Phillips, *A List of Geographical Atlases in the Library of Congress,* 4 vols. (Washington, D. C., 1909–1920); Clara Egli LeGear, *A List of Geographic Atlases in the Library of Congress,* Vol. 5 (Washington, D. C., 1958), which updated the Library of Congress's holdings between 1920 and 1958; Clara Egli LeGear, *United States Atlases; A List of National, State, County, City and Regional Atlases in the Library of Congress,* 2 vols. (Washington, D. C., 1950–1953), for Ohio, see I, 243–56; II, 159–69; Richard W. Stephenson, *Land Ownership Maps; A Checklist of Nineteenth Century United States County Maps in the Library of Congress* (Washington, D. C., 1967), for Ohio, see pp. 32–36.

9. A typewritten list of Ohio maps in the National Archives is at the Ohio Historical Library. See also, National Archives, *Civil War Maps in the National Archives* (Washington, D. C., 1964).

10. Christian Brun, *Guide to the Manuscript Maps in the William L. Clements Library* (Ann Arbor, Michigan, 1959).

The Mapping of Ohio

CHAPTER ONE

The Ohio Country

Soon after the first Europeans set foot in North America, explorers, traders and trappers followed the many rivers and narrow animal and Indian trails westward through the deep, pristine forests, penetrating the region around the Great Lakes Basin and the vastness of the upper Mississippi and Ohio valleys. As these early adventurers became familiar with the topography of these regions, their discoveries appeared upon maps, and slowly a picture of the territory west of the Alleghenies and north of the Ohio River was pieced together. The earliest maps were based on information gathered by energetic French pioneers, who had endured the hardships of that almost endless wilderness in expanding their trade and spreading their religion. By the middle of the eighteenth century, when the final struggle between England and France commenced for domination of a worldwide empire, which would include the Ohio Valley, the British had begun to probe the interior of the continent west of the Alleghenies, and their maps reflected this westward movement.

The information about the Ohio Valley contained on the many general maps of North America that were made by American, British and French map makers between 1650 and 1778 illustrated both the patterns of exploration as well as the growth in knowledge of the western territory. By the time of the American Revolution, the Ohio Country was a familiar portion of the maps of North America. These maps, made from first-hand reports, located major waterways, Indian villages, trading posts and deposits of natural resources. Many European adventurers and cartographers, including Samuel de Champlain, Louis Joliet, Jean Baptiste Louis Franquelin, Herman Moll and Guillaume Delisle, made maps that illustrated both the French and British empires in North America.[1] Although their work is significant to the story of the exploration of the Ohio Country, others, building on new evidence and each other's resources, made maps that represent milestones in the history of the mapping of Ohio.

The first map to illustrate all five of the Great Lakes, including Lake Erie, the last of the lakes to be discovered, was "Amerique Septentrionale Par N. Sanson d'Abbevile, Geog. du Roy A. Paris Chez l'Auteur et chez Pierre Mariétte rue S. Lacques a l'Espérance 1650" (Plate I). This map made by Nicolas Sanson d' Abbeville, the royal engineer and geographer to the French monarchy between 1647 and 1667, included the outline of Lake Erie nineteen years before Louis Joliet, a French trader and explorer,

Plate I: Sanson, "Amerique Septentrionale," (1650), Yale University. One of the earliest maps to illustrate Lake Erie.

was credited with being the first European, on record at least, to have paddled over its surface in a canoe. The presence of Lake Erie had been suggested earlier by Samuel de Champlain, the Frenchman who led the way for the French across North America, when he indicated on his map that a waterway existed between Lake Huron and Lake Ontario. Sanson, in order to gather information for his map, had obviously pored over the official records in the French Archives, and undoubtedly had consulted with Catholic priests and trappers who had returned to visit their native soil after long and hard treks across the regions of the Great Lakes Basin. The famous French cartographer did not label Lake Erie as such, but hinted its name when he noted that to the southeast there lived the N. du Chat, or Cat Nation, as the Erie Indians were called. It was not until six years later, in 1656, on another of his maps of North America, that Sanson specified the lake as Erie, Du Chat. It is difficult to speculate about the significance of all the rivers that appeared to flow into Lake Erie, as early cartographers commonly dressed up their works with unidentified rivers and streams; yet the appearance of the rather wide ribbon of water that apparently drained what is now northern Ohio and entered into Lake Erie about the center of its southern shore is certainly curious. It is probable that Sanson learned from his talks with those who had visited the region that the Indians spoke of a large river that flowed to the south of Lake Erie; with only that amount of information, the map maker allowed his imagination to do the rest. Certainly what Sanson placed on his map in 1650 represented either advanced knowledge of the lake region before actual exploration or some very fine guesswork on his part.

One of the major problems that French cartographers had during the remainder of the seventeenth and the first part of the eighteenth century was to determine the exact location of the Ohio River and its tributaries. It was the popular belief that the Ohio was a tributary of the Wabash River, and that it extended beyond the famous Forks of the Ohio to the headwaters of the Allegheny River. Of course, such mistakes could not be corrected until each river was traced to its source, and this took some time. How much reliance was placed on the knowledge of the geography of the Ohio Valley possessed by the Indians who roamed freely over this vast area is not certain. It is known, however, that in 1668 several Seneca Indians described to René-Robert Cavelier, Sieur de la Salle, that south of Lake Erie there flowed a magnificent river that eventually reached the sea. About seventy years later, George Croghan, the famous Pennsylvania trader, was being guided by Indians in his penetration of the Old Northwest. To Richard Peters he wrote on August 8, 1748, shortly after the signing of the Treaty of Lancaster, that several of the western Indians had given to him valuable geographic information about the Ohio Valley. "By their information," Croghan wrote, "it appears that the River Ouabache [Wabash] takes its rise from a Lake at a small distance from the West end of Lake Erie. . . . The principal Deputy of the Twightwees [Miami] laid down with Chalk the Courses of the Mississippi, of Ouabache, and of Ohio, marking the Situation of their own Towns, of Lake Erie, and of two Forts that the French have on the Mississippi."[2] It should be kept in mind that no matter how much recognition Europeans and Americans claimed for discovering and mapping these western regions, the Indians were familiar with their own surroundings and knew the topography of the Ohio Valley quite well; it remained the task of the white man merely to locate and map this region.

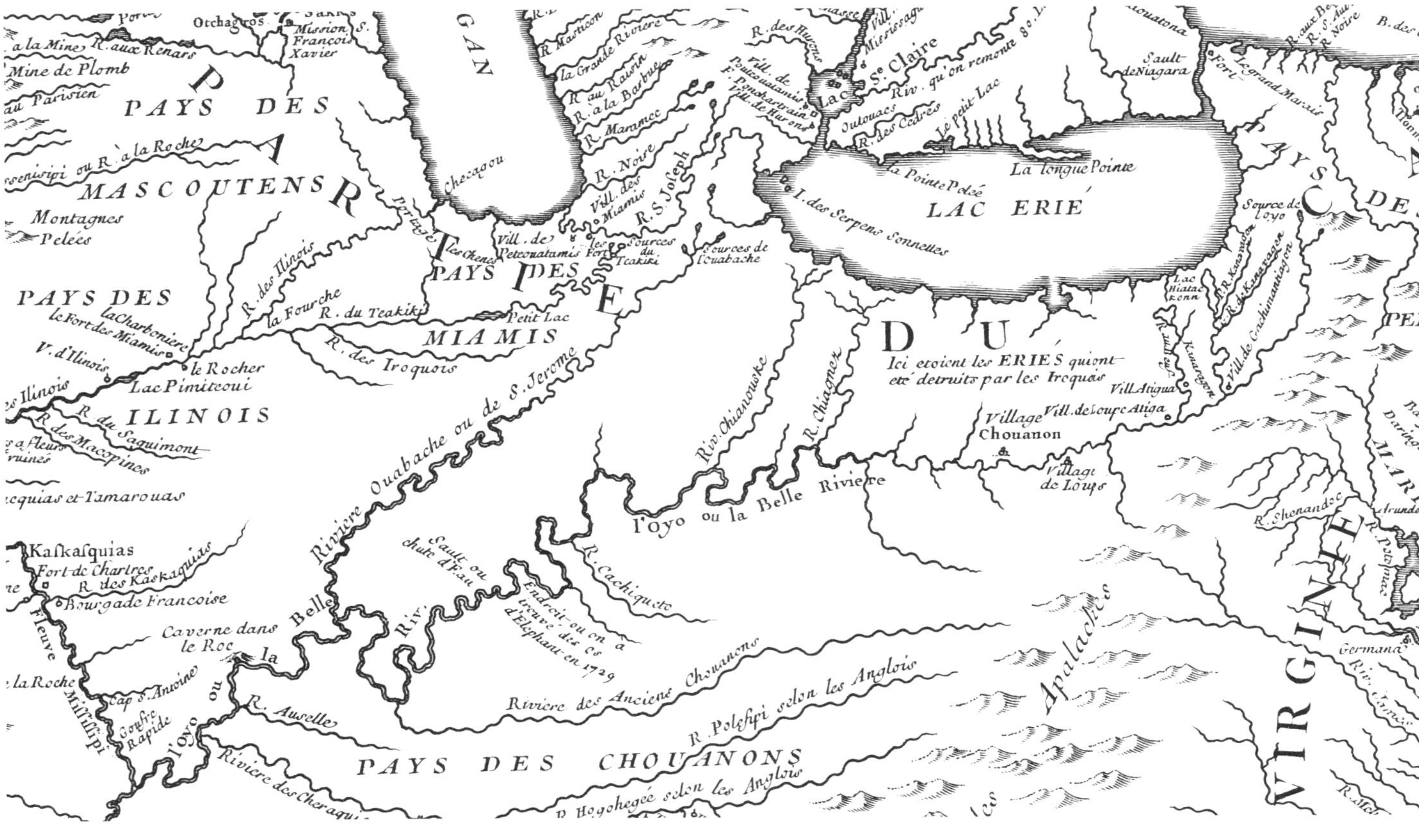

<table><tr><td>

Plate II: Bellin, "Carte de la Louisiane," (1744), printed in Charlevoix, *Histoire et Description Generale de la Nouvelle France,* (1744). This map accompanied the first history written about North America.

</td></tr></table>

During the second quarter of the eighteenth century great progress was made in the cartography of North America. Through the effort of Jacques-Nicolas Bellin, a French geographer and cartographer, one of the most accurate descriptions of the Ohio Country to that time was included on his map, *"Carte de la Louisiane cours du Mississippi et pays voisins . . .,"* (Plate II) which appeared in Volume II of Pierre Francois Xavier de Charlevoix's *Histoire et Description Generale de la Nouvelle France avec le Journal Historique d'un Voyage fait par ordre du Roi dans l'Amérique Septentrionnale.* Charlevoix, a Jesuit priest, explorer and historian, had visited North America between 1720 and 1722, and during that time had personally traveled over a portion of the Great Lakes and down through the Illinois Country to New Orleans on the Gulf of Mexico. His *Histoire*, which appeared in 1744, was the first descriptive work of the interior of the new continent; because of the very nature of the volumes, along with the author's masterful style, it proved to be quite popular. Bellin's map, which appeared as the frontispiece with this work, was in part based on both Charlevoix's notes and his first-hand knowledge of the country. In addition, however, Bellin, who was the Premiere Geographer of the Marine and of the Depot des Cartes et Plans, also had access to the official records of all the earlier explorations. The combined efforts of these two men allowed the French cartographer to detail a classic map of the French empire in North America.

Bellin's *"Carte de la Louisiane . . ."* is significant to the story of the mapping of the state of Ohio because of the interpretation of the location of Lake Erie and the Ohio River. His outline of Lake Erie became the norm copied by almost every European map maker during the remainder of the eighteenth century; in fact, it was not altered much until the United States government conducted its official surveys of the Great Lakes in 1841. Charlevoix, during his journey from Montreal to New Orleans, had entered the eastern end of Lake Erie through the Niagara River on May 27, 1721, and sailed along the northern shore westward toward the Detroit

River. In his *Journal Historique,* which was appended to his famous *Histoire*, he explained:

> Lake Erie is a hundred leagues in length from east to west. Its breadth from north to south is thirty leagues, or thereabouts. The name it bears is that of an Indian nation of the Huron language, which was formerly seated on its banks, and who have been entirely destroyed by the Iroquoise. Erie in that language signifies Cat, and in some accounts this nation is called the Cat nation. This name comes, probably, from the quantity of these animals formerly found in this country. They are no larger than ours and their skins are reckoned very valuable.[3]

In an effort to put as much information on his map as possible, Bellin noted on his map just below the lake's southern shoreline, *"Ici etoient les ERIES qui ont été detriute par les Iroquois."*

On June 5, Charlevoix and his party reached the western end of the lake, and he recorded in his *Journal*: "Towards four o'clock in the afternoon we perceived the land on the south shore, and two little islands which lie very near it. These are called Rattlesnake islands, and we are told they are so infested with these reptiles that the air is infected with them."[4] It is probable that the Frenchman either saw West Sister Island and Pelee Island, or the Bass islands north of Catawba Point. Although he never set foot on any of these islands, Charlevoix obviously had learned about them from the Indians or other Frenchmen who had. He was correct to call them *I. des Serpens Sonnettes* because as late as the first part of the nineteenth century this group of islands was famous for the great number of rattlesnakes that crawled over them almost unmolested.

This particular map by Bellin was also one of the first popular works to show the headwaters of the Wabash River and the suspected source of the Ohio in their proper location, and to illustrate that the former was a major tributary of the latter. His detailed map work along the upper Ohio no doubt reflected his reliance on sources other than those furnished by Charlevoix, who penetrated no farther south than the northern shore of Lake Erie while on his voyage. It is known that Bellin used the journal of Chaussegros de Léry, an engineer who directed the construction of the fortifications at Niagara in 1726, and who himself journeyed down the Ohio in 1729. Bellin indicated that a waterway between Lake Erie and the Allegheny River existed via a portage to *Lac de Kanarongon*, or Lake Chautauqua, through that lake and then down the *Kanaragon,* or Conewango Creek. The *Vill. de Gachisnanragon* shown at the junction of the Conewango and the Allegheny was Paille Coupee, or De-ga-syo-noh-dyah-goh, or Cut Straw. Farther down the Allegheny River, Bellin showed the *R. au Beuf,* or *Rivière aux Boeuf,* or French Creek, as it was known by the English. A little beyond French Creek was *Vill. de Loupe Atigua* [*Attigue*]. The French called the Delaware "Loups," and this particular village located by Bellin was, with little doubt, the rather large Delaware settlement called Attigue which apparently was located on both sides of the Allegheny near the mouth of the Kiskiminetas River. The *Village de Loups* along the south side of the river was another large Delaware settlement, and was later known as Shannopin's Town. It was near the confluence of the Allegheny and the Monongahela rivers where Pittsburgh now stands. Several miles beyond, on the west bank of the Ohio, Bellin noted the *Village Chouanon.* This Shawnee village, actually located on the east bank of the Ohio and reputed to have been the largest Indian village on the Upper Ohio, was Chinigue or Logstown, the important trading center for the French and the English west of the Allegheny Mountains.

The problem of the identification of the rivers that Bellin showed to flow entirely within the state of Ohio is one for speculation. Between *Village Chouanon* and *R. Chiagnez*, two major rivers, both without names, are shown. As already mentioned, sometimes the early cartographers placed rivers on their maps in order to fill in space rather than to illustrate geographic information. Bellin's *Riv. Chiagnez* was the Muskingum River, while his *Riv. Chianouske* was the Scioto. These were the names by which the French knew the main streams that drained Ohio's interior and served as important links to the Great Lakes to the north.[5]

Further evidence that the French cartographer had reference to these two rivers is seen by his location of their source just south of Lake Erie. On another Bellin map, *Carte Des Lacs Du Canada Dressée sur la Manuscrits du Depost des Cartes, Plans et Journaux de la Marine et sur le Journal du R P de Charlevoix*, which appeared as the frontispiece in the third volume of Charlevoix's *Histoire*, he showed all the Great Lakes, but only that portion of the state of Ohio above the fortieth parallel. The headwaters of both the "*Riv. Chiagues*" and "*Riv. Chianouske*" are shown south of Lake Erie, and Bellin noted that after a portage, each "*tombe dans la belle Rivière.*" Just above the southern shore of Lake Erie, Bellin admitted, "*Toute cette coste n'est presque point connue.*" It is probable that while the existence of the various rivers that drained the northern half of Ohio and emptied into Lake Erie was not known to him, he was assured by his sources that several large rivers passed through the region and emptied into the Ohio River. He no doubt assumed that these north-south routes, rather than being a combination of streams and portages, were indeed extended waterways.

As already mentioned, his "*Carte de La Louisiane . . .*" was the first popular map to locate the approximate source of the Ohio and Wabash rivers and to indicate their proper relationship to each other. It is obvious that Bellin did not know of the Monongahela River and referred to the Allegheny as *l'Oyo.* He located the source of the *Rivière Ouabache ou de S. Jerome,* or Wabash southwest of the western end of Lake Erie. The location of this river, of course, was no doubt familiar to Bellin since the Erie-Wabash-Ohio-Mississippi route between Canada and New Orleans had been used by French missionaries and traders by the end of the seventeenth century. The Allegheny-Ohio-Mississippi route, on the other hand, was denied to the French because of the hostilities of the Iroquois and was not used until later.

It is apparent that Bellin corrected Charlevoix. The French explorer believed, as did many, that it was the Wabash and not the Ohio that emptied into the Mississippi River, but Bellin clearly named *l'Oyo ou la Belle Riv.* at its mouth, while he illustrated that the *Rivière Ouabach ou de S. Jerome* was its tributary. It is particularly interesting to compare the information Bellin illustrated on his map with that described by Charlevoix in his *Journal.* Charlevoix stated:

> On the fifteenth [November 15, 1721], the wind changed to the north, and the cold continued to increase. We advanced four leagues to the southward, and then found that the river run four leagues more towards the north. Immediately after this uncommon winding we passed on the left the fine river Ouabache, by means of which there is a passage as far as the country of the Iroquois, when the waters are high. This river, at its entrance into the Mississippi, is not less than a quarter of a league in breadth. There is not, in my opinion, a place in all Lousiana more proper for a settlement than this, nor where it is of great importance to us to have one. The whole country watered

> by the Ouabache and Ohio, which runs into it, is extremely fertile consisting of vast meadows, which feed thousands of buffaloes. Besides it communication with Canada is as easy as that by the river of the Illinois, and the passage much shorter.[6]

This comparison of Charlevoix's writing with that of Bellin's strongly suggests that the map maker knew the Ohio Valley better than did the explorer. Bellin's map recorded the accumulated geographic knowledge of the Ohio Valley that the French had gained through exploration between 1720 and 1744, and it is probable that he had a greater familiarity with the French Empire in North America than any other European cartographer before 1755. In fact, George Washington used Bellin's map when he was ordered by Virginia's royal governor, Robert Dinwiddie, to reconnoiter the activities of the French at the Forks of the Ohio in 1755 on the eve of the French and Indian War.

Washington's reliance on French maps to help guide his way to the Forks indicated that the British were not familiar enough with the Ohio Valley during the first half of the eighteenth century to map it for themselves and were consequently content to borrow from French intelligence. Justin Winsor discussed Washington's journey in his *Mississippi Basin* and judged that despite the fact that Bellin's map was already ten years old by 1753, it was "still much better than any which the English had yet produced."[7] By 1755, however, the British were engaged against the French in a worldwide struggle for empire, and the Ohio Valley was part of the prize. Although many maps appeared after the outbreak of hostilities in 1754, two wonderful and competent maps of North America were produced in 1755. Both helped to satisfy the curiosity aroused in Europe about the American theater of war, and to illustrate British claims in the vast region west of the mountains.

The first of these maps, "A map of the British and French dominions in North America, with the roads, distances, limits, and extent of the settlements," (Plate III) was compiled and drawn by John Mitchell in 1750 and originally issued in England in 1755. Mitchell, a physician and botanist rather than a map maker, lived in America for at least twenty years, although he apparently never traveled extensively except between his home in Urbanna (Middlesex County, Virginia) and Philadelphia. He prepared his map at the request of the Board of Trade, and had access to the records and maps in that office and to "those of several private People"[8] in addition to those kept by the British Admiralty. He also used the materials and journal recorded by Christopher Gist, the famous explorer, during his trip through the Ohio Country in 1750–51, when he toured west of the mountains as an agent for the Ohio Company of Virginia, consulted Charlevoix's *Histoire*, along with Bellin's maps, and had access to John Patten's account of his journeys north of the Ohio during his captivity by the Indians between 1750 and 1752.[9] Mitchell's map was probably the most famous one ever produced of North America. It was used by Parliament to prepare the Quebec Act of 1774, consulted at the peace negotiations that concluded the American Revolution, involved in Ohio's border dispute with Michigan in 1836, considered in the negotiation of the Webster-Ashburton Treaty of 1842; and played an important part in many national and international boundary disputes, even as late as 1932. There were many editions of the so-called Mitchell map, including several in French, Dutch, Italian and Spanish.[10] The map reproduced in this volume is the second impression of the third edition, which was

(Overleaf) Plate III: Mitchell, "North America," (1755), Library of Congress, Washington, D.C. Mitchell's map was used in the settlement of international disputes between 1774 and 1932.

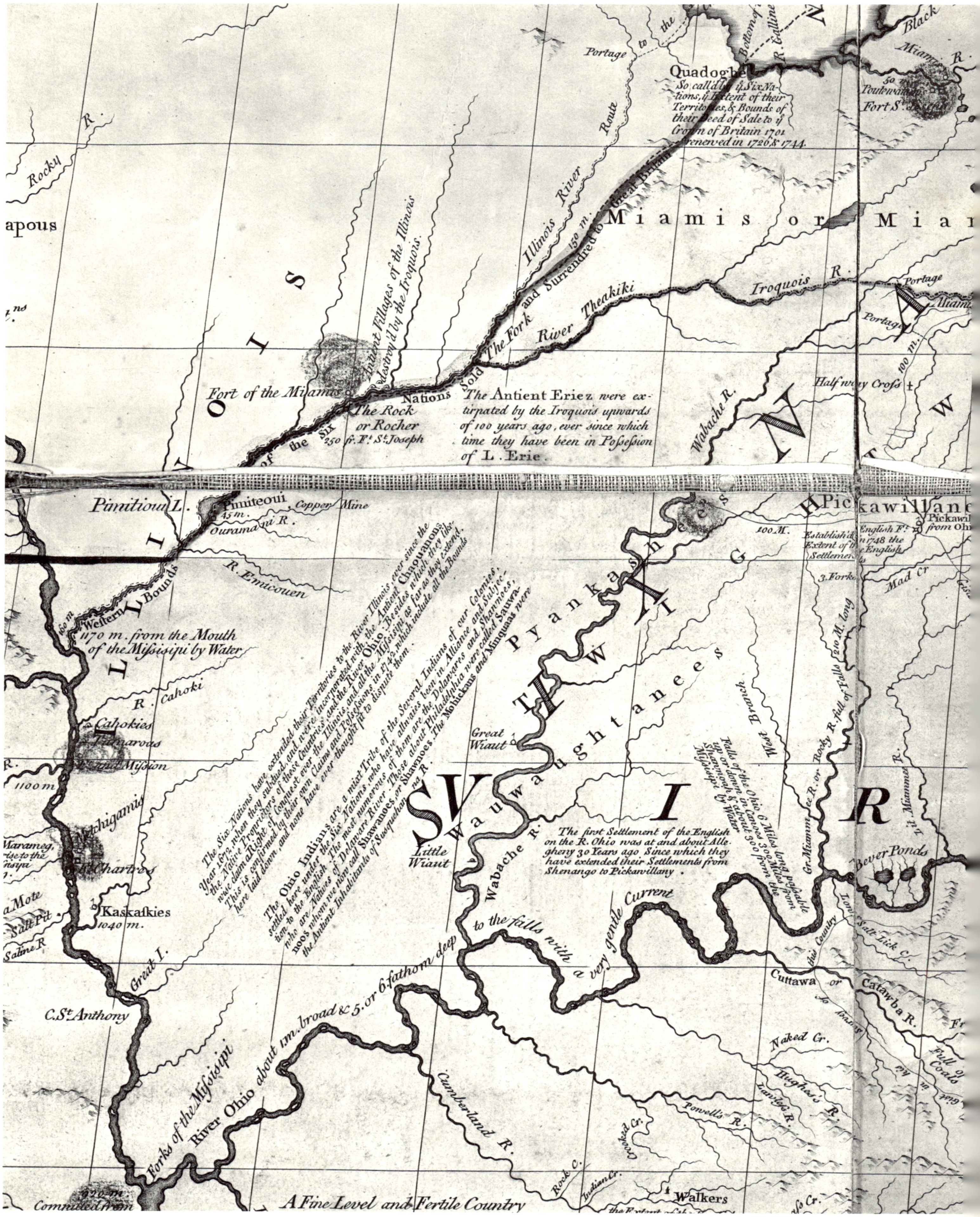
Portage
Quadoghe
So call'd by y Six Nations, y Extent of their Territories, & Bounds of their Deed of Sale to y Crown of Britain 1701 renewed in 1726 & 1744.
Route
Illinois River
Miamis or Miami
Rocky R.
apous
Black
Miami R.
Fort St Joseph
Antient Villages of the Illinois destroy'd by the Iroquois.
The Fork and Surrendered to Great Britain
Iroquois R.
River Theakiki
Portage
Portage
Half way Cross
100 m.
Fort of the Miamis
The Rock or Rocher
Nations Sold
250 m. Ft St Joseph
The Antient Eriez were extirpated by the Iroquois upwards of 100 years ago, ever since which time they have been in Possession of L. Erie.
Wabache R.
Pimitiou L.
Pimiteoui
Copper Mine
Ouramani R.
R. Emicouen
Western Bounds
1170 m. from the Mouth of the Misisipi by Water
Pickawillanes
100 M.
Mad Cr
3 Forks
R. Cahoki
Cahokies
Tamaroas
Mission
Michigamis
Chartres
Kaskaskies
1040 m.
Salt Pit
Saline R.
Great I.
C. St Anthony
The Six Nations have extended their Territories to the River Illinois ever since the Year 1672, when they subdued, and were incorporated with, the Antient Chaouanons, the Native Proprietors of these Countries, and the River Ohio. Besides which they lay claim a Right of Conquest over the Illinois, and all the Missisipi as far as they extend. This is confirmed by their own Claims and Possessions in 1742, which include all the Bounds here laid down, and none have ever thought fit to dispute them.
The Ohio Indians are a mixt Tribe of the Several Indians of our Colonies, settled here under the Six Nations who have always been in Alliance and Subjection to the English. The most numerous of them are the Delawares and Shawnoes, who are Natives of Delaware River. Those about Philadelphia were called Sauwanoos whom we now call Shawanoes or Shawnoes. The Mohickons and Minquaas were the Antient Inhabitants of Susquehanna R.
Pyankashees
Twightwees
Great Wiaut
Little Wiaut
Wauwaughtanees
Wabache R.
West Branch
Falls of the Ohio 6 Miles long passable up or down in Canoes 300 Miles from Shannoah & about 500 from the Missisipi by Water
Rocky R. full of Falls 210 M. long
Gr. Miammee R.
Little Miammee R.
Bever Ponds
The first Settlement of the English on the R. Ohio was at and about Allegheny 30 Years ago. Since which they have extended their Settlements from Shenango to Pickawillany.
to the falls with a very gentle Current
River Ohio about 1 m. broad & 5 or 6 fathom deep
Forks of the Misisipi
Cuttawa or Catawba R.
Naked Cr.
Hughes's R.
Powell's R.
Crooked Cr.
Cumberland R.
Rock C.
Indian Cr.
Walkers
A Fine Level and Fertile Country

LAKE ERIE
OR
OKSWEGO
Long Point
Bare Pt.
Rattle Snake I.
Vill. of Hurons
Vill. of Utawas
The Strait
Canahogue Bay
Sandoski
Canahogue
The seat of War, the Mart of Trade, & chief Hunting Grounds of the Six Nations, on the Lakes & the Ohio
The Forts lately usurped by the French
The Places called Licks, and Lick Creeks are Salt water, which afford plenty of Salt to Man & Beast in those Inland Parts. The Resort of all sorts of Game Huntsmen, Traders & Warriors, especially by the Salt Ponds.
Salt Ponds
L. Otsanderket
White Woman's Cr.
Salt Lick Cr.
OHIO INDIANS
Tuscaroraas
Legs T.
Mohickans
Kittanning
Mingos
Shenango or Cheningue an English Settlem.
Shawnoes
Pict's
Delaware T.
Hockhocken or Margaret's T.
Traders
Delawares
Maguck
Harriskintom
Shawnoes
Hockhocken R.
Muskingum or Elk R.
Muskingum English Factory
Sawanoos
Monongahela R.
Little Conhaway R.
Conhaway
Wood R.
New R.
Green Briar R.
Cow Pasture
Calf Pasture
North R.
James River
Bedford C.
Pr. Edward C.
Albemarle
Goochland C.
Cumberland C.
Chesterfield C.
Amelia C.
Prince George C.
Dinwiddie Co.
Brunswick C.
Henrico C.
Hanover C.
Louisa C.
Spotsylvania C.
Orange C.
Culpepper C.
Prince William C.
Fairfax C.
Alexandria
Frederick C.
Winchester
Shanando River
Blue Ridge
Pennsilvania
Tuscarora
Kittatinny
Shippensburg
Cumberland
Virginia
Maryland

printed by Jeffreys and Faden and is part of the large map collection held by the Library of Congress.

Mitchell's map was one of the first to illustrate exact locations of important and strategic sites west of the mountains through the use of mathematical computation. His calculations were based on the information supplied in the journals kept by Gist, who almost daily recorded the approximate direction and distance that he had traveled. Mitchell apparently used Fort Cumberland as a fixed reference point and consequently was able to compute Gist's daily advance down the Ohio River. By such a system, Mitchell approximated the location of *"Ft du Quesne,"* Loggs Town, the Muskingum River, "Shannoah," Pickawillany, Hockhocken, "Sandoski," the Miami River and Lake Erie. Although Mitchell's figures later proved to be inaccurate, he did produce the most informative British map of North America to that time and his map proved to be not only the most important one in American history but one that is of great significance to the story of the mapping of Ohio.

Comparison of the geographic knowledge of the Ohio Valley displayed on Mitchell's map with that shown on Bellin's quickly demonstrates the increased interest and activity in that region on the part of the British at the time of their final conflict with France for control of North America. Despite France's claim to the whole of the Ohio Valley, which had recently been reinforced by Pierre Joseph Celoron de Blainville's excursion down the Ohio River in 1749, Mitchell tried to illustrate on his map that the British too laid claim to the fertile and rich Ohio River basin through the rights of discovery and occupation. Mitchell, for instance, claimed that English traders had established a post at Pickawillany on the Great Miami River in 1748, a year before Celoron visited that region, and a notation on his map just above the Ohio River between the "Miammee R." and the Wabash River explained that for the past thirty years English settlements had spread from the Allegheny throughout the Ohio Country, so that by 1755, there were British posts from "Shenango to Pickawillany." Probably his most striking claim was that England controlled the Valley through its alliance with the Iroquois. Mitchell reasoned that since 1672 the Iroquois claimed by right of conquest all the territory to the Illinois River. Because the Six Nations "who have all waise been in Alliance and Subjection to the English" were England's allies in North America, Britain could therefore claim control of the Ohio Valley.

Mitchell noted the great "Traders Road" over which Gist traveled part of the way from Shannopin's Town at the Forks of the Ohio to Pickawillany on the Great Miami River. Gist traveled from the Forks to Hockhocken. From there he cut south to Maguck and Shannoah Town, down the Ohio to the Great Miami and then up that stream to the large Indian Village and trading center at Pickawillany. The route Gist took on his return trip from Pickawillany to the Indian villages at the mouth of the Scioto was marked by Mitchell who labeled it "Gist's Route & Traders Road." Because few Englishmen had visited those regions in the northwestern portions of the state, little information was available. It is curious that Mitchell referred to the Ohio River as the "Splawacipiki R.," which was his phonetical spelling of what the Shawnee called that great waterway.

It is of particular interest to identify the various sites first mapped by Mitchell within Ohio because, in most cases, these same locations were later selected as village sites by white surveyors and settlers as they worked their way across the state. It is necessary to rely on Gist's *Journals*, much as did

Mitchell when he produced his famous map, for information about the exciting Ohio Country.

The first village that Gist visited in Ohio was "Tuscaroraas" located by Mitchell on the west bank of Elks Eye Creek. Gist explained that this village was "a Town of Ottaways, a Nation of French Indians. . . . There are not above six or eight families belonging to this Town."[11] This small village was, in fact, situated between the Tuscarawas River and the Big Sandy Creek, just above the present site of Bolivar in Tuscarawas County. Several years later, by 1764, it was known as King Beaver's Town and occupied by Delaware Indians who had moved from Fort Duquesne before the French evacuated that post in 1758. Gist then proceeded to Muskingum, or "T of Owendoes," or Conchake, a fairly large village established by Wyandots in 1748 a little above the Forks of the Muskingum on the east bank of the Tuscarawas River, close to the present site of Coshocton. "The Town of Muskingum consists of about one hundred Families," Gist noted, and estimated that England probably had the support of about half the braves of the village.[12] Mitchell indicated on his map that an "English Fort" was established here on the banks of the Muskingum. Even though he tried diligently to substantiate England's claim to this region with whatever evidence he could muster, Mitchell no doubt stretched the fact and implied that the small trading post established by George Croghan, a famous Pennsylvania trader who was quite familiar with the vast land north and west of the Ohio himself, was indeed a fort. The goods that Croghan had stored here were confiscated by the French in 1753. This site was abandoned by the Indians shortly before 1760, probably because the French lost their influence throughout the Ohio Valley when their principal post, Fort Duquesne at the Forks of the Ohio, fell to the British in 1758.

"Hockhocken" was a Delaware village located along the Hocking River at Lancaster. Gist mentioned that "it was a small Town with only four or five Delaware Families." This village was also called Margaret's Town, so named for French Margaret or Margaret Montour, the daughter of Madame Montour, the famous interpreter for the Five Nations and mother of Andrew Montour,[13] who also accompanied Gist on his journey through Ohio in 1750 and 1751. The village was located near the Hocking River, which was originally called Hocking Hocking, a Delaware term that meant the site of arable land.

Gist proceeded to Maguck, "a little Delaware Town of about ten families by the N Side of a plain or clear Field about 5 M[iles] in Length NE & SW & 2 M[iles] broad, with a small Rising in the Middle, which gives a fine Prospect over the whole Plain, and a large Creek on the N Side of it called Sciodoe Creek." This particular village was located on the famous Pickaway Plains in Pickaway County just south of Circleville. Several miles below Maguck was "Harrickintoms T." or Hurricane Tom's Town, which consisted of about "five to Six Delaware Families on the SW Sciodoe Creek." Mitchell mistakenly placed it on the east side of the Scioto, for it actually was situated below the city of Chillicothe, nearly opposite the mouth of Salt Creek.[14] The last Delaware town in the Scioto Valley was Lower Delaware Town, a small village "of about twenty Families," along the east bank in Clay Township, Scioto County, founded in approximately 1738. Its head man was Windaughalah, who for years had significant influence among the Indians that inhabited Ohio. He was prominent at the negotiations of the treaties of Fort McIntosh and Fort Finney; and Buckongahelas, the famous chief of the Delawares, was his son. Gist recorded that his party was treated kindly by the old chief and that "he ordered a Negro Man that belonged to him [Windaughalah] to feed our Horses well."[15]

The principal community of Indians in Ohio visited by Gist was Shannoah Town, Shawnoah, or Lower Shawnee Town, a center for Indian trade in the Ohio Valley. Although the largest concentration of lodges was at the present site of Alexandria on the west bank of the Scioto opposite Portsmouth, a smaller community also existed along the south bank of the Ohio in Green County, Kentucky. "The Shannoah Town is situate upon both Sides of the River Ohio, just below the Mouth of Sciodoe Creek," Gist recorded. It "contains about 300 Men," he continued, and "there are about 40 Houses on the S Side of the River and about 100 on the N Side, with a Kind of State-House of about 90 Feet long, with a light Cover of Bark in wch they hold their Councils." In 1758, George Croghan was there when the Ohio River flooded almost fifty feet above its normal level and destroyed that portion of Shannoah Town on the north side. That same year the entire community relocated to the famous Pickaway Plains below the Delaware village of Maguck. This new settlement became known as Lower Shawnee Town, Upper Chillicothe, Old Chillicothe or Pluggy's Town and it was from there that the Shawnee raided into the Virginia settlements prior to the American Revolution. The old village along the Ohio still appeared on later maps as Old Lower Shawnee Town even though it was no longer occupied.

From Shannoah, Gist set out on land across the present counties of Scioto, Adams, Highland, Fayette, Madison, Clarke, Champaign and Logan for Pickawillany, the major village of the Twightwees, or Miamis, in Ohio, located opposite the mouth of Loramie's Creek along the west bank of the Great Miami River in Miami County. This village was founded before 1748. The agent for the Ohio Company of Virginia reported that Pickawillany "consists of about 400 Families, & daily increasing, it is accounted one of the strongest Indian Towns upon this Part of the Continent." He was apparently impressed by these Indians and remarked: "They are accounted the most powerful People to the Westward of the English Settlements, & much superior to the six Nations [Iroquois] with whom they are now in Amity." Mitchell noted on his map that an English fort had been established there in 1748 and that it represented "the Extent of the English Settlements" west of the mountains. As mentioned earlier, Mitchell tried to use whatever evidence he could to illustrate that England did have legitimate claims to the Ohio Valley. This fort, although claimed to be British, was in fact one built by the Indians and known as La Demoiselle Fort, named for La Demoiselle or Old Britain, the famous chief of the Miamis who was put to death by the French in 1752. The fort was frequently used by English traders for shelter and as a supply depot. While there, Gist and his party helped "to bring Loggs to line the Inside" in an apparent attempt to strengthen the structure.

Southwest of Lake Erie, near the headwaters of the "Miamis River," or Miami of the Lakes, or Maumee River, Mitchell located an Indian village and a fort. This village, known as Kekionga, or Kiskakon, which was composed of about forty to fifty cabins, was located near the junction of the St. Joseph and St. Mary's rivers where Fort Wayne, Indiana, now stands. The fort, first called Fort St. Phillippe and later renamed Fort Miamis or "the fort of the Miami," was built by the French in 1721.[16] In 1747, because of an Indian uprising fostered by the British against the French, this post was pillaged, only to be restored the following year. No doubt, Mitchell claimed this post had actually been under English control because the Miamis, friendly to the British, had gained control of this site for a short while.

Farther east, along the southern shore of Lake Erie, Mitchell indicated that a fortification called "Sandoski," situated on the western bank of the "R. Blanc," had been "usurped by the French, 1751." The French called what is now known as the Sandusky River the White River, and it appeared as such on many of their maps as "R. Blanc." The fortification Mitchell showed was probably Fort Sandusky, built by the French in 1750–51 south of the Portage River on the north side of Sandusky Bay near Port Clinton. It was abandoned in 1752–53 even though French traders continued to maintain posts along the south side of Sandusky Bay for several years after that. As noted by Mitchell, Sandusky Bay was known as "Bottom of the Lake," probably because it was believed to have been that point where the southern shore of Lake Erie dipped to its most southerly point.

The large area eastward from Sandusky and south of Lake Erie to the Cuyahoga, including the upper Muskingum Valley, was known as the White River Country[17] or Canayahaga or Canahogue and was described by Mitchell as "the seat of War, the Mart of Trade, & chief Hunting Grounds of the Six Nations, on the Lakes & the Ohio." The British map maker had taken much of his information about northern Ohio from the work of the French cartographer, Robert de Vaugondy. Mitchell's note was a further attempt to strengthen England's claim to the Ohio Country through its alliance with the Iroquois. Vaugondy had shown on his map an Iroquois village along what is now the Cuyahoga River. Mitchell copied this but did not indicate that George Croghan had established a small trading post at the Seneca village in 1744 as the Pennsylvania trader began to thrust British trading operations into a region that was once solely under French influence. In his own journals of 1760–61, Croghan mentioned that at the mouth of the "Gichawaga Creek," or the Cuyahoga River, there was a fine harbor. He was forced by the French to abandon his post there in 1750, only to continue his economic interests in the Twightwee village along the Great Miami.[18]

It is apparent that Mitchell attempted to place on his map every scrap of valuable information about the region west of the mountains that he had available to him. Salt, for instance, was so necessary to life that he indicated two sources of salt in Ohio. The first, the "Salt Ponds," or springs, he located close to the southern shore of Lake Erie. His location of these so-called ponds is erroneous because he no doubt had reference to the salt springs in Wetherfield Township in Trumbull County east of the Cuyahoga. Augustus Porter, the principal surveyor for the Connecticut Land Company, visited these springs in 1796 and recorded in his journal: "At this place we found a small piece of ground, say two or three acres, and a plank vat of sixteen or eighteen feet square, and four or five feet deep, set in the ground, which was full of water, and kettles for boiling salt; the number we could not ascertain, but the vat seemed to be full of them. An Indian and a squaw were boiling water for salt, but from appearances with poor success."[19] The most famous salt licks in Ohio's history were located in Jackson County on Salt Creek, which emptied into the Scioto. Gist crossed Salt Creek a little to the east of the Scioto River and made particular note of the location of these valuable deposits. It is likely that he did not actually visit these springs but that he recorded what others told him about them. "About 1 M up this Creek [Salt Creek] on the S Side," he recorded, "is a very large Salt Lick, the Streams which run into this Lick are very salt [*sic*], & tho clear leave a blueish Sediment: The Indians and Traders make salt for their Horses of the Water, by Boiling it; it has at first a blueish Colour, and somewhat bitter Taste, but upon being dissolved in fair Water and boiled a second Time, it becomes tolerable pure Salt."[20] Of interest also was Mitchell's notice of the "Long Reach 25 m." on the Ohio River east of the Muskingum. Between Sardis in Morgan County and Beavertown in Washington County, the Ohio River straightens out for approximately sixteen miles in nearly a straight line in a southwesterly direction. After following such a winding river, the Long Reach appeared most unusual to those who viewed its majestic natural beauty in the mid-eighteenth century. George Croghan noted in his journal in 1765 that there "the banks continue high; the country on both sides, level, rich and well watered."[21]

A second important map of North America detailing the Ohio Country was "A general Map of the Middle British Colonies in North America" by Lewis Evans. Like Mitchell's, this map appeared in 1755. Evans, a Philadelphian, was a geographer and surveyor who had earned a fair reputation in the American colonies as a map maker on the eve of the French and Indian War. This particular map, published by Benjamin Franklin's printing house and accompanied by a pamphlet, *Geographical, Historical, Political, Philosophical and Mechanical Essays: The First, Containing an Analysis of a General Map of the Middle British Colonies in America,* was the first of at least ten editions that appeared between 1755 and 1807. One of these later editions was published by Thomas Pownall, a close associate of Evans, to whom the first edition was dedicated. It is, however, the Pownall edition of the Evans map, published in 1776 (Plate IV), that is significant to the study of the mapping of Ohio.

Pownall, during his administrative career in the American colonies, had been the lieutenant-governor of New Jersey, and later governor of Massachusetts and South Carolina. He had a great interest in the region west of the mountains and was able to persuade Franklin to propose a plan for the creation of two western colonies to serve as "buffer colonies" for the British settlements against French intrigues in the West. Following the end of the Seven Years War, he proposed that an English Atlantic community should be welded into a solid commercial dominion. He sympathized with the American colonies; in 1774, during the American Revolution, Pownall, as a Member of Parliament, proposed a peace plan that fell mostly on deaf ears.

To accompany his reprint of Evans's map, Pownall also published *A Topographical Description of Such Parts of North America As Are Contained in the (Annexed) Map of the Middle British Colonies, etc., in North America,* which included most of the descriptive material that Evans had used in his *Analysis.* The former colonial governor, however, expanded the text to include new data about North America and especially material concerning that region north and west of the Ohio River. For instance, at the end of his pamphlet, Pownall printed part of the journal kept by Captain Harry Gordon, chief engineer for the Western Department, who in 1766 journeyed down the Ohio and produced the first survey map of that river,[22] and the entire diary maintained by Christopher Gist when he toured Ohio in 1750. Based on Gordon's fairly accurate map of the Ohio River, Pownall corrected Evans's original map work and relocated parts of the Ohio and

(Opposite page) Plate IV: Pownall, "Middle British Colonies," (1776), to accompany his *Topographical Description . . . of North America* (1776). Pownall's map, made over the plates used by Lewis Evans, indicated the increase in geographical information between 1755 and 1776.

LAKE ERIE
Ohio R.
Cayahoga R.
Kentucke R.
G. Salt Lick R.
G. Kanhawa R.
L. Kanhawa R.
Big Sandy C.
Mineami R.
Shawane T.
The Falls
PIQUES
TAWIXTAWIS
WIANDOTS

the Scioto, Little Miami and Big Miami rivers. On his revised map, however, he chose to retain the originality of the first edition and did not erase Evans's interpretation of the Ohio Country, but rather sketched in the course of the Ohio and its major tributaries as determined by Gordon and described by Gist. Evans had been convinced more than was Mitchell that the Ohio was a relatively simple meandering stream. "The [Ohio] River," he stated, "from hence [the Forks] downward, is agreed by all who have gone down it, to be in general pretty strait, nor can its Curves be indeed considerable where it is confined in a Manner by a Chain of little Hills, from the last mentioned Fork to 10 Miles Below the Falls."[23] The Pownall edition was probably the first major map of North America to illustrate the Big Bend in the river above the Scioto. Because of the method selected by Pownall to revise this map, it is easy to see the changes that were made and the effect that new information about the Ohio Country made available to map makers had on the mapping of the Northwest. Each new journey by traders and trappers across the mountains furnished additional knowledge about the vast and rich region, and new maps, elaborately produced, were obsolete almost before the ink was dry. As Pownall himself explained:

> None of the Parts of the Map West and North of the Ohio are presumed to be other than such a Sketch as shall give a general Idea. Every new Map may correct the last before it, and yet be no more than a Sketch at best. We must wait for Observations and Surveys in our future Knowledge of this Country, in order to give an actual Map. There is none such yet: nor are there any Materials as yet from which any such Map can be compiled, whatever may be pretended. In Justice to Mr. Lewis Evans's Industry, I will venture to say none as yet can give a better Idea of those Parts than this Map has done, not even those done by the French while they had possession and commanded in these Parts.[24]

The information about Ohio that appeared on Evans's original map was similar to Mitchell's, yet Evans's was more precise and detailed. Evans claimed that his sources for the western portion of his map were traders who had traveled the country. "The Routs across the Country, as well as the Situation of the Indian Villages, trading Places, the Creeks that fall into Lake Erie, and other affairs relating to Ohio and its Branches," Evans explained in his *Analysis,* "are from a great Number of Informations of Traders and other, and especially of a very intelligent Indian called The Eagle, who had a good Notion of Distances, Bearings, and delineating."[25] Evans had access to a journal kept by William Franklin, the son of Benjamin Franklin, when the former accompanied George Croghan on one of his missions to the Ohio Country in 1747–1748.[26] The journal was used to help sketch the Ohio from the famous Forks, or from the junction of the Allegheny and the Monongahela rivers, to the Scioto. Unfortunately, Franklin's journal has not been preserved over the years. Besides the Indian, The Eagle, and young Franklin, Evans mentioned that Joseph Dobson, an agent of the famous Baynton, Wharton and Morgan trading firm of Philadelphia, had supplied him with "an Account of the Distance from Creek to Creek as they fall in [the Ohio], and of the Islands, Rifts, and Falls all the Way from the Fork to Scioto [Scioto River]," while Alexander McGinty and Alexander Lowrey, both traders for many years in Ohio, furnished important material from the Scioto to the Falls at Louisville, Kentucky.[27] Conspicuously missing from Evans's list of informants were Gist and Croghan. Apparently, the map maker did not avail himself of the knowledge of this region possessed by these two famous explorers.

Pownall used a great deal of the information in Evans's *Analysis* in his own *Topographical Description* to describe what is now Ohio. Since water transportation was the principal mode of travel through the wilderness, the descriptive material concentrated on Ohio's major streams and their immediate surroundings. The Muskingum, Hocking, Scioto, Little Miami, and Great Miami were each discussed, and the following paragraphs from Pownall's pamphlet indicate the extent of information known about this rich region at the beginning of the American Revolutionary War.[28]

> "Muskingum is a fine gentle River, confined within high Banks that prevent its Floods from damaging the surrounding Land." It is 250 Yards wide at its Confluence with the Ohio. "It is passable with large Batteaux to the Three Logs [Legs], and with small Ones to a little Lake at its Head, without any Obstruction from Falls or Rifts. From hence to Cayahoga [Cuyahoga], is a Portage a Mile long. Cayhahoga [Cuyahoga], the Creek that leads from this Portage to Lake Erie, is muddy and middling swift, but no where obstructed with Falls or Rifts. As this has fine Land, wide extended Meadows, lofty Timber, Oak and Mulberry fitted for Shipbuilding, Walnut, Chesnut [*sic*], and Poplar for domestic Services, and furnishes the shortest and best Portage between Ohio and Lake Erie; and its Mouth is sufficient to receive good Sloops from the Lake: It will in Time become a Place of Consequence. *Muskingum,* though so wide extended in its Branches, spreads all in most excellent Lands, abounding in good Springs and Conveniences, particularly adapted for Settlements remote from Marine Navigation, as Coal, Clay, and Freestone. In 1748 a Coal Mine, opposite Lamenshikola [Sandy Creek, Stark County] Mouth, took Fire, and kept burning above a Twelve-month, where great Quantities are still left. Near the same Place is excellent Whetstone; and about Eight Miles higher up the River is Plenty of white and blue Clay for Glass Works and Pottery. Though the Quantity of good Land on Ohio, and its Branches, is vastly great, and the Conveniencies attending it so likewise; we may esteem that on Muskingum the Flower of it all.
>
> "Hockhocking is passable with Batteaux Seventy or Eighty Miles up; it has fine rich Land, and vast grassy Meadows, high Banks, and seldom overflows. It has Coals about Fifteen Miles up, and some Knowls of Freestone.
>
> "Sioto [Scioto] is a large gentle River, bordered with rich Flats, which it overflows in the Spring; spreading then above Half a Mile in Breadth, though when confined to its Banks it is scarce a Furlong wide." The Latitude of its Mouth 38° 22′. I have marked the Error of its being placed too high in the Map. Muskingum is in Evans's Map placed in its general Run much too far to the West; I have in some Measure corrected it in this Edition. "If it floods early, it scarce retires within its Banks in a Month, or is fordable in a Month or Two more. The land is so level, that in the Freshes of Ohio the Back-water runs Eight Miles up. Opposite the Mouth of this River is the Lower Shawane Town, removed from the other Side, which was One of the most noted Places of English Trade with the Indians. This River, besides vast extents of good Land, is furnished with Salt on an Eastern Branch, and Red Bole on Necunsia Skeintat. The Stream is very gentle, and passable with large Batteaux a great Way up, and with Canoes near 200 Miles to a Portage near the Head, where you carry over good Ground Four Miles to Sanduski. Sanduski is a considerable River, abounding in level rich land, its Stream gentle all the Way to the Mouth, where it will receive considerable Sloops. This River is an important Pass, and the French have secured it as such; the Northern Indians cross the Lake here from Island to Island, land at Sanduski, and go by a direct Path (Great Warriors' Path) to the Lower Shawane Town, and thence to the Gap of Ouasioto (Cumberland Gap), in their Way to the Cuttawas Country. This will, no Doubt, be the Way that the French will take from *Detroit* to *Moville* [Mobile], unless the English will be advised to secure it, now that it is in their Power.
>
> "Little Mineami [Miami] River is too small to be gone far with Canoes. It has much fine Land, and some Salt Springs; its high Banks, and middling Current, prevent its overflowing much the surrounding Land.

"Great Mineami [Miami] River, Assereniet, or Rocky River [French name], has a very stone Channel, a swift Stream, but no Falls. It has several large Branches, passable with Canoes a great Way; one extending Westward towards the Quiaaghtena River [Wabash River]; another [Loramic Creek] towards a Branch of Mineami [Maumee] River (which runs into Lake Erie) to which there is a Portage, and a Third [Great Miami River] has a Portage to the West Branch of Sanduski; besides Mad Creek, where the French have lately established themselves. A Vein of elevated Land, here and there a little stony, which begins in the Northern Part of the Peninsula, between the Lakes Erie, Huron, and Michigan, extends across the Lake Mineami [Maumee] River, below the Fork, and Southward along the Rocky River, to Ohio; and is the Reason of this River's being stony, and the Grounds rising a little higher than the adjacent Plains. It is, like all the Land on this River, very rich, and would scarce have been perceived, had not the River worn the Channel down to the Rocks which lie beneath."[29]

It is easy to see that Evans's work of the Ohio Country was more detailed than was Mitchell's. Between the Muskingum and the Hocking River, Evans located "Wanduxales old T." This village was built by the Delaware chief Windaughalah and, although it is shown opposite the Little Kanawha River, it actually was situated at the mouth of the Little Hocking. Windaughalah abandoned this site in approximately 1738 and moved his village to the east bank of the Scioto just above Lower Shawnee Town. Gist visited this village in 1751 and Mitchell showed it on his map as "Low. Delaware T." Below the mouth of the Hockhocking River, in the Ohio River was located Beatu's Island. This was probably Buffington Island. A little farther down the Ohio was "Le Torts Falls, a Rift." James Le Tort was a famous Shamokin trader from the Susquehanna, who, as early as 1728, was trading with the Delaware and Shawnee along the Ohio. His prominence apparently was recognized, for at one time there was a Le Tort's Creek, Le Tort's Island and Le Tort's Rapids. Today, Le Tort's Creek is probably Old Town Creek in Meigs County. The rift in the Ohio known as Le Tort's Rapids was removed from this river as a navigational hazard years ago. It was described by an army engineer who surveyed the Ohio between Pittsburgh and Louisville in 1835 to determine obstacles to river navigation. He reported:

With regard to the rapids called Le Tart's falls, the river here passes over a rocky bottom, with a descent and current far greater than is found at any other point above the falls [Falls of the Ohio at Louisville]. The bottom is free from dangerous projections, and the concentration of the water at this place, by the more prominent parts of the rocky bed on each side, renders the depth over these rapids much greater than upon most of the shoals in the river; the current, which is here five and three-fourths miles per hour, at a moderate stage, can, in general, be stemmed by steamboats of sufficient power; but for keel boats, and the lighter class of steamers, permanent warping fixtures are here necessary.[30]

Only Letart Island remains from the original sites. Letart Falls in Meigs County, however, as well as Letart, West Virginia, directly across the river, remains as a tribute to this adventurer. Both names are a corruption of Le Tort.[31]

"Kiskiminetas old T.," just below "The big Bent" in the Ohio on Evans's map was known as Kiskiminetas 2 and located on the west side of the Ohio, approximately eight miles above the mouth of the Great Kanawha. The original Kiskiminetas was located north of the Forks of the Ohio along the Allegheny. It is believed that about 1738 the Indians from

Kiskiminetas 2, along with the inhabitants from the Shawnee town shown south of the Ohio, moved to the mouth of the Scioto and there built Lower Shawnee Town.[32] It was this village that Gist visited and described in his journal of 1750–51.

One of the most curious notations made by Evans on his map, one that also appeared on Mitchell's, was "The Cross," or "Half way Cross," located between what Evans called "The Eng. Tawixtior T." on the Great Miami and the "Fr. Tawixtwi T." shown on the Maumee. This was not merely an important crossroads, as may be suspected by the number of trails shown leading through this site; it had a deeper meaning to the French. According to John Patten, a Pennsylvania trader who was familiar with the Ohio Country and who had been captured by the French at Fort Miami in 1750, it was here, at the headwaters of the St. Mary's River, probably in Auglaize County, "a place about 65 miles from the Miamis Town, . . . [that] the French have erected a large Wooden Cross, to be worshipped by their Traders who pass this way."[33]

At Sandusky Bay near the western end of Lake Erie, Evans indicated the location of one fort and three village sites. The exact location of the French fort built there in 1751 caused much speculation among antiquarians until the turn of the nineteenth century, when the journals kept by Lieutenant Joseph Gaspard Chaussegros de Lery, an engineer in the French army, were discovered at Laval University in Quebec, Canada. De Lery, on one trip from Detroit to Presqu' Isle in 1754, and another from Detroit to Fort Duquesne in 1755, visited the Sandusky Bay region and determined the location of these sites. Included with his journal was a very rough diagram of Fort Sandusky, the first post built in Ohio by Europeans.[34] "F. Sandusky," as Evans correctly noted, was on the west side of Sandusky Bay. "I imagined that some vestiges still remained of the Fort the French had built in 1751, and which was afterwards evacuated," De Lery wrote in his journal. "To find it I followed the shore on the north side of the said lake [Lake Otsandoske or Sandusky Bay], which runs East and West. After proceeding about three leagues, I found a clearing, where I landed at noon and discovered the ruins of the old Fort."[35] This post, abandoned in 1753, was located about one mile east of present Port Clinton, and at the south end of a portage trail across Marblehead Peninsula. The "Wiandots" village along the east bank of the small river leading into Lake Erie was one of several small Wyandot or Huron villages located around the bay area. "Junundot," shown on the southeast side of Sandusky Bay was either Ayonontout, a Wyandot village, or the village of Chief Nicolas who rebelled against French influence in 1747 and moved his village to the headwaters of the Muskingum in 1748 and founded Conchake.[36] It is difficult to distinguish which particular village Evans wanted to portray, since his map was not that precise and the word Junundot meant Wyandot Village. The third village, shown at a fork on the Sandusky River south of the bay, was probably Sunnyendeand or Junqueindundeh and located on present Picherel Creek.[37]

As already mentioned, the region around what is today the Cuyahoga River or, as the French called it, the White River Country, was an important center of trade for both the French and the British. Again, Evans's map was much more detailed than Mitchell's, and because Pownall did not revise this particular section, the American map maker portrayed the situation as it was on the eve of the French and Indian war. The most important site on the Cuyahoga was the "French House," a small trading post established there in 1742 by Sieur du Saguin, a French trader from Detroit, who held much influence with the Indians in the White River region. The Cuyahoga River, in fact, was known to the French as Saguin River, named for the Detroit trader while the English used the Indian name, Cayahoga, for the same river. When that region was surveyed by members of the Connecticut Land Company near the end of the eighteenth century, they mistakenly believed that a separate Saguin River existed. Evans properly called the first river east of the Cuyahoga the Elk or, as it was known to the French, Rivière La Biche. Cartographers for a while believed that there was an extra river between the Cuyahoga and the Elk and called it Saguin or the Chagrin. As time passed and more was learned of that region, the name Elk was lost and Chagrin became the common name for the river.[38] The "Tawas" Evans located on the east side of the Cuyahoga was the Ottawa village just north of Tinker's Creek in what is now Independence Township, Cuyahoga County, where David Zeisberger and John Heckewelder settled their Moravian Indians in 1786.[39] The "Mingoes T." south of the "French House" seemed to refer to one of several Seneca or Mingo villages that were situated along this important waterway.

Two other Indian towns, Mohicon John's Town and White Woman's Town, appeared on Evans's map. "Mohicons," which the map maker located on the east side of a branch of the Muskingum north from Muskingum Town, referred to Mohicon John's Town, which was probably located near the forks of Jerome and Muddy Creeks in Mohicon Township in Ashland County.[40] In 1761, Major Robert Rogers noted in his journal that he had "crossed Maskongam Creek [Jerome Creek], running south, about twenty yards wide. There is an Indian Town about twenty yards from the Creek, on the east side, which is called Mingo Cabbins."[41] Several years later Thomas Hutchins, an army engineer and surveyor, said that this village "consisted only of a few houses."[42] White Woman's Town was located nearly opposite Killbuck Creek in Coshocton County. The village was established sometime before 1750 and was named for Mary Harris, a woman who, while not yet ten years old, was taken from her home in Deerfield, Massachusetts, on February 29, 1704, by some French and Indians from Canada. Gist, who visited here in 1751, recorded that on January 15, he left the Muskingum River and went "to the White Woman's Creek, on which is a small Town; this White Woman was taken away from New England, when she was not above ten Years old, by the French Indians; She is now upwards of fifty, and has an Indian Husband and several Children—Her name is Mary Harris; she still remembers they used to be very religious in New England, and wonders how the White Men can be so wicked as she has seen them in the Woods."[43] In 1756 she was living in Cahnawaga, an Indian Village near Montreal, and one of her sons was reportedly an officer in the French service.[44]

On November 1, 1778, two years after Pownall published his edition of the Evans map, Thomas Hutchins, an officer in the British service between 1757 and 1779, produced his "New Map of the Western Parts of Virginia, Pennsylvania, Maryland and North Carolina; Comprehending the River Ohio, and all the Rivers, which fall into it; Part of the Mississippi, the Whole of the Illinois River, Lake Erie, Parts of the Lakes Huron, Michigan, &c. And all the Country bordering on these Lakes and Rivers" (Plate V), along with *A Topographical Description of Virginia, Pennsylvania,*

Maryland, and North Carolina, Comprehending the River Ohio, Kenhawa, Sioto, Cherokee, Wabash, Illinois, Mississippi, &C. The Climate, Soil and Produce, whether Animal, Vegetable, or Mineral; The Mountains, Creeks, Roads, Distances, Latitudes, &c. and of every Part, laid down in the annexed Map. Hutchins's work was printed in London and was sold in the United States by Robert Aitkin of Philadelphia for one guinea. The map and *Topographical Description* was reprinted several times, both in England and the United States. A French edition appeared in 1781.[45]

Hutchins's map of the western country is important to the history of the mapping of Ohio because it was one of the first produced by a map maker who had personally visited and surveyed parts of the country himself, and because it served as a guide to so many hardy pioneers who first ventured west of the mountains.[46] The map maker had been north and west of the Ohio River at least five times between 1760 and 1768 and probably many more times not recorded. His first visit was an official trip for the British from Fort Pitt to Presqu' Isle in 1760;[47] the second was his tour of the Indian villages for the English department of Indian Affairs in 1762;[48] the third was with Colonel Henry Bouquet's expedition to the Muskingum in 1764;[49] the fourth was along with George Croghan in Captain Harry Gordon's surveying expedition of the Ohio River in 1766;[50] and his fifth was when he was assigned to the garrison at Fort Charta in the Illinois Country in 1768.[51] To produce this map, Hutchins relied on Lewis Evans's "A General Map of the Middle British Colonies in North America" and his *Analysis,* along with the materials provided by Croghan in his journals. A great amount of his map work dealing with the western country, however, rested on his own observation and surveys. As he himself explained:

> Those parts of the country lying westward of the Allegheny mountains, and upon the rivers Ohio and Mississippi, and upon most of the other rivers; and the lakes (laid down in my Map) were done from my own Surveys, and corrected by my own Observations of latitudes, made at different periods preceding, and during all the campaigns of the last war (in several of which I acted as an Engineer) [French and Indian War] and since in many reconnoitring tours, which I made through various parts of the country, between the years 1764 and 1775.
>
> It is fit also, that I should take notice, that in the account which I have given of several of the branches of the Ohio, and Alleghany rivers I have adopted the words of the late ingenious Mr. Lewis Evans, as I found he had properly described them in the Analysis to his Map of the Middle Colonies.[52]

Hutchins seemed to have detailed more around the Forks of the Muskingum than either Mitchell or Evans, possibly because he was more familiar with that area than with any other. Not only had he visited that area in 1764 with the Bouquet expedition, but he was author of a manuscript, "A Description of part of the Country Westward of the River Ohio, with the Distances Computed from Fort Pitt to the several Indian Towns by Land & Water,"[53] a description of the various Indian and trader's trails and villages located in the central and eastern part of Ohio. He also wrote "The Rout from Fort Pitt to Sandusky, and thence to Detroit" in 1764 which described the Indian trail and villages from the famous Forks of the Ohio, through northern Ohio to Sandusky Bay.[54] Certainly no man knew that region along the Muskingum or, in fact, the Ohio Country, better than did Hutchins.

At the famous "crossing place" on the Tuscarawas stood Beaver's Town, located at the present site of Bolivar in Tuscarawas County. This

(Overleaf) Plate V: Hutchins, "New Map of the Western Parts," (1778), in Hutchins, *Topographical Description* (1778). Hutchins's map was drawn mainly from first-hand observations and personal explorations.

Portage 15 Miles
A Number of Small Lakes
Here are many Small Lakes
Miami River
R. Portage
Portage 1¼ Mile
Sandusky Lake
Sandusky
Junqueindunde
Wyandot Town
River
Sandusky River
Miami Village
Miami Fort
Riviere a Boite
R. a Lang Lois
Ecor a Amelins
From the Miami Fort to the Lower Shawanoe Town the Country is Level & rich abounding in Streams of Water, and fine Timber; and in many Places the Road directs its Course through extensive Meadows.
Ball Eagle Creek
Ottawas
Mohickon John's
Road from the lower Shawanoe Town to Sandusky
The English Tawixtwi or Picque Town taken in 1752 by the French.
Rich Level Land
Mad River
Great Miami or Rocky River
Rapid and Stony but no Falls
Lick Town
Waukat
A Spring
Lower Shawanoe T.
Indian
Lead Mine
Miami River
Little Miami
Sioto River
Hurricane Toms
Salt Springs
Creek 10 Yards Wide
Creek 8 Yards Wide
Crooked Creek
Big Bones
Creek 10 Yards Wide
Elk River 30 Yards Wide
Creek 15 Yards Wide
Wanduxales
Old Lower Shawanoe T.
Creek 15 Yards Wide
River 30 Yards Wide
Great Salt Spring
The Path to the

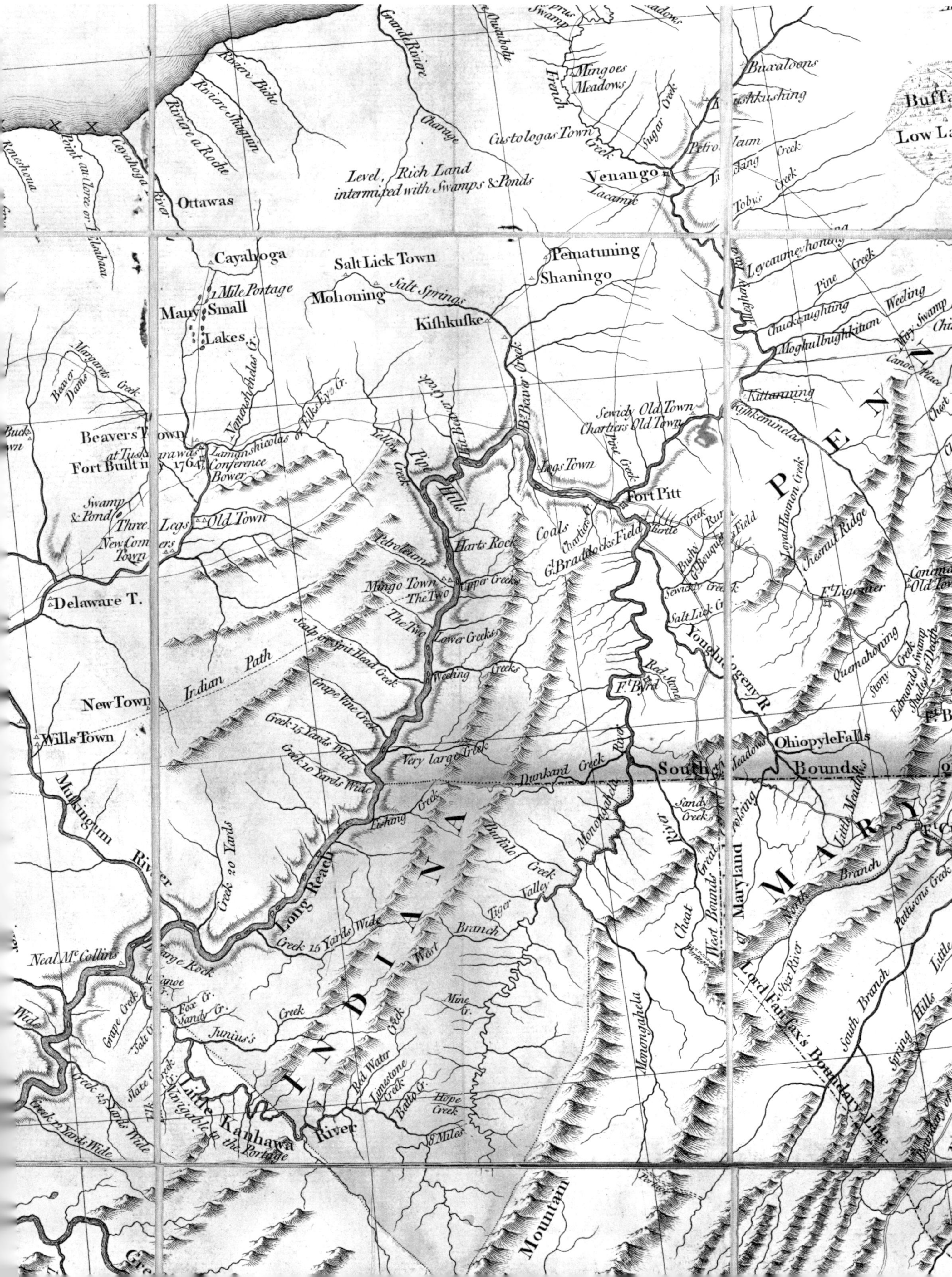
Grand Riviere
Mingoes Meadows
French
Buxaloons
Kuskuskushing
Buffal
Low La
Riviere Biche
Riviere Shagrun
Riviere a Roche
Cayahoga River
Ottawas
Change
Custologas Town
Sugar Creek
Level, Rich Land intermixed with Swamps & Ponds
Venango
Tobys Creek
Cayahoga
Salt Lick Town
Pematuning
Shaningo
1 Mile Portage
Many Small Lakes
Mohoning
Salt Springs
Kishkuske
Pine Creek
Weeling
Beaver Dams
Margarets Creek
Beavers Town
Fort Built in 1764
Conference Bower
Kittanning
Sewicly Old Town
Chartiers Old Town
Legs Town
Fort Pitt
Coals
Three Legs Old Town
New Comers Town
Delaware T.
Harts Rock
Mingo Town
The Two Upper Creeks
The Two Lower Creeks
Indian Path
Weeling Creeks
New Town
Wills Town
Grape Vine Creek
Creek 15 Yards Wide
Creek 10 Yards Wide
Very large Creek
Muskingum River
Dunkard Creek
Monongahela River
Fishing Creek
Buffalo Creek
Creek 20 Yards
Long Reach
INDIANA
Tiger Valley
West Branch
Creek 15 Yards Wide
Neal McCollins
Large Rock
Fox Cr.
Sandy Cr.
Juniuss
Mine Cr.
Red Water Creek
Limestone Creek
Hope Creek
Little Kanhawa River
Navigable to the Portage
8 Miles
Creek 25 Yards Wide
Creek 12 Yards Wide
Gt. Braddocks Field
Turtle Creek
Chestnut Ridge
Loyal Hannon Creek
Ft. Ligonier
Salt Lick Cr.
Youghiogeny R
Quemahoning
Red Stone
Ft. Byrd
Ohiopyle Falls
South Bounds
Sandy Creek
Cheat River
Maryland
MARY
PENN
West Bounds of Maryland
Lord Fairfax's Boundary Line
Styx River
South Branch
North Branch
Spring Hills
Monongahela
Mountain

village, named for Chief King Beaver, the leading chief of the Delaware, appeared on Mitchell's map as "Tuscaroraas," and was also known as Delaware Town. Major Robert Rogers was there for nearly four days in January 1761, and in his journal reported: "This Indian town stands on good land, on the west side of the Mashongam [Tuscarawas] River, and opposite to the town, on the east side, is a fine river [Big Sandy Creek] which discharges itself into it. The latter is about thirty yards wide, and the Maskongam about forty; so that when they both join they make a very fine stream, with a swift current running to the southwest. There are about 3,000 acres of cleared ground round this place. The number of warriors in this town is about 180. All the way from the Lake Sandusky [Sandusky Bay] I found level land and a good country."[55]

The fort that Hutchins depicted below the mouth of the Big Sandy, below Beaver's Town on the west bank of the Tuscarawas River, was built by the men under Colonel Henry Bouquet's command on October 15, 1764, during his expedition from Fort Pitt to the Forks of the Muskingum to overawe the Ohio Indians following the Pontiac uprising. After initial contact with the Indians in the upper Muskingum Valley was made and they were requested to meet with Bouquet, it was recorded in the official account of the expedition that the Colonel "ordered a small stockade fort to be built to deposite provisions for the use of the troops on their return; and to lighten the convoy."[56] Hutchins, in his description of the trail from Pitt to Sandusky stated that "about two miles below Tuscarawas [Beaver's Town], on a very high bank, with Muskingum at the foot of it, which is about 100 yards wide at this place, with a fine, level country at some distance from its banks, producing stately timber, free from underwood, and plenty of food for cattle, was erected a small stockaded fort, to deposit provisions for the use of the troops on their return, and to lighten the convoy."[57] The "Conference Bower" Hutchins placed along the west bank of the Tuscarawas "at some distance from the camp [fort]." This bower apparently was either a leafy shelter or one built from hides or cloth, possibly as illustrated in the cartouche in the lower right corner of Hutchins's map of the Bouquet expedition. On October 17, Hutchins wrote in the historic account of that expedition: "The Colonel with most of the regular troops, Virginia volunteers and light horse; marched from the camp to the bower erected for the congress. And soon after the troops were stationed, so as to appear to the best advantage, the Indians arrived, and were conducted to the bower. Being seated, they began, in a short time, to smoak their pipe. . . ."[58]

The "Lamanschicolas or Elks Eye Cr." which emptied into the Tuscarawas was mentioned by Evans in his *Analysis* and identified as the Big Sandy Creek. "Memenshehelas Cr." which emptied into Lamanshicolas is the present Nimishillen Creek that flows into the Big Sandy at Sandyville in Sandy Township, Tuscarawas County. Hutchins mentioned that where the trail to Sandusky crossed this creek it was "about 50 feet wide at the ford, which is a little above where it empties itself into the aforesaid branch [Big Sandy] of Muskingum."[59] Directly to the north from Beaver's Town, Hutchins believed that the source of the Tuscarawas was located among "Many Small Lakes" and explained that "the country is fertile, interspersed with swamps and many small Lakes."[60] Beyond this point northward to the Cuyahoga River and the Indian village, Cayahoga, was the "1 Mile Portage." Hutchins stated that between the Tuscarawas River "and a little Lake [Summit Lake] communicating with Cayahoga river, there is a level carrying place a Mile and a half over. Just below where the stream of this Lake joins the Cayahoga, there is a fall, which occasions a portage of about 25 yards, after passing of which there is good navigation for Canoes to where this river discharges itself into Lake Erie. . . ."[61] Cayahoga, also known as Canayiahagen, Canayahaga, Kichaga, Cayahawge, Gichaga, was located on the west side of the Cuyahoga River and situated at the north end of the portage from the Tuscarawas. "This town," Hutchins wrote, "is computed to be between 35 and 40 miles from Tuscarawas [Beaver's Town.]" This village probably is what Henry Howe, in his *Historical Collections of Ohio,* called Old Portage, which he located three miles north of the present city of Akron.[62]

Eastward from the Cuyahoga River lay the Mohoning River, an important branch of Beaver Creek. On the west side of the Mohoning was situated Salt Lick Town near what Mitchell noted on his map as the "Salt Ponds" and Hutchins called "Salt Springs." Salt Lick Town was established by the Delaware about 1758 and was located approximately one and a half miles south of the river and nearly one mile west of the present city of Niles. In 1758, John McCullough, who was held captive by the Delaware from 1756 until his release in 1764, lived in Salt Lick Town. He explained in his *Narrative* that in 1758 he moved from Beaver "to where they were settling a new town, called Kseek-he-vong, that is, a Place of Salt, a place now well known by the name Salt Licks, on the West Branch [Mohoning] of Beaver; where we lived about one year."[63] Down the river from Salt Lick Town was Mohoning Town, which was probably situated above the present site of Newton Falls in Trumbull County. McCullough was taken here also, and he wrote that in 1759 "we moved to a town about fifteen miles off [from Salt Lick Town], called Mohoning, which signifies, a lick."[64] It should be noted that Hutchins mistakenly placed "Rivière a Roche," or Rocky River, on the east side of the Cuyahoga rather than the west, and that he illustrated both the "Rivière Shaguin" or Chagrin and "Rivière Biche" as two separate streams when, in fact, those two names were applied to the same stream.

To the northwest from Beaver's Town, Hutchins noted some "Beaver Dams" along "Margaret's Creek." Margaret's Creek was the present Sugar Creek, which empties into the Tuscarawas at Dover in Tuscarawas County, and the beaver dams were probably along Apple Creek in either East Union Township or Salt Lick Township, both in Wayne County. Hutchins explained that these dams were located "Where there is a brook, the streams of which, when stopped by the beavers, occasioned a pond on the left; the remains of the Dam is still [1764] visible."[65] Farther to the northwest, along White Woman's Creek, or what is the Mohican River, Hutchins located an old Ottawa fort. "The remains of a Fort built by the Ottawas, at which there is a pleasant running brook, is ten miles from Mohickon John's Town,"[66] he wrote. This site was undoubtedly the same place noted by Lieutenant Joseph Gaspard Chaussegros de Lery who passed through this region along the historic Conchake Path, which ran from Sandusky Bay to Conchake, in 1755 on his trip from Detroit to Fort Duquesne. "We passed a place on the bank of the said river [Mohican] where some Hurons had taken refuge after the treacherous deed they committed at Detroit," de Lery wrote. "It is called the Fugitives' Camp."[67] At Sandusky Bay, Hutchins positioned a fort called Sandusky on the southern shore at the eastern end of the bay. Here, the "British built a Block-house, surrounded with a Stockade,"[68] Hutchins wrote. This fort was built in 1760 but was captured and destroyed in 1763 during Pontiac's uprising.

Below Beaver's Town, along the east bank of the Tuscarawas just south of the mouth of Stillwater Creek, was Three Legs Old Town. This site was abandoned by 1764 and little of its history, other than its location, is known. New Comers Town, or Gekelemukpechnuk, below Three Legs Old Town, was built sometime before 1762, by the Delaware chief, Nettawatwees or Nettaulwaleman, and was located on the site of the present town of Newcomerstown in Coshocton County. In 1772, David McClure, a Congregational minister, toured the Indian Villages along the upper Muskingum Valley. He visited New Comers Town and recorded his experiences in a diary published in 1899. "This town is called New Comers Town by the English and stands on the west bank of the Muskingum [Tuscarawas], containing about sixty houses, some of logs, and others of bark of trees, fastened by elm bark to poles stuck in the ground and bent over at the top," McClure recorded. "There are nearly 100 families," he continued, and "it is the principal Town of the Delaware Nation, and the residence of the king and the greater part of the Councillors. . . . Eight or ten acres around the Town are cleared. On the opposite side of the River is a large corn-field, in rich low ground. It is inclosed within one common fence, and each family has its divisions to plant. Some of the houses are well built with hewed logs, with stone chimneys, chambers and cellars."[69] New Comers Town was abandoned in 1775, three years after McClure was there, and the Delaware established a new village at the Forks of the Muskingum. They called the village Goshachgunk, the Delaware name for Conchake, the same Wyandot town that Mitchell noted as Muskingum or "T. of Owendoes" on the north side of the Tuscarawas, just above the Forks. The town of Coshocton is now situated on this old Delaware village.

The "Delaware T.," a little farther down the Muskingum, was approximately three miles below the Forks and situated on the east bank. Hutchins reported that "Mowheysinck Town [Delaware Town] is Situate about 100 yards from the River [Muskingum], on this side, and has upwards of 35 Persons in it, 15 of which are Warriors. Their Cornfields are better than half A mile below the Town, close on the Opposite side of the River."[70] This village was also known as Bullet's Town as shown on Hutchins's map of the Bouquet expedition in 1764. Across the river, on the west bank, was Waukautaumeka. Hutchins explained that this village was "Situate on A very high Bank, about 100 yards from the Muskingum River. There is About 40 Houses in this Town and near the same Number of warriors, and upward of 90 women and Children. Their Cornfields are near 3/4 of a Mile from the Town, on the same side, up the [Wakatomika] Creek."[71] Waukautaumeka was near the present site of Dresden, Cass Township, Muskingum County. Farther down the Muskingum, at the mouth of the Licking River, was located a village which Hutchins did not label on his map. In his description of the Ohio Country, however, he remarked that the path from Waukautaumeka led "to a small Delaware Town, the path to which in some places touches near A few Ridges which are not Steep or Difficult. This Town is Situated in A fork where A Creek [Licking River] Empties into the Muskingum, and has 8 Houses in it, and about 12 Warriors and 30 women & Children. Their Cornfields are close to the Town. Muskingum & The Creek that Runs into it are both Fordable here."[72] This village site was located near the city of Zanesville. Below the

Licking River, on the east side of the Muskingum was "Will's Town," situated north of the mouth of Salt Creek, near the present site of Duncan Falls, Salt Creek Township, Muskingum County. "The path still Continues through level Land, free from underwood, 3 Miles, to Will's Town, Se-Key-unck, or the Salt Licks, 50 Miles from the Ohio," Hutchins noted. "Will's Town has 35 Houses in it and About 45 Warriors, & 80 Women & Children. The Houses are close together, and their Cornfields in sight of the Town. The Muskingum is not Fordable opposite the Town."[73]

To the east of Lower Shawanee Town, Hutchins located a lead mine, and in his *Topographical Description* he commented that "Salt Springs, as well as Iron Ore, and rich Lead Mines are found upon the River Ohio. One of the latter," he said, "is opened on a branch [Kinnikinnick] of the Sioto River, and there, the Indian natives supply themselves with a considerable part of the lead, which they use in their wars, and hunting."[74] This mine was never identified after white settlement of the region and may have been either a rumor or some rock formation containing a limited supply of lead that had been pushed southward by the glacier and was quickly exhausted by the Indians. Almost at the head of the Scioto on the map, he indicated "Lick Town." This was a small Indian village known as Big Salt Lick Town and was situated near Columbus. Along the north bank of the Ohio, half way between the Hocking and the Muskingum rivers, Hutchins located "Neal McCollins [McCollen]". He explained that "15 Miles from the Canoe place [Roxbury in Morgan County] is A Shelving Rock, under which Neal McCollin, a Delaware Indian, Built a Cabbin."[75]

The discussion in this chapter about the maps of North America that contained details of the Ohio Country mentioned but a few from among a great body of cartographic material made between 1632 and 1778. The works of European cartographers such as Louis de Hennepin, Jean Baptiste Louis Franquelin, Herman Moll, John Senex and others, although not directly important to the mapping of Ohio, cannot be ignored in any comprehensive study of the delineation of the United States. The maps used in this study plotted French and English exploration and documented the growth of geographic discovery and knowledge as it related to the Ohio Valley during more than a century. Each piece by Sanson, Bellin, Mitchell, Evans, Pownall and Hutchins represented an important development in the preliminary drawing of the Ohio Country and the subsequent mapping of the state itself. By the eve of the Seven Years War, a certain body of geographic knowledge of the Ohio Valley was a familiar part of most of the maps of North America. Despite the amount of activity west of the mountains by both the French and the English prior to 1754, however, it was not until the years between the Seven Years War and the American Revolution that a general sketch of Ohio was compiled. The task of mapping the western frontier was left to American surveyors and cartographers, and it was not until the land was settled, the Indian threat removed, and a system of efficient internal transportation evolved that full and accurate maps of Ohio were produced.

1. One of the earliest writers to pay attention to cartographic evidence as a historical resource was Francis Parkman, who described a collection of French maps in the *Depot des Cartes de la Marine et des Colonies* in Appendix I, "Early Unpublished Maps of the Mississippi and the Great Lakes," in his *LaSalle and the Discovery of the Great West,* first published in 1869. Probably the most important work written in the nineteenth century about the cartography of the Ohio Valley during the exploration period was by Justin Winsor in Volume IV of his *Narrative and Critical History of America,* which appeared in 1887. Winsor built on the work done by Parkman, and many early maps, traced by the author himself, were reproduced in this volume along with lengthy critical analyses of each piece. A more recent monograph, Louis Karpinski's *Bibliography of the Printed Maps of Michigan: 1804–1880,* which was published in 1931, is indispensable to the overall story of the mapping of the Great Lakes region. Although most of this volume dealt with the maps of Michigan, he ranged much wider than that in his informative survey in Chapter III, "Significant Maps of the Great Lakes Area." In 1959 the University of Pittsburgh Press published *Early Maps of the Ohio Valley* by Lloyd Arnold Brown. This volume is a collection of fifty-four manuscript and printed maps that appeared between 1673 and 1783, and even though the author did not directly concern himself with the historical relationship of one map to another, he did place each map in its historical perspective.

2. Albert T. Volwiler, *George Croghan and the Westward Movement* (Cleveland, 1926), p. 65, fn. 109.

3. Charlevoix, *Histoire et Description Generale de la Nouvelle France avec le Journal historique d'un Voyage fait par ordre du Roi dans l'Amérique Septentrionnale* III (Paris, 1744), 253.

4. Charlevoix, *Histoire et Description Generale,* III, 255.

5. Charles A. Hanna, *The Wilderness Trail or the Ventures and Adventures of the Pennsylvania Traders on the Allegheny Path* (New York, 1911), II, 126–27.

6. Charlevoix, *Histoire et Description Generale,* II, 240.

7. Justin Winsor, *The Mississippi Basin: The Struggle in America between England and France,* 1697–1763 (Boston, 1895), p. 306. This does not mean that the British were totally ignorant of the Ohio Country. Individuals, such as George Croghan and Christopher Gist, were in Ohio as early as 1743 and were familiar with the country; however, their knowledge of the area, prior to 1755, at least, was not recorded on any map that was put into general use. See Nicholas B. Wainwright, *George Croghan, Wilderness Diplomat* (Chapel Hill: Institute of Early American History and Culture at Williamsburg, Univ. of North Carolina Press, 1959), pp. 3–68.

8. Letter from Thomas Penn to Richard Peters, February 21, 1755, quoted in Lawrence Henry Gipson, *Lewis Evans* (Philadelphia, 1939), p. 61, fn. 14.

9. Gipson, *Evans,* pp. 60–61; see Reuben Gold Thwaites, ed., *Collections* of the State Historical Society of Wisconsin, XVIII (Madison, 1908). pp. 113–14.

10. For a detailed discussion of the various editions of the Mitchell map, see Hunter Miller, ed., *Treaties and Other International Acts of the United States of America* (Washington, D.C., 1933), pp. 328–51.

11. William M. Darlington, *Christopher Gist's Journals* (Cleveland, 1893), p. 36.

12. *Ibid.,* p. 37.

13. See Chapter VIII, "Andrew Montour, 'The Half Indian'," in Hanna, *The Wilderness Trail,* I, 223–46.

14. In 1965, archaeologists Olaf H. Prufer and Ellen Andors began excavation of what they called the Morrison Village Site, a village identified as belonging to the Madisonville Phase of the Fort Ancient Tradition, a late prehistoric Indian group. This village was the same one that Gist identified in 1751 as the Delaware's Hurricane Tom's Town. While he called it Delaware, Hurricane Tom's Town was actually a mixed village, with both Delaware and Shawnese living together. Modern archaeological research has uncovered sufficient evidence to indicate that the Fort Ancient people were the Shawnese in the Ohio Valley and that Hurricane Tom's Town is a site that provided the bridge between the prehistoric and historic. See Olaf H. Prufer and Douglas N. McKenzie, eds., *Studies in Ohio Archaeology,* (Cleveland, 1967), pp. 216–17; and Olaf H. Prufer and Orrin C. Shane, III, *Blain*

Village and the Fort Ancient Tradition in Ohio, (Kent, Ohio: Kent State University Press, 1970), pp. 263–65.

15. Darlington, *Gist's Journals,* p. 43.

16. Paul Woehrmann, *At the Headwaters of the Maumee,* Indiana Historical Society Publications (Indianapolis, 1971), Vol. 24, 2. Also see Bert Anson, *The Miami Indians* (Norman: University of Oklahoma Press, Oklahoma, 1970), pp. 34–35. Discussed also in Erminie Voglin, "Report of Aboriginal Use and Occupancy of Royce areas 87 and 88," The Great Lakes-Ohio Valley Research Project of Indiana University, m.d., [1956], p. 8, unpublished ms., copy in possession of author.

17. Hanna, *The Wilderness Trail,* I, 332–33.

18. Volwiler, *George Croghan,* p. 35. See also, Nicholas B. Wainwright, *George Croghan, Wilderness Diplomat* (Chapel Hill, 1959), pp. 6–15.

19. Charles Whittlesey, *Early History of Cleveland, Ohio* (Cleveland, 1867), p. 159.

20. Darlington, *Gist's Journals,* pp. 42–43.

21. Clarence Walworth Alvord and Clarence Edwin Carter, eds., *The New Regime: 1765–1767, British Series,* II, *Collections* of the Illinois State Historical Society, Vol. XI (Springfield, Illinois, 1916), p. 129.

22. Gordon's journal was reprinted in Alvord and Carter, *The New Regime: 1765–1767,* pp. 290–311; see also Beverley W. Bond, Jr., ed., *The Courses of the Ohio River* (Cincinnati, 1942).

23. Gipson, *Lewis Evans,* p. 155.

24. Thomas Pownall, *A Topographical Description of Such Parts of North America As Are Contained in the (Annexed) Map of the Middle Colonies in North America* (London, 1776), p. 124. For comments about the map, see Charles A. W. Pownall, *Thomas Pownall* (London: H. Stevens, Son & Stiles 1908), pp. 278–82.

25. Gipson, *Lewis Evans,* p. 154; Thomas Pownall, *A Topographical Description,* p. 126.

26. Thomas Pownall, *A Topographic Description,* p. 126; Volwiler, *George Croghan,* p. 67; Gipson, *Lewis Evans,* p. 57.

27. Thomas Pownall, *A Topographical Description.* p. 125; Gipson, *Lewis Evans,* p. 154.

28. In this excerpt, Evans's observations are enclosed within quotation marks while Pownall's remarks are not.

29. Thomas Pownall, *A Topographical Description,* pp. 143–46 passim; Gipson, *Lewis Evans,* pp. 172–74 passim.

30. "Report from the Engineer Department to the Secretary of War," Document #1, *Senate Documents,* 24th Congress, 1st Session (1835), Series #279, p. 184. A map, "Letart's Islands and Rapids," was included. Fortescue Cuming, in the journal of his tour through Ohio in 1808, recorded, "Eight miles below Old Town Creek we were carried through Le Tart's falls at the rate of six knots an hour, but the rapids, which it ought to be called more properly than falls, is not more than half a mile long." Fortescue Cuming, *Sketches of a Tour to the Western Country,* Vol. IV of *Early Western Travels: 1748–1846,* ed. Reuben Gold Thwaites (Cleveland, 1904), 139.

31. Hanna, *The Wilderness Trail,* I, 168; II, 142.

32. Hanna, *The Wilderness Trail,* I, 168, 267; II, 142.

33. Thwaites, *Collections* of the State Historical Society of Wisconsin, XVIII, 113.

34. Lucy E. Keller, "Old Fort Sanduski of 1745 and the 'Sandusky Country'," *Ohio Archaeological and Historical Society Publication,* XVII (1908), 357–430.

35. Hanna, *The Wilderness Trail,* II, 168–76.

36. Hanna, *The Wilderness Trail,* I, 320; II, 170–74. Portions of Major Robert Rogers's journals kept during his trip to Lake Erie in 1760 are reprinted in Hanna. Rogers mentioned crossing Sandusky Bay and coming to "A town of the Windot Indians, where we halted to refres." *Ibid.,* p. 184. [William Smith], *Historical Account of Bouquet's Expedition Against the Ohio Indians in 1764* (Cincinnati, 1868), p. 150.

37. Voglin, "Report . . .," pp. 16–18.

38. Hanna, *The Wilderness Trail,* I, 335–36; Whittlesey, *Early History of Cleveland,* p. 213.

39. Whittlesey, *Early History of Cleveland,* p. 137; Lucius V. Bierce, *Historical Reminiscences of Summit County* (Akron: T. & H. G. Canfield, 1854), p. 120; Hanna, *The Wilderness Trail,* I, 333–34.

40. Hanna, *The Wilderness Trail,* II, 187.

41. *Ibid.,* pp. 183–85.

42. *Ibid.,* p. 206.

43. Darlington, *Gist's Journals,* p. 41.

44. *Ibid.,* pp. 114–15; Stephen W. Williams, *Memoir of the Revolution: John Williams* (1853), p. 121.

45. See Frederick Charles Hicks, *Thomas Hutchins: A Topographical Description of Virginia, Pennsylvania, Maryland and North Carolina* (New York, 1907). For a list of Hutchins's western maps, see pp. 55–58.

46. *Ibid.,* p. 56. David Summers, student at Ohio University from Blue Licks, Kentucky, found a Hutchins map of the "Western Parts" among the possessions of one of his ancestors who used the map on his westward migration from Albermarle County, Virginia, to Kentucky. Rufus Putnam used Hutchins's famous map, among others, to plan the western settlements of the Ohio Company and it was referred to by him in his letter to George Washington of June 16, 1783, when Putnam sought the General's support for the Newburgh Petition. Archer B. Hulbert, ed., *Ohio in the Time of the Confederation,* Vol. III of the *Ohio Company Series* in the *Marietta College Historical Collections* (Marietta, Ohio, 1918), 59–60.

47. "A Journal of a march from Fort Pitt to Venango—and from thence to Presqu' Ile, 1760," *Pennsylvania Magazine of History,* II, 149–53.

48. For a reprint of the journal kept during this trip, see Hanna, *The Wildernerness Trail,* II, 362–67; see also "A Manuscript Map of a Tour from Fort Cumberland to Fort Pitt, 1762, with a Description of the County and the names of the different Indian Tribes," Henry E. Huntington Library.

49. [William Smith], *Historical Account of Bouquet's Expedition Against the Ohio Indians in 1764,* (Cincinnati, 1868).

50. Bond, *The Courses of the Ohio River,* "Extracts from the Journal of Captain Harry Gordon," in Pownall, *A Topographical Description,* pp. 153–66.

51. "Journal from Fort Pitt to the Mouth of the Ohio In the Year 1768," Indiana Historical Society *Publication,* II, 417–21. See list of Hutchins's publications in Hicks, *Thomas Hutchins,* pp. 53–66.

52. Hicks, *Thomas Hutchins,* pp. 71-72.

53. This manuscript is in Vol. I of the Thomas Hutchins Mss., Pennsylvania Historical Society, Philadelphia, Pennsylvania. It is also partially reprinted in Hanna, *The Wilderness Trail,* II, 192–202.

54. This is included among the Thomas Hutchins's manuscripts in the Pennsylvania Historical Society's collection and was reprinted by Hanna, *The Wilderness Trail,* II, 202–08. Part of this record was reprinted in [Smith], *Bouquet's Expedition.*

55. Hanna, *The Wilderness Trail,* II, 185. For many years the Tuscarawas River was called the Muskingum.

56. *Bouquet Expedition,* pp. 51–52.

57. Hutchins, "Pitt to Sandusky," Hutchins Mss., Pennsylvania Historical Society; Hanna, *The Wilderness Trail,* II, 204. According to Smith in his *Bouquet's Expedition*, p. 60, this was not the only stockade built during Bouquet's thrust to the Muskingum. A second was built on October 25, 1764, near the Forks of the Muskingum.

58. [Smith], *Bouquet's Expedition,* p. 51.

59. Hutchins, "Pittsburgh to Sandusky," Hutchins Mss., Pennsylvania Historical Society; Hanna, *The Wilderness Trail,* II, 203–04.

60. Hutchins, "Pittsburgh to Sandusky," Hutchins Mss. Pennsylvania Historical Society; Hanna, *The Wilderness Trail,* II, 205.

61. Hanna, *The Wilderness Trail,* II, 205.

62. Henry Howe, *Historical Collections of Ohio,* II (Cincinnati: C. J. Krehbiel & Co. 1900), p. 626.

65. Hutchins, "Pittsburgh to Sandusky," Hutchins Mss., Pennsylvania Historical Society; Hanna, *The Wilderness Trail,* II, 205.

66. Hutchins, "Pittsburgh to Sandusky," Hutchins Mss., Pennsylvania Historical Society; Hanna, *The Wilderness Trail,* II, 206.

67. De Lery's *Journal,* quoted in Hanna, *The Wilderness Trail,* II, 206. In 1747 some Hurons and other Indians under Chief Nicolas, who had built a fort at Sandusky Bay, killed several French traders at Detroit. After this incident, Nicolas and his followers moved from Sandusky and apparently established here for a short time a small fortification for fear of French retaliation.

68. Hutchins, "Pittsburgh to Sandusky," Hutchins Mss., Pennsylvania Historical Society; Hanna, *The Wilderness Trail,* II, 206.

69. David McClure, *Diary of David McClure, Doctor of Divinity, 1748–1820* (New York, 1899), p. 61; Hanna, *The Wilderness Trail,* II, 310–11.

63. John McCullough's narrative is cited in Hanna, *The Wilderness Trail,* I, 351; also see Hutchins, "A Description of part of the Country Westward of the River Ohio," Hutchins Mss., Pennsylvania Historical Society.

64. Hanna, *The Wilderness Trail,* I, 350.

70. Hutchins, "A Description"; Hanna, *The Wilderness Trail,* II, 189.

71. Hutchins, "A Description"; Hanna, *The Wilderness Trail,* II, 194.

72. Hutchins, "A Description"; Hanna, *The Wilderness Trail,* II, 195.

73. Hutchins, "A Description"; Hanna, *The Wilderness Trail,* II, 195.

74. Hicks, *Thomas Hutchins,* p. 32.

75. Hutchins, "A Description"; Hanna, *The Wilderness Trail,* II, 198.

CHAPTER TWO

Mapping the Prehistoric Earthworks

The tumuli, fortifications, effigies and earthworks that appear in every county in Ohio and throughout the Ohio Valley have fascinated people and excited their imaginations since the first explorers set foot north of the Ohio River and looked upon the ancient monuments. These works of prehistoric man have been the subject of descriptive and speculative literature, learned treatises and archaeological investigation since 1772, when David Zeisberger, the Moravian missionary, first observed the mounds in the Tuscarawas Valley.[1] Knowledge about the architects of these great prehistoric monuments even today is limited and somewhat speculative; but modern research techniques and investigations have provided a fuller understanding of the daily social, economic and religious life of these once mysterious prehistoric people. Since these works generously dot the Ohio countryside, most numerously in the Miami, Scioto, Hocking and Muskingum river valleys, and so much energy was spent by some of the most learned men in the country during the nineteenth century to describe these monuments and to synthesize a history of the first inhabitants of the Ohio Valley, some of the maps and plans produced by these surveyors and scholars of the prehistoric mounds should be included as part of the history of the mapping of Ohio.

There is little doubt that the study of archaeology in the United States during the major part of the nineteenth century was concentrated around the antiquities located in the state of Ohio and in the Ohio Valley, and that it gave to American men of learning a topic for consideration that promoted American scholarship independently of European initiative. During the first half of the century, those mounds were discussed and described both in articles that appeared in learned eastern journals, and in popular accounts of western travels. Among these, one of the most important appeared in 1820, when the American Antiquarian Society, founded in Boston in 1812 and interested in the study of American archaeology, published Caleb Atwater's "Description of the Antiquities Discovered in the State of Ohio and other Western States" as Volume I, *Archaeologia Americana,* in its *Transactions and Collections.* Atwater's work, complete with diagrams, was the first systematic study made of the prehistoric earthworks in Ohio. Just before mid-century, in 1848, the Smithsonian Institution produced *Ancient Monuments of the Mississippi Valley; Comprising the Results of Extensive Original Surveys and Explorations* in Volume I of *Smithsonian Contributions to Knowledge.* This comprehensive survey of the western antiquities was prepared by two Ohioans, Ephraim George Squier and Edwin Hamilton Davis, and is considered a classic in archaeological literature. Two years later, in 1850, this particular work was followed by the third volume of *Smithsonian Contributions to Knowledge*, in which Charles Whittlesey published his "Description of Ancient Works in Ohio" as a supplement to the Squier and Davis survey.[2] Between 1851 and 1900, 684 separate articles and books were produced that discussed the state's earthen monuments.[3] During these years the Archaeological Association of Ohio, founded in 1875, from which later developed the Ohio Archaeological and Historical Society, encouraged investigations into the state's mounds, and *The American Antiquarian,* "A Quarterly Journal Devoted to Early American History, Ethnology and Archaeology" and to the "study of Historic and Pre-historic Antiquities," was first published in Cleveland in April 1878. Certainly the western mounds must have been considered of some consequence to have occupied the serious thoughts of so many of the country's leading intellectuals during the century, and to have had not only a fair amount of material printed about them, but to have had two of the nation's leading learned societies devote their first scholarly volumes to the subject. Although earth mounds are found around the world, the study of the almost endless number located along the major waterways that criss-crossed the state, or situated high on the tops of prominent bluffs that commanded large areas of picturesque southern Ohio, offered to American scholars and to men of learning a stimulus for academic investigation that would be solely American.

The first mention of a prehistoric site in Ohio was made by the Moravian missionary David Zeisberger, who helped to found several Moravian villages among the Delaware Indians along the Tuscarawas River between 1772 and 1798. In his journal kept during his visit to Ohio in 1772, Zeisberger observed: "Long ago, perhaps more than a century ago, Indians must have lived here, who fortified themselves against the attack of their enemies. The ramparts are still plainly seen. We found three forts in a distance of a couple of miles. The whole town must have been fortified, but its site is now covered with a thick wood."[4] Later, in 1779 and 1780, the Moravian missionary wrote a history of the North American Indians in which he offered the idea that the mounds were used for underground dwellings. "Here and there," Zeisberger said, "traces may be found, particularly along the Muskingum [Tuscarawas], in which region one may yet see many places, where embankments, still to be seen, were thrown up around a whole town. Here and there, furthermore," he continued, "near the sites of such towns there are mounds, not natural, but made by the hand of man, for in those days the natives carried on great wars with one another. . . . At the top of these mounds there was a hollow place, to which the Indians brought their wives and children when the enemies approached and attacked them, the men ranging themselves round the mound for defensive action. . . . On the above named hills they always had great blocks lying all about, in order that should the enemies attempt to storm the heights these might be rolled upon and among them so as to keep them off. In such attacks both sides usually lost many men, which were often buried in one pit and a great mound of earth raised above them, such as may even now be seen bearing in these days great and mighty trees."[5]

In 1775 the first plan of an Ohio earthwork was published, along with its description, in the *Royal American Magazine* (Plate I). This plan of the famous works at Circleville in Pickaway County was made on October 17, 1772, and was obtained perhaps from a trader by the Reverend David Jones who visited among the Indians in Ohio that year and the next. Jones sent his sketch and description along with the following letter to the editor of the *Royal American Magazine:*

> I Herewith inclose to you a curious piece of fortification work, found in the Shawanese country. Who the constructors of it were, the Indians who live in neighborhood to it, have no tradition upon which we may depend. I was not at the spot myself, which was about two days journey to the south from the furthest Indian town I visited. The gentleman who gave me the account and plan of it, went some distance out of his way in his travels through the wilderness, on purpose to see it. From personal acquaintance I could rely on it. There are a number of this kind of forts, that are to be seen in the country of the Ohio, though I saw none so large and complete as this. They are mostly built near a river or run of water, and in one or two instances the walls run in a circular form, and appear to come together at the bank of the river, leaving only a covered way to go to the water.
>
> The present Indian inhabitants were not the builders, and they can give no satisfactory account who were. By the largeness of the trees growing in the fort, they must be of ancient standing and consequently erected long before the Shawanese and Delawares removed into the country west of the Ohio; for their removal into that country happened since the English took possession of the sea. It seems probable that the former inhabitants were constantly at war, nation against nation, and erected these forts as their *dernier reforts* in times of danger. They were doubtless better skilled in the arts of fortification and architecture than the present inhabitants.
>
> Those forts, or whatever they were designed for, are not all built alike, for some are circular, some in the form of a half moon, and others square, and contain within the walls from an half to eight or ten acres. This fort is situated towards the head waters of the river Sciota, which empties into the Ohio on the North West.[6]

Jones himself recorded:

> North of this town [Chillicothe] are to be seen the remains of an old fortification, the area of which may be fifteen acres. It lies near four squares, and

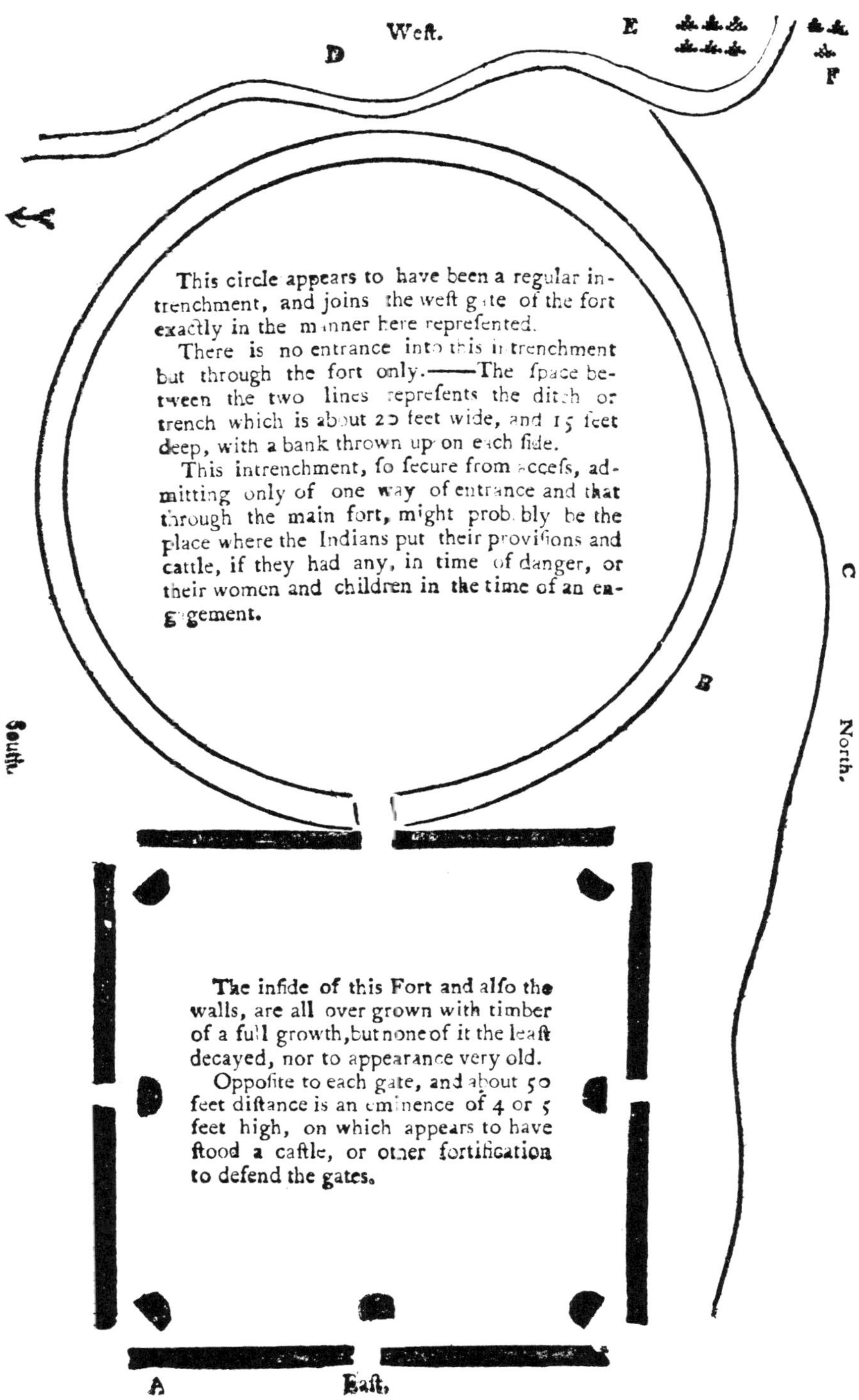

Plate I: "A Plan of an Old Fort," (1772), printed in *Royal American Magazine* (1775). This sketch of the circular works, at what is now Circleville in Pickaway County, was the first plan of an Ohio earthworks published in North America.

appears to have had gates at each corner, and in the middle likewise. From the west middle gate, went a circular entrenchment including about ten acres, which seems designed to defend on all quarters. This circle included a spring. Mr. [John] Irwine told that another exactly in this form is to be seen on the river Sciota, the banks of which remain so high as to intercept sight of men on horseback. 'Tis evident to all travellers that this country has been inhabited formerly by a martial race of mankind enjoying the use of iron, for such entrenchments, as appear in various places, could not have been made otherwise; but of this part of antiquity we shall remain ignorant.[7]

The first surveyed plan of an ancient monument was prepared by Captain Jonathan Heart, who was stationed at Fort Harmar at the mouth of the Muskingum River and was later killed while a member of Governor Arthur St. Clair's ill-fated expedition against the Indians in November

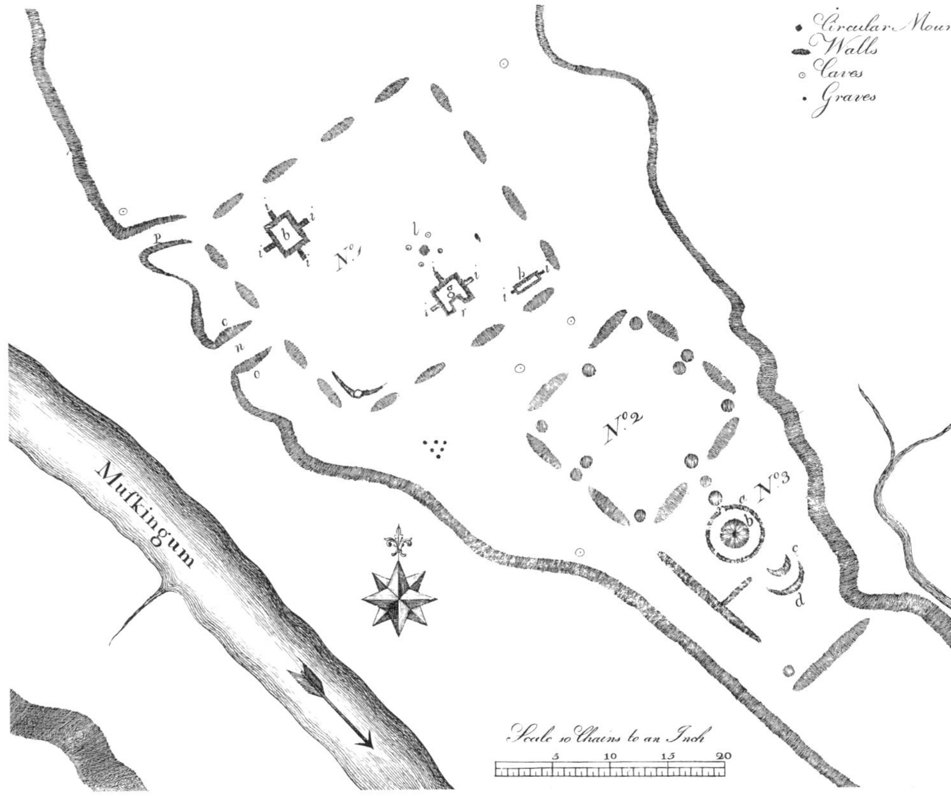

Plate II: Heart, "Plan . . . of Ancient Works," (1787), appeared in the *Columbian Magazine* (1787). Jonathan Heart's drawing was the first made from actual surveys of Ohio earthworks.

1791. Heart's plan and description were printed in the *Columbia Magazine* in May 1787[8] (Plate II). For clarification, Heart in his description divided the Marietta works into three categories: No. 1 was the town; "No. 2, the fortification, and No. 3, the pyramid, and some detached works." He further explained:

> The town is about one quarter of a mile square, surrounded with a line of walls of earth, from six to ten feet high, and from twenty to forty feet thick, the walls on each side are divided into four nearly equal parts, by three openings in each, directly opposite to each other, and openings also at each angle of the town; the openings in the center of the walls are the largest, particularly that on the side next to the Muskingum, from which opening a covered way, 120 feet wide, leads by a gradual descent 120 yards to the low grounds, where it is probable the Muskingum then run—this covered way is guarded on each side with walls, commencing at 20 yards distance from the walls of the town, and gradually become more elevated as the covert way descends, and near their termination towards the Muskingum are 30 feet high, being nearly on a plane with the walls of the town; the covert way is sloping from the center towards the walls, like the pavement of a street, as tho' there might have been canals all joining the walls, to carry the water from the town—at the north-west corner of the town is an oblong mount, 74 by 44 yards square, and six feet high, the top a perfect plane, a regular ascent leads to the top of it, at the center of each side, directly opposite each other, and in a line with the opposite openings in the walls of the opposite sides of the town—near the south-wall of the town is a similar mount, 50 by 40 yards square, with this difference only, that instead of an ascent to go up, on the side next the wall of the town, is a hollow way 10 feet wide, leading 20 feet towards the center of the mount, and then a gradual ascent to the plane of the top—a little back is a small circular mount, with four small caves at equal distances and opposite each other—at the south-east corner is also an oblong mount, 36 by 18 yards, but less conspicuous—at the south-west corner is a semicircular parapet, guarding the opening with a circular mount on the parapet, opposite the angle of the town.
>
> No. 2. The fortifications are also nearly square, with openings only in the center of the opposite walls, and at the four angles, each opening is

guarded with a circular mount 10 feet high, the openings on the east and west sides of the fortifications have two of those mounts back of each other—between the towns and fortifications are some large caves, (1) mounts, graves, etc.

No. 3. The pyramid is a circular mount, a little oval, 50 feet high, 390 in circumference, surrounded with a ditch 5 feet deep, and 15 wide, a parapet outwards, 759 feet in circumference, an opening in the parapet towards the fortifications. There are other walls, mounts and caves less conspicuous, and perhaps independent of those particularly mentioned, [which] might be considered as works of nature, but in connection with other parts are proof of art and design. The trees growing on the pyramid and different parts of the works are large, in some instances there are white oak trees of near four feet diameter, growing from a rise of earth, evidently made by the decay of a more ancient growth—the soil as well as the timber are the same in appearance in every part of the works, as in the common wilderness—The graves mentioned between the town and fortifications are small mounts of earth, from some of which human bones have been taken—in one were found bones in the natural position of a man, buried nearly east and west and a quantity of ising-glass (2) on his breast—in the other graves the bones were irregular, some calcined by fire, others only burnt to a certain degree, as to render them more durable; in others the mouldered bones retained their shape without any substance, others were partly rotten, and partly the remains of decayed bones—in most of the graves were found stones evidently burnt pieces of charcoal, Indian arrows, and pieces of earthen ware, which appear to be a composition of shells and a cement.

Nothing is yet found which can lead to a discovery, when, or by whom those works were constructed or the design of the different parts; the accounts of the Indians are irregular and inconsistent, and carry more fable than appearance of tradition—but the uniform regularity, and prodigious extent of the works, as well as their former height (if we may calculate from their present appearance, and their probable antiquity) are convincing proofs that they were constructed by a people not only numerous, but well acquainted with the art of fortification and defence, and added a beautiful uniformity to usefulness in the construction of every part.

Explanation of the Plate.

No. 1. The town b, g, h, mounts of earth i, i, i, etc., ascents leading to the top of the mounts, r. a hollow way, leading also to the top of the mount, l, a circular mount with four caves, m, a semicircular parapet, with a circular mount at m, p, a covered way 120 feet wide, 120 yards long, with walls 30 feet high at o, o–n, a covered way.

No. 2. The fortifications.

No. 3. The pyramid, a, the parapet and ditch 759 feet in circumference, 5 feet deep, and 15 feet wide, with an opening at r,—b, a circular mount 50 feet high, and 390 feet in circumference; c, a semicircular redoubt, d, a ditch and parapet.

The few published references to the Ohio mounds that had appeared before 1785 aroused the curiosity of many easterners. This inquisitiveness was further encouraged by descriptions of these mounds in the frequent letters from soldiers who were stationed at Fort Harmar, which found their way into print in eastern journals. Many of the leading intellectuals of the late eighteenth century viewed these western monuments with fascination. They discussed in correspondence the prehistoric cultures of the Ohio Valley, and on occasion, their letters were printed. As early as 1783 Ezra Stiles, President of Yale University, presented an address to the General Assembly of Connecticut in which he discussed the western antiquities and speculated about the origin of the North American Indian.[9] Several years later, a series of letters between Stiles and Noah Webster, the famous lexicographer, was published in 1787 and 1788 in several issues of the *American Magazine* in which they debated the origin of the prehistoric Indians and their works.[10] Winthrop Sargent, the president of the Congress under the Articles of Confederation and secretary of the Northwest Territory, busied himself with the study of the earthworks he found in the Ohio country, and on March 27, 1787, sent to James Bowdoin, Governor of Massachusetts and the first president of the American Academy of Arts and Sciences, a description and a plan of the Marietta ruins. Sargent's "Plan of the Ruins of an ancient Town or fortified Camp near the Confluence of the Ohio and Muskingum Rivers presented to the American Academy of Arts and Sciences by their most obedient Winthrop Sargent,"[11] which was not brought to the Academy's attention until 1850. Finally published in 1855, it was more finely drawn than Heart's, and minor discrepancies between the two are apparent. Descriptions and plans of the Ohio works continued to be sent to the East where scholars and men of science and learning discussed and debated their origin and purpose. As settlement began to move rapidly westward at the beginning of the nineteenth century, and the Ohio Country became more accessible, the amount of material about the earthworks grew, and during the first half of the century most of the major mounds and earthworks in Ohio were meticulously surveyed.

Of all the mounds in Ohio, those that were located high on top of prominent terrain features commanding a view of the valleys below most fired the imagination of the inquisitive. The first diagram of one of these fortified hilltops drawn to scale was a "Plan of an Ancient Fortification on the East Bank of the Little Miami River, about four miles above the mouth of Todd's Fork, and 30 miles North East from Cincinnati, in the county of Warren, State of Ohio," published in 1810 (Plate III).[12] This fairly accurate survey of what is now called Fort Ancient was the first made of this earthwork. Without doubt, it was studied with interest while the following description, which accompanied the diagram, was carefully read:

Plate III: "Plan of an Ancient Fortification," (1810), published in *Port Folio* (1810). The first published view of the famous Fort Ancient in Warren County.

EXPLANATION OF THE PLATE

The fortification stands upon a plain nearly horizontal, about two hundred and thirty-six feet above the level of the river, between two branches, that have very steep and deep banks. The walls made of earth are represented by dotted lines; the gates are marked by spaces: the plain extends eastward along the state road, leading from Lebanon to Chilicotha [*sic*], nearly level, about half a mile. The fortification on all sides, except near the north end where the road runs through is surrounded with precipices almost the shape of the wall. The wall on the inside varies in height according to the shape of the ground on the outside, being generally from eight to ten feet, but on the plain it is about nineteen and a half feet high on the inside and out, on a base of four and a half poles [16½ feet or a rod]: in a few places it appears to be washed away in gutters from twenty to sixty feet deep, made by water collecting on the inside. At twenty poles east of the gate through which the state road runs, are two mounds, ten feet eight inches high, the road running between them nearly equidistant from each. From these mounds are gutters running nearly north and south to communicate with the branches on each side. North-east from the mounds on the plain, are two roads, about a pole wide, elevated about three feet, and run nearly parallel about a quarter of a mile, and then form an irregular semicircle round a small mound, as represented in the annexed plan.

Near the south end of the fortification, on the south-west side are three circular roads, about forty poles in length, cut out of the side of the precipice between the wall and the river, perhaps for the purpose of annoying boats or canoes. There is no appearance of regular stone work, though some loose

> stones seem to have been collected in places on the side next to the river. Within the fortification are a kind of basons dug several feet deep, having circular banks as if intended for some kind of subterraneous habitations.

The published plan of Fort Ancient was merely one small fragment of a growing body of literature that appeared in print about the mounds found west of the mountains. Interest was encouraged by Thaddeus Harris's *The Journal of a Tour into the Territory Northwest of the Allegheny Mountains; Made in the Spring of the Year 1803,* which appeared in 1805.[13] His account of the land and people in Ohio was popular in New England and the Middle Atlantic states and his description of the earthworks found at Marietta and elsewhere in Ohio was read with care. Several years later, in 1810, there appeared *Sketches of a Tour to the Western Country Through the States of Ohio and Kentucky; a Voyage down the Ohio and Mississippi Rivers and A Trip through the Mississippi Territory and part of West Florida. Commenced at Philadelphia in the Winter of 1807, and Concluded in 1809 by Fortescue Cuming.* In this travel account, the mounds along the Muskingum, Scioto and Miami rivers, and others were observed by Cuming. These two works were eagerly read not only because of the material they contained about the earthworks themselves, but because an anxious eastern reading public was keenly interested in any information about that vast and untouched region west of the mountains as well.

Concern for both the preservation of the western mounds and the gathering and storage of data about them occupied the attention of a small group of American intellectuals and scholars across the country. Determined in their efforts, this group formed the American Antiquarian Society in Boston on November 19, 1812, with the expressed design "to discover the antiquities of our own continent; and, by providing a fixed and permanent place of deposit, to preserve such relics of American antiquity as are portable, as well as to collect and preserve those of other parts of the globe."[14] This Society encouraged the investigation of the prehistoric earthworks and requested that the information gathered should be sent to their library in Worcester, Massachusetts. In Ohio, during the years following the close of the War in 1812, interest in the Society's activities was intense indeed, and investigations of the mounds and earthworks were conducted by individuals wherever they were located.

It was no accident that the first volume published by the American Antiquarian Society was Caleb Atwater's "Description of the Antiquities Discovered in the State of Ohio and other Western States." Atwater, the author of the first history of Ohio, was a lawyer by profession, but a researcher, collector and writer by avocation. He was described by Henry Howe as "a queer talker, [who] . . . appeared to me like a disappointed unhappy man . . . [whose] life appears to have been a struggle with penury."[15] He gathered information about the earthworks in the Ohio Valley either from personal observations or from correspondents who sent to him descriptions, relics and surveys of these mounds. "I continue to receive, by every mail, specimens of minerals, and drawings of ancient works, accompanied by descriptions of them; specimens of something either curious or valuable relative to the natural history or antiquities of the country," he stated in a letter to Isaiah Thomas, president of the Society.

"The objects themselves are numerous all over this great secondary region," he continued. "It is indeed nothing but one vast cemetery of the beings of past ages."[16]

His work was a labor of both scientific and scholarly inquiry into the mysteries of the prehistoric past and, without doubt, one of love. Years later he recalled: "To collect that information, I spent much time, money, and labor."[17] Atwater, however, was not alone with his labor, for many Ohioans contributed materially to the preparation of the volume. For instance, Samuel P. Hildreth of Marietta, a physician whose intellectual curiosity ranged wide from meteorological observations to the writing of history, provided facts concerning the earthworks found at the mouth of the Muskingum. John Johnston, the United States Indian agent at Piqua, assisted the author by furnishing data to him about the Ohio Indians in general, while Roswell Mills, the county surveyor for Perry County, surveyed the fortifications in that county for Atwater.

Others contributed to Atwater's study as well. Samuel Williams, and probably E. P. Kendrick, surveyed the earthworks for the plan "Ancient Works, on the North fork of Paint Creek near Chillicothe: Ohio" (Plate IV) in South Union Township in Ross County. At the time this particular plan was made, the works were located on lands owned by a Mr. Ashley and Colonel John Evans, a "man of energy and enterprise" who moved from Baltimore County, Maryland, to Ross County in 1800. Atwater, who personally visited each site during the preparation of his study, described these ancient earthworks as follows:

> Five miles and a half from Chillicothe, on the above mentioned stream (North Fork of Paint Creek), these works are situated on a beautiful piece of what we call second bottom. The area of the largest enclosure contains about one hundred and ten acres. On the northeast and west side of it, is a wall, with an intrenchment or ditch on its outside. It is generally twelve feet from the bottom to the summit of the wall, which is of earth. The ditch is about twenty feet wide, and the base of the wall the same. There is no ditch on the side next [to] the river. The small work, on the east side, contains about sixteen acres, and the walls are like those of the larger work, but there is no ditch. The largest circular work, which consists of a wall and ditch like those

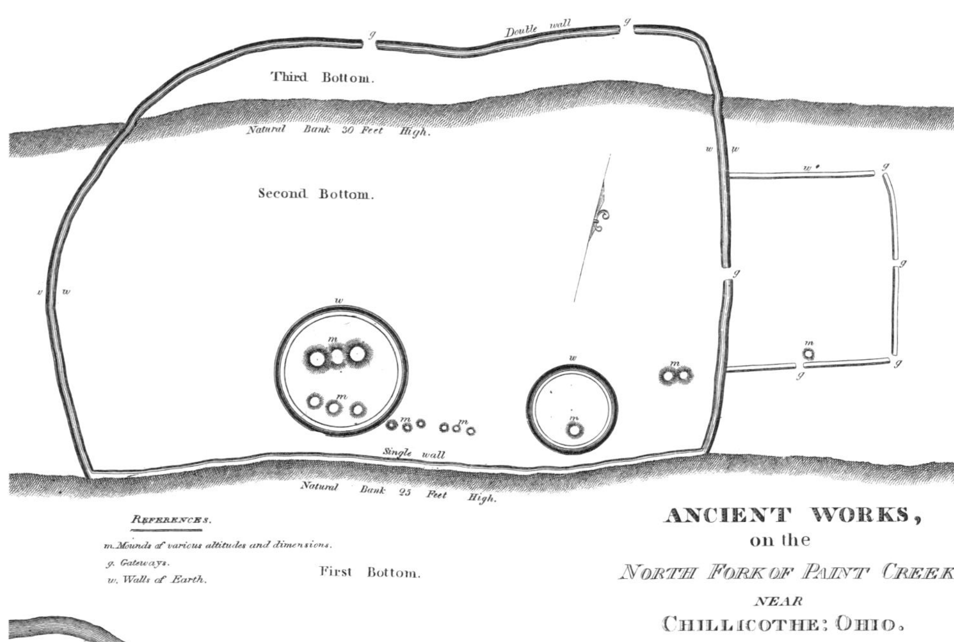

Plate IV: Atwater, "Ancient Works . . . near Chillicothe," (1820), in the first volume of the American Antiquarian Society *Transactions* (1820). Atwater's study of the Ohio earthworks was the first organized attempt to gather data about the state's prehistoric monuments.

> already described, is a sacred enclosure, including within it six mounds, which have been used as cemeteries. By examining the drawing and measuring them by the annexed scale, a correct idea of their dimensions may be easily obtained. The observation applies to the gateways in the outer wall.[18]

This mound group was excavated several times between 1846 and 1922, and in the latter year, William C. Mills, curator of the Ohio State Archaeological and Historical Society, identified the builders of these earthworks as a distinct prehistoric culture which he designated Hopewell, named for Captain M. Cloud Hopewell, then owner of the farm on which the works were located.[19] Atwater concluded that "the immense labor, and the numerous cemeteries filled with human bones, denote a vast population near this spot in ancient times." Today, most of the original Hopewell site is gone, having fallen victim to agricultural progress and the need for advanced productivity. All that remains of the site is a small portion of the north double wall and slight evidence of the bases of several of the circular mounds.

At Circleville, Guy W. Doan helped Atwater to survey the works there. Together they produced the plan "Ancient Works at Circleville: Ohio" (Plate V) which represented those earthworks found along the east bank of the Scioto River just south of the mouth of Hargus Creek. Atwater believed that this site was a fortification and speculated that perhaps it was one of the most northerly found in the Scioto Valley. Actually, he explained that there were two forts there. The first, a round one, had a diameter of sixty-nine feet, while the second, a square, enclosed fifty-five square rods. Before the white man laid out and began to build the settlement of Circleville on this same site in 1810, the two walls of the circular works measured, from the bottom of the ditch to the top of the walls that separated them, at least twenty feet. When Atwater surveyed the works in 1820, he complained that the walls were reduced to only "about five or six feet in height now; on the inside, the ditch is, at present, generally not more than 15 feet. They are disappearing before us daily, and will soon be gone."[20] The walls of the square earthwork, that were still standing when

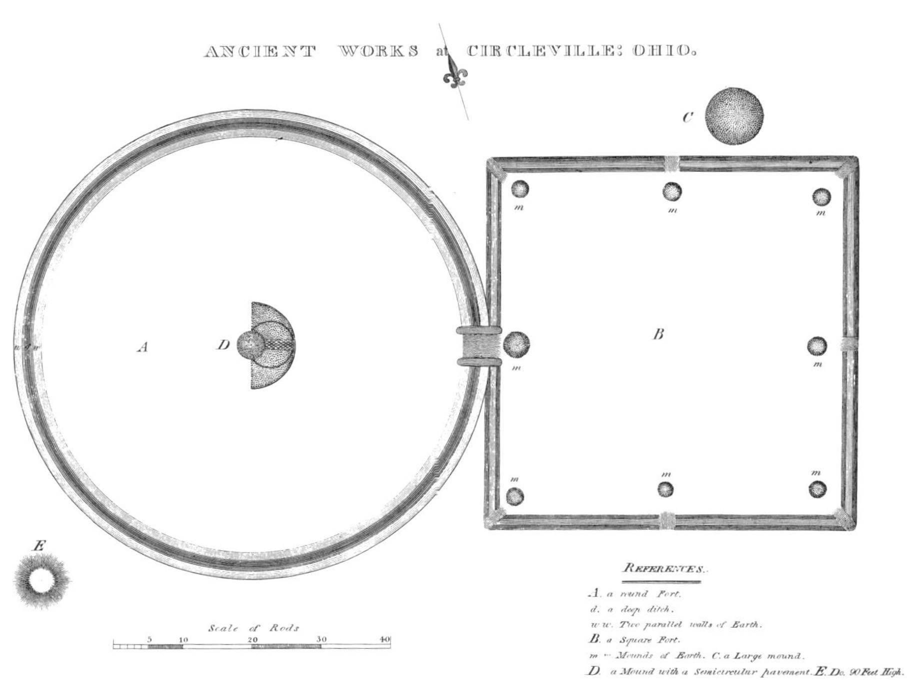

Plate V: Atwater, "Ancient Works at Circleville," (1820), from *Transactions* published by the American Antiquarian Society (1820). By 1820, Atwater warned of the destruction of Ohio's earthworks.

Atwater observed them, measured about ten feet high. While the community of Circleville was being built, the soil from the walls of the square work was used to manufacture bricks in order to construct the square and rather unimaginative homes and shops of the early settlers of Ohio. Workmen, while tearing into the mounds, found ashes, calcined stones, sticks and "vegetable mould," Atwater reported.

The eight mounds situated around the interior of the square fort, opposite the openings in the walls, were "perhaps four feet high, 40 feet perhaps in diameter at the base, and 20 or upwards at the summit." These were located two rods distant from the opening and, Atwater explained, "were in a perfectly straight line, and exactly parallel with the wall." The mound marked "E" to the southwest of the circle in Atwater's drawing was actually located on a small rise of ground which, with the mound, reached a height of ninety feet. This earthwork, Atwater said, "must have been the common cemetery, as it contains an immense number of human skeletons, of all sizes and ages."[21] In 1834 it was removed in order to provide a right of way for one of the city's streets. Within the mound workmen found many skeletons, some well preserved. One was described as having a cranium with "A high forehead and large and bold features, with all the phrenological marks of daring and bravery. Poor fellow, he died overwhelmed by numbers; as the fracture of the right parietal bone by the battle-axe and five large arrows sticking in and about his bones, still bear silent, but sure testimony."[22]

The most unusual mound in this group was the one that had stood in the center of the circle but which had been destroyed several years before Atwater made his survey. This one, he stated, had been ten feet high and several rods wide at its base, with a level summit thirty feet across. Wrapped around its base, on the eastern side, in a semicircle, was a pebble pavement, approximately six rods wide. Leading from this pavement up to the summit was a "raised way . . . like a modern turnpike." Atwater had witnessed the removal of this mound. He reported that it had been the final resting place for two individuals, apparently of some note. Along with the skeletons were found a number of large spear points, the remains of a wood-handled knife, some mica and what he called "a plate of iron."

"The present town of Circleville covers all the round and the western half of the square fort," Atwater stated. "These fortifications, where the town stands, will entirely disappear in a few years." Today, as he predicted, all of this significant Hopewellian earthwork has disappeared except for a very small fragment, hidden in an alleyway on the southeastern side of the town. Troubled and saddened by the destruction of the prehistoric works that he witnessed daily throughout Ohio, Atwater cried:

> Why were these wantonly destroyed? 'They were rude.' Were they not venerable on account of their high antiquity and simplicity? Are the modern Turks and Arabs, who trample on the busts of ancient heroes, the moss grown and prostrate columns of ancient temples, baths, palaces and theatres, the only barbarians? 'But those who buried in tumuli, worshipped the manes of the heroes there deposited.' And were not the Greeks and Romans also idolaters? And, have not all the civilized nations of Europe joined in condemning those who wantonly violate the sacred repositories of the dead, in those countries where the arts once flourished? It is true that the citizens of the ancient republicks [*sic*] enjoyed not the Christian religion; that commerce, and even agriculture, had made no great progress among them. To defend their country, and extend their conquests, were the great objects which they constantly kept in view. Soldiers became heroes; and these, after

> death, were exalted to gods. The love of military glory was constantly connected with the love of country. Having but few objects of pursuit, their passions were more intensely fixed and treated as such. They put to the sword, or sold as slaves, their prisoners of war. But what makes us to differ from them, unless it be an acquaintance with Christianity?[23]

The classical example of nineteenth-century archaeological literature published about the mounds and earthworks in Ohio was prepared by Ephraim George Squier and Edwin Hamilton Davis, and appeared as the first volume of Smithsonian Institution's *Smithsonian Contributions to Knowledge*. This work, the first comprehensive study made of the western antiquities, took two and a half years to prepare, engaged the energy and patience of many men interested in archaeological study, and represented a compilation of thought about the prehistoric Indians of some of the finest minds of the first part of the century. They concluded that nearly ten thousand mounds and between one thousand to fifteen hundred enclosures were located throughout the state. As with Atwater's work published twenty-eight years earlier, many men of science and learning contributed to the final preparation of this volume. Because of its importance to the history of archaeological studies generally, and because it deals primarily with those earthworks found in Ohio specifically, several examples of its map work must be included among the historic maps of Ohio.

After careful planning, it was decided by Squier and Davis that the study of the ancient earthworks should commence with those found in the Scioto River and Paint Creek valleys, centered around Chillicothe in Ross County. The "Map of a Section of Twelve Miles of the Scioto Valley with its Ancient Monuments" (Plate VI) by Squier was the result of that particular study. The numerous mounds and earthworks found in these two valleys, estimated by Squier and Davis at one hundred enclosures and five hundred mounds, identified the Scioto Valley region as one of the most important cultural centers of the mound-building Indians in North America. Both authors, however, cautioned that although the works found there were many, "it is proper to observe, to prevent misconception, that there are few sections of country of equal extent which embrace so large a number of ancient works. The fertile valley of the Scioto river was a favorite resort of the ancient people, and was one of the seats of their densest population."[24]

At the very top of the map, designated "A", were the Dunlap's Works, a rhomboidal earthwork, located in Union Township, Ross County, 6 miles north of Chillicothe. The parallel walls that approached the enclosure from the southeast were 1,130 feet long. They stopped within 60 feet of an opening 120 feet wide in the eastern wall. To the west of the works, and upon a small rise, was situated an outwork, 80 feet wide and 200 feet long, that overlooked the entire complex. About half a mile south of the Dunlap's Works was located a number of mounds identified today simply as Mound #4. Included in this group was a truncated mound, 15 feet high with a base diameter of 100 feet, which was one of the first explored by Squier and Davis. They noted that of all the mounds in this section of the valley, this group was the only one not subject to flooding. In fact, it was pointed out that during the high water in 1832, this particular mound gave refuge to an entire family and nearly 100 head of livestock. Both these mounds and the Dunlap's Work have succumbed to the farmer's plow.[25]

Across the Scioto River from Mound #4, situated on the table land along the east bank, were located the Cedar Bank Works, or "B", in

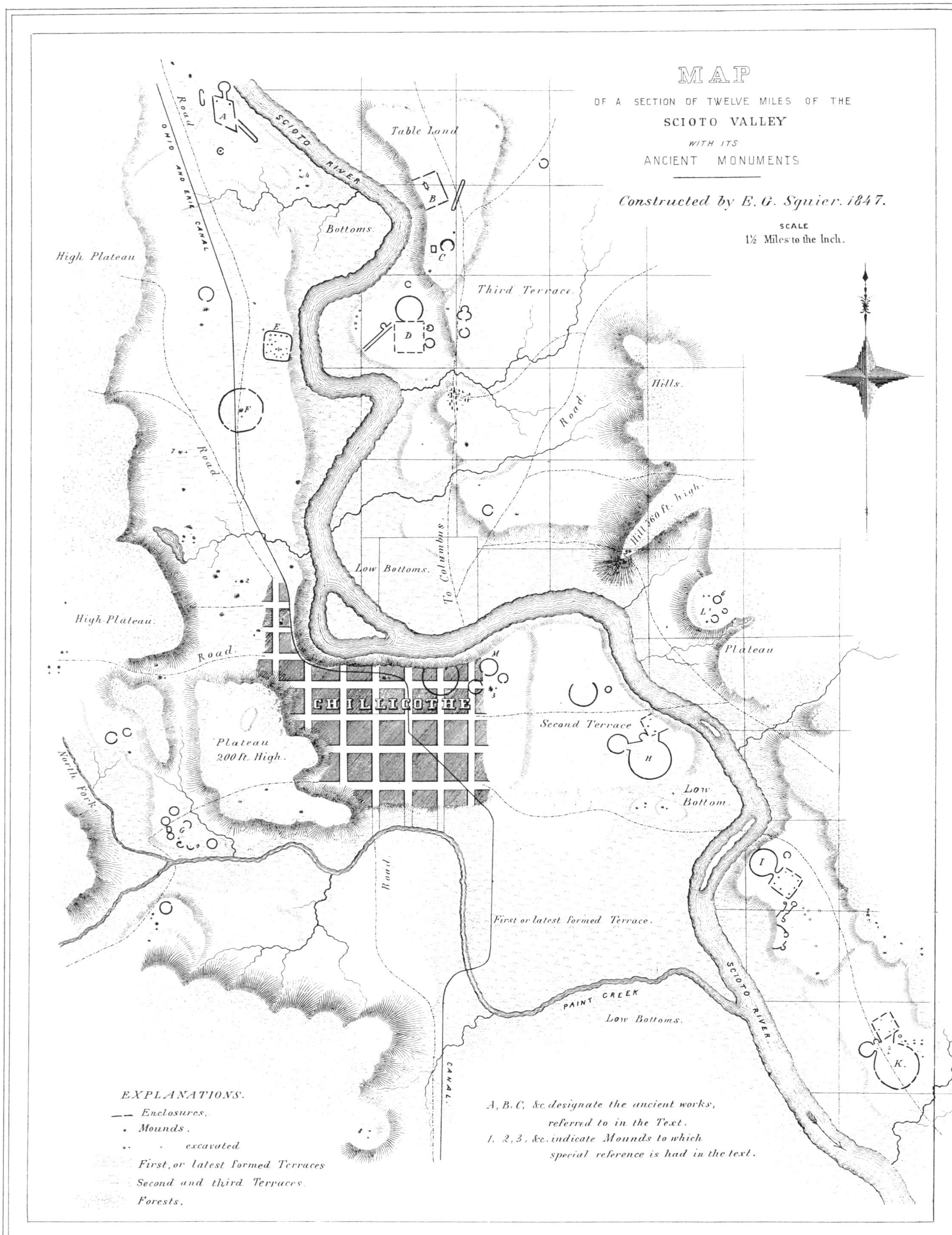
MAP
OF A SECTION OF TWELVE MILES OF THE
SCIOTO VALLEY
WITH ITS
ANCIENT MONUMENTS
Constructed by E. G. Squier. 1847.
SCALE
1½ Miles to the Inch.
Table Land
Bottoms.
High Plateau
Third Terrace.
Hills.
Road
OHIO AND ERIE CANAL
SCIOTO RIVER
Hill 160 ft. high.
Low Bottoms.
To Columbus.
High Plateau.
Plateau
CHILLICOTHE
Second Terrace
Plateau 200 ft. High.
North Fork
Low Bottom.
First or latest formed Terrace.
PAINT CREEK
Low Bottoms.
CANAL.
EXPLANATIONS.
Enclosures.
Mounds.
excavated
First, or latest formed Terraces
Second and third Terraces.
Forests.
A, B, C, &c. designate the ancient works, referred to in the Text.
1, 2, 3, &c. indicate Mounds to which special reference is had in the text.

GreenTownship 5 miles above Chillicothe. Three walls of the square-like enclosure were 6 feet high and 40 feet at their base. The side facing the river had no wall, but rather was protected by a natural bluff that rose some 70 feet high from the river shore, "so steep as to admit of no ascent." The eastern wall was 1,400 feet long while both the north and south walls were 1,050 feet long. A ditch 5 feet deep and 40 feet wide surrounded the entire enclosure except along the river side. Inside the enclosure, opposite the gateway in the north wall, was an elevated square 250 feet long and 150 feet wide. At both ends, graded ways 30 feet wide led from the enclosure's floor to the top of the square. Three hundred feet from the enclosure's southeastern corner were single parallel walls, 70 feet apart, 870 feet long.[26] At the time of the Squier and Davis survey, this earthwork already had been partially destroyed by the construction of the Chillicothe and Columbus turnpike. As late as 1941, however, some of the walls of the Cedar Bank Works still remained; today most of the components are still visible to the trained observer.

South of these works, about one-third of a mile down river, was a small group of earthworks the most prominent of which was indicated by "C" on the map. Here on a terrace that overlooked the river below was located a truncated pyramid and a small circle. The pyramid, or square mound, was 120 feet wide at its base and about 9 feet high, while the circle was 250 feet in diameter with an opening 30 feet wide along its south side. The location, according to Squier and Davis, appeared to have been artificially smoothed. Within the circle, along the wall opposite the opening, was a semicircular interior wall that paralleled the exterior one. A ditch separated the two. The two early archaeologists were puzzled as to the purpose of these particular works but speculated that they were either for games of some sort or for religious ceremonies.[27] These works, too, are now gone.

Below them, but 4 miles above Chillicothe, were the Hopeton Works in Springfield Township shown as "D". Situated at the base of an elevated plain, these earthworks were different from the others of similar configuration because the circle extends into the rectangle rather than standing separately. The circle itself enclosed nearly 20 acres and its wall, although farmed for many years even before the Squier and Davis survey, still averaged nearly 5 feet high. The rectangle, which had 12 gateways, also enclosed almost 20 acres. Its clay walls were 12 feet high and 50 feet thick at the base. From the rectangle's northwest corner, running southwesterly toward the river, were parallel walls 2 feet high, 150 feet apart and nearly half a mile long.[28] Today, after almost a century and a half of farming, only a very small segment of the rectangle of the Hopeton Works is visible.

Opposite the Hopeton Works, on the west bank of the Scioto River, was located a series of earthworks, identified as "E" and "F" and the small circular configuration to the northwest of "E". Of these works, Mound City, or "E", was, according to Squier and Davis, "in many respects the most remarkable in the Scioto Valley." Here in 1848 within a three-foot-high wall which enclosed 13 acres, 29 mounds were located on "the beautiful level of the second terrace, . . . [which was] still covered with the primitive forest."[29] All the mounds were excavated and determined to be places of burial. Until June 1917 the site had been put to agricultural use but, at that time, the government located Camp Sherman, a training camp for recruits during World War I, on this spot. Most of the mounds that had not been reduced through agriculture over the years, were leveled except for the largest one. This one, 17 feet high and 100 feet wide at the base, which was

(Opposite page) Plate VI: Squier and Davis, "Scioto Valley," (1848), in the first volume of the Smithsonian Institution's *Smithsonian Contributions to Knowledge* (1848). This sketch illustrates how numerous were the mounds and earthworks throughout the Scioto Valley.

saved from destruction by the camp's commander at the request of the Ohio State Museum, was located in the center of the camp. The earth carried by the prehistoric Hopewells so long ago to construct the mounds in which to bury their dead, along with the bones of their honored heroes and their purest artifacts, was scattered throughout the camp to become part of Camp Sherman's grounds and roads and drill fields. Although the large center mound was saved for posterity, it nevertheless helped to house part of the camp's water pumping system.

Following the armistice, in 1920, William C. Mills, curator for the Ohio Archaeological and Historical Society and well-known archaeologist, secured permission from the War Department to conduct a study of the Mound City group while many buildings from the military camp still remained. He and his team were able to locate only twelve of the mounds originally identified by Squier and Davis in 1846. Mills's excavation during 1920–21 revealed some exquisite Hopewell artifacts and added greatly to former studies.[30] On March 2, 1923, President Warren G. Harding proclaimed this site the Mound City Group National Monument. It was dedicated "to conserving the scenic, scientific, and historic heritage of the United States for the benefit and inspiration of its people."[31]

In Chillicothe itself were located Chillicothe Works #1, which consisted of three circular earthworks enclosing a little over an acre each. These mounds, two of them located on Water Street and the third just to the north, were part of a large complex that had been obliterated by the time Squier and Davis conducted their survey in 1846. To the east of Chillicothe, but on the west side of the Scioto, was situated the Chillicothe Works #2 shown as "H" on Squier's map. Near this site in 1796, the first settlement, called Prairie Station, was built. That year the first crops were planted in the bottom land along Paint Creek and the Scioto, and apparently the earthworks were not spared from the first farmer's plow. When the works were surveyed, the investigators noted that "the ground had been in cultivation for more than forty years" and consequently, "the walls are much reduced, although distinctly traceable at this time [1846]."[32] Portions of the square had been eroded, probably by the action of the river, and the circle too had been partially destroyed if, Squier and Davis conjectured, the works indeed had ever been finished. All traces of this earthwork have since disappeared.

Below the Chillicothe Works #2, along the east bank of the Scioto River, set upon "a bold bank," rising 75 or 80 feet above the water, was situated the High Bank Works, identified as "I" on the map. Similar in configuration to several other of the Hopewellian earthworks found in the valley, the square in this particular group was more nearly an octagon that enclosed approximately 18 acres. Squier and Davis explained that the earthen walls of the octagon were "very bold" and measured between 11 and 12 feet high and 50 feet wide at their base. The circle's wall, only 4 to 5 feet in height, enclosed nearly 2 acres, the same as the Hopeton Works. From the southernmost angle of the octagon to the circular works along the edge of the high bank ran parallel walls. The right wall, the shorter of the two, was 1,400 feet long and lay between two circular enclosures that surrounded nearly an acre and a half each. The circular works along the high bank to the south of the main group consisted of two circles, the large one enclosing nearly 7 acres. This group had been partially destroyed by erosion and land fall caused by flooding.[33] By 1846, when the plan for the High Bank Works was made, the dimensions of the works had already been diminished through agricultural use. Today, the circle and the square can still be traced on the gound.

The last earthwork that appeared at the bottom of this map of the ancient monuments near Chillicothe in the Scioto Valley was called by Squier and Davis merely the "Ancient Works" and specified as "K". Later this work was identified as the Harness Group, named for the owners of the farm on which it was located, Edwin and Daniel Harness. Located along the Scioto's east bank and on the river's third bank, it was "a very fair type of a singular series occurring in the Scioto valley—all of which have the same figures in combination, although occupying different positions with respect to each other, viz. a square and two circles." The square enclosed almost 27 acres. Its walls averaged 4 feet in height and were broken by 8 gates each 30 feet wide. Small mounds 4 feet high stood within the enclosure 40 feet from, but directly in front of, each side gateway. Nearly 52 acres were enclosed by the larger circle, while the smaller one surrounded approximately 11½ acres. The wall of the great circle was interrupted by 12 gateways. Within the circular enclosure itself were situated 5 mounds, the largest of which was 160 feet long, close to 20 feet high and 80 to 90 feet wide. It was later known as the Edwin Harness Mound. Squier concluded that "the whole work appears to have been partially finished or constructed in great haste. The mounds at the gateways, and those outside of the walls, were formed by carelessly scooping up the earth at their base, leaving irregular pits near them."[34]

The mounds at this site were first explored by Squier and Davis in 1846, as they had investigated most of the other mounds they surveyed. As a result, a massive—and indeed a spectacular—collection of Hopewellian artifacts was gathered. To help defray the expenses encountered during their two-and-a-half-year study, these Hopewell pieces were placed on the market for sale. No individual or institution within the state of Ohio, or for that matter, in the country would meet the surveyors' price. The collection was boxed and stored in Washington. During the American Civil War, the Englishman William Blackmore, a lawyer, anthropologist and member of the prestigious council of the British Anthropological Society, purchased the Squier and Davis materials. In 1863 he gave the entire collection to the city of Salisbury, England, where it became the nucleus of the Blackmore Museum. It is now part of the holdings of the British Museum and, although some of the major pieces from this collection are displayed today, the great bulk of it remains boxed and stored away. It is ironic, indeed, that the fruits of one of the earliest and most important archaeological projects in the United States were destined to be housed in England.

Despite this tragic loss, however, artifacts from the Scioto Valley have remained in Ohio. In 1885, Frederick Ward Putnam explored this site, and the 278 specimens taken during his excavation are at the Peabody Museum at Harvard University.[35] Not long before Putnam began his work, several boys from a local school that stood nearby, under the supervision of a Mr. Wilson, probably the boys' teacher, dug into the Harness Mound. No doubt determined that learning was a personal and exciting experience and that certain topics should be studied first hand, Wilson and his gang turned up some of the best finds to come from this particular group. These artifacts were later either acquired or purchased from the students by Daniel R. Harness and given to the Peabody Museum. Eleven years later, in 1896, Warren King Moorehead, a graduate of Denison University at Granville, Ohio, who studied under Putnam and also served as Curator of the Ohio Archaeological and Historical Society between 1894 and 1897, conducted excavations again at this site. In 1903 and 1905, the Harness mounds were investigated for the third time, this time by William C. Mills. As a result of the last two excavations, 1,200 artifacts were recovered and are now part of the prehistoric Indian collection of the Ohio Historical Society's Center in Columbus.[36]

Only portions of the Harness group have survived the progress of civilization. Squier and Davis had noted in 1846 that part of the works had been reduced in height at that time by cultivation. Putnam, in 1885, mentioned that all the small mounds connected with the earthworks had either been levelled or reduced by the farmer's plow. Later, the right of way of the Cincinnati, Hamilton and Dayton Railroad passed through the works, as did the road from Chillicothe to Jackson, Ohio. Edwin Harness constructed his home and a large barn over the small circle. Today, only portions of the smaller circular earthwork remain as evidence of this once great prehistoric site, and it, too, is threatened by highway construction.

To the southwest of Chillicothe, high upon a beautiful plain on a third terrace overlooking the junction of North Fork of Paint Creek and Paint Creek, was located the Junction Group ("G" on the map). This group consisted of 4 circles, 3 crescents, 2 squares and 4 mounds. Of these 13 monuments, the square at the extreme eastern end of this group was most prominent. Although closely resembling a circle, its obtuse-angled sides, which were 3 feet high, formed a square that enclosed a little less than an acre. It was surrounded by a ditch that formed a square 160 feet to a side, and was entered through a gateway 25 feet wide on the south wall. To the west of this square, and commanding the entire group, was a mound 7 feet high and 45 feet at its base, which was situated at the end of a crescent-shaped earthwork. "That they were not designed for defense is obvious," Squier speculated, "and that they were devoted to religious rites is more than probable. They may have answered a double purpose, and may have been used for the celebration of games, of which we can have no definite conception."[37]

Another example of the profusion of earthworks found in the proximity of the Scioto River Valley near Chillicothe was displayed by the "Map Exhibiting a Section of Six miles of the Paint Creek Valley with its Ancient Monuments" constructed by Squier in 1847 and also published in the first volume of *Smithsonian Contributions to Knowledge* (Plate VII). Within this section, however, aside from the two Hopewell earthworks that are similar to those found along the banks of the Scioto, Squier mapped a fortified hill position located atop a lofty bluff and a very unusual stonework.

The earthworks to the left of the map marked "A", but unnamed by Squier, were in fact the Seip group, which contained the famous Seip

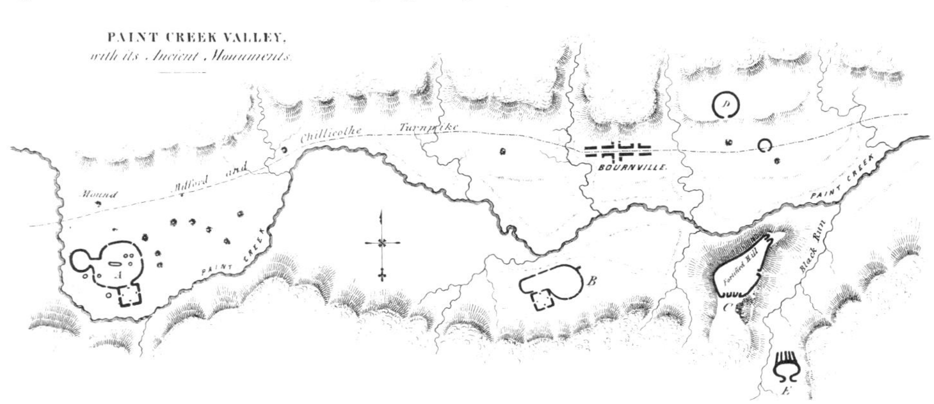

Plate VII: Squier and Davis, "Paint Creek," (1847), published in *Smithsonian Contributions to Knowledge* (1848). Thousands of prehistoric sites dot the valleys of Ohio's major waterways.

Mound. This group, the largest prehistoric earthworks in the Paint Creek Valley, was located along the north bank of Paint Creek in Paxton Township, three miles east of Bainbridge. It was situated on land owned by John Woodbridge of Chillicothe when Squier and Davis made their surveys in 1847; but in 1909, when the group was excavated by William C. Mills for the Ohio Archaeological and Historical Society, the works were owned by the Seip brothers, whence its name derives. Gerard Fowke, in his extensive *Archaeological History of Ohio* published in 1902, mistakenly identified this earthwork as the Pricer Group.[38] Pricer, whose farm was adjacent to the Seips', owned part of the large elliptical mound located almost in the center of the larger circle, and his name was given to that mound. As early as 1847, several of the small mounds and circles that accompanied the larger works had already "been so much reduced by the plough as to be traced with difficulty."[39] By 1909, the group had been further reduced by cultivation, but "when the site is freshly plowed, the old walls [of the circle] are easily traced by a slight elevation." The walls of the large square, which once enclosed 27 acres, however, were gone.[40] Twenty-two years later, in 1931, all traces of the walls were gone except for a one-hundred-foot segment which was included as part of a barnyard and subsequently saved from cultivation.[41] Today, only the Seip Mound remains and is part of the Ohio Historical Society's State Memorial system.

Further down, but on the right bank of Paint Creek opposite the small village of Bourneville, the Baum Group (indicated as "B") was located. "It is but another combination of the figures composing the work belonging to this series," Squier and Davis explained, and "in structure, it differs in no material respect, except that the walls are higher and heavier." This site was situated on land owned by Baum, one of the pioneer farmers in the Paint Creek Valley who cleared away giant black walnut, oak, sycamore and ash trees from the land in order to farm the valley's rich floor. In 1846, when this site was surveyed, a large portion of the southern wall of the great circle had been obliterated because of water flowing through the works from the hills to the south.[42] The square was intact. In 1902, Gerard Fowke commented that from a spring in the hills below this earthwork "a perennial stream [flows] toward the junction of the square and the larger circle. This and surface drainage have carried down sufficient of the loose material from above to obliterate all traces of the wall and excavated areas at this point."[43]

The most impressive and spectacular work of all the earthworks found along Paint Creek is that of Spruce Hill, a stone fortification site, located on "the summit of a lofty, detached hill . . . [that] is not far from four hundred feet in perpendicular height; and is remarkable, even among the steep hills of the West, for the general abruptness of its sides, which at some points are absolutely inaccessible."[44] The stone wall, though now deteriorated by the ravages of time and weather and man, followed along the edge of the summit at an average height of probably 8 feet. It measured on line nearly 2¼ miles in length and enclosed over 140 acres. At both ends of its northeast and southwest axes were gateways protected by walls nearly quadruple the thickness of the other portions of the embankments, which guarded natural avenues of approach to the summit from the valley below. The summit itself was a wide and fertile plain containing a large pond of fresh water that covered 2 acres nearly centered within the enclosure.[45] Spruce Hill was one of several fortified hilltops found in Ohio, including Fort Hill in Highland County and Fort Ancient in Warren

County, which are now both historical sites owned and preserved by the State of Ohio. Like these other fortified summits, it was built by the ancient Hopewells and appears to have been a site of occupancy.[46] Gerard Fowke judged that "this fort is probably the largest area in the world surrounded by an artificial wall made entirely of stone."[47]

One of the most unusual stone works that existed in the state was the Black Run Stone Works, letter "E" on the map, located to the east of Stone Run, a small tributary of Paint Creek. The identity of its configuration and the purpose for which it was constructed are unknown. At the time that the survey was made, it was in a terrible state of disrepair, even though the general outline of the works could be ascertained. The wall, which was probably 8 feet high, formed an ellipse, of which the conjugate axis was 170 feet and the transverse 250. An opening or gateway 50 feet wide faced the south. The walls at either end of this opening curved outward and back along the enclosure's wall for 60 feet. To the north of the main wall, and 10 feet from it, were 5 single stone walls, 20 feet wide and 10 feet apart, slightly angled, that extended northward for 100 feet. The westernmost single wall, as well as the western portion of the main wall of the enclosure, was nearly gone, as many of the larger stones found their way into the foundations, hearths and chimneys of the dwellings of the early settlers of the Paint Creek Valley. It is believed that the figure was intended to resemble a bear's claw. In 1878 Stephen D. Peet of Astabula, who was at one time corresponding secretary of the American Antiquarian Association and secretary of the Archaeological Association of Ohio, became editor of the *American Antiquarian,* a journal published in Cleveland and devoted to the study of early American history, ethnology and archaeology. He conjectured that this stone work was really 2 intertwined snakes, while the 5 walls represented its tails or rattles.[48] Squier and Davis, however, had simply commented that "the purposes of the strange work are entirely inexplicable: its small size precludes the idea of a defensive origin. It is the only structure of the kind which has yet been discovered in the valleys, and it is totally unlike those found on the hills."[49]

Most of these important and magnificent prehistoric earthworks, like those in the Scioto Valley, have been carelessly destroyed by those who have not had an appreciation of their value. As late as 1959, one of the very few remaining sites mapped by Squier in the Paint Creek Valley fell victim to progress. That year, workers were clearing ground for a new Route 50. Just east of Bourneville, below the circular works marked "C" on the map, stood the fairly large Bourneville Mound. A bulldozer cut into the mound and its operator was surprised to find that a quantity of bones lay within. Other workers appeared and, with their shovels, soon uncovered 7 or 8 skeletons, a copper breast plate, several copper ear spools, 4 drilled grizzly bear teeth, and other relics connected with prehistoric people and their burial sites. "The men had put these artifacts in one pile, and the bones in another pile. Then the foreman came along and said they would have to go ahead with the work. The men each grabbed something out of the pile of artifacts, and a carryall scooped up the bones in a load of earth and hauled them away to a dump."[50]

Of course, the Scioto Valley was not the only noteworthy location of ancient earthworks in Ohio. For years those at Marietta had been studied, and since the first settlement there in 1788, as already noted, several plans of the works there had been drawn. This particular map, "Marietta Works.

(Opposite page) Plate VIII: Squier and Davis, "Marietta Works," (1848), included in their study in volume one of *Smithsonian Contributions to Knowledge* (1848). This plan of the earthworks at Marietta was made by Charles Whittlesey.

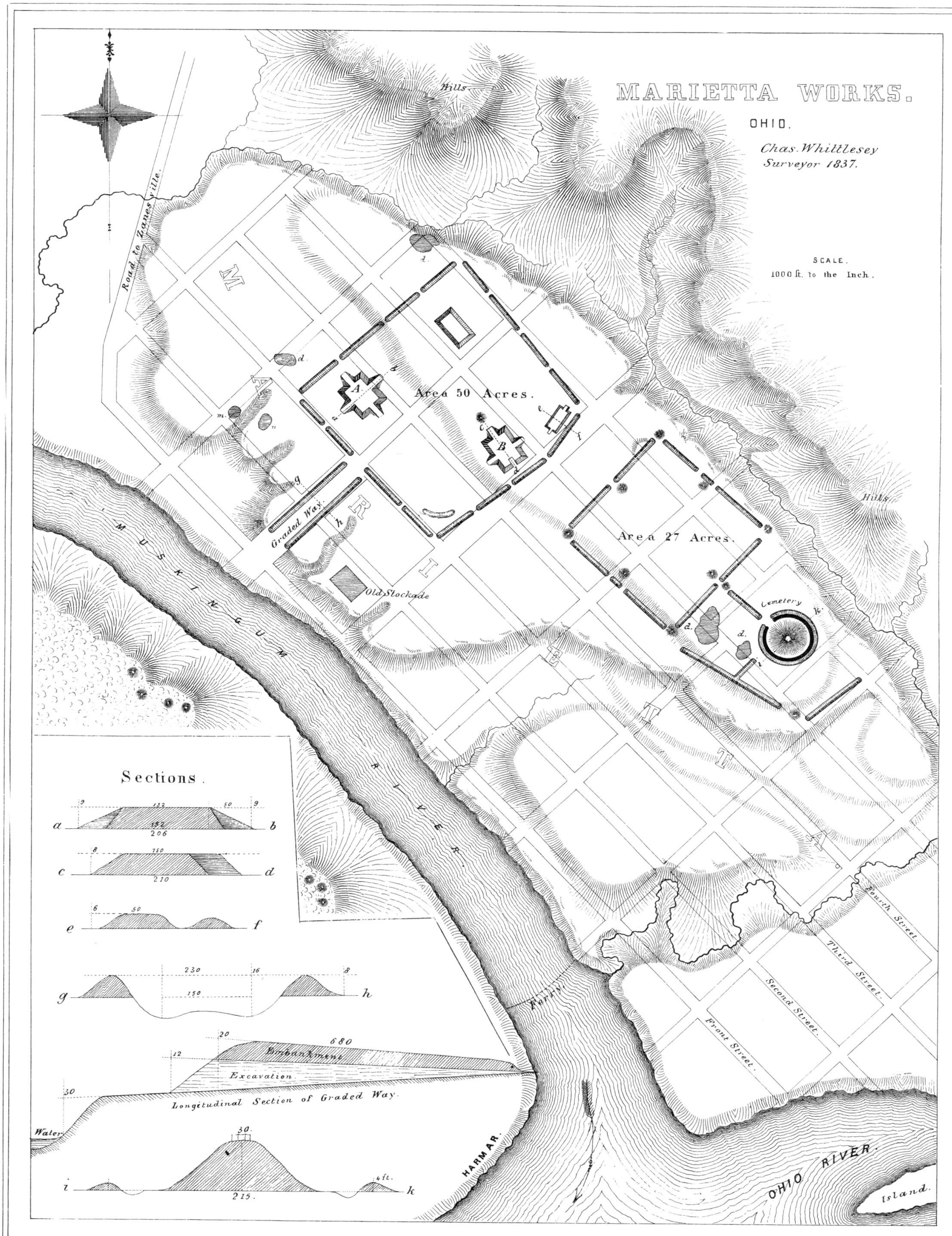
MARIETTA WORKS.
OHIO.
Chas. Whittlesey
Surveyor 1837.
SCALE.
1000 ft. to the Inch.
Hills.
Road to Zanesville.
Area 50 Acres.
Area 27 Acres.
Graded Way.
Old Stockade
Cemetery
MUSKINGUM RIVER.
OHIO RIVER.
HARMAR.
Ferry.
Island.
Front Street.
Second Street.
Third Street.
Fourth Street.
Sections.
Embankment
Excavation.
Longitudinal Section of Graded Way.
Water.

Ohio." by Charles Whittlesey, which appeared in *Ancient Monuments of the Mississippi Valley* (Plate VIII) was, according to Squier and Davis, an accurate and true representation of the works located along the east bank of the Muskingum River. Whittlesey, from Cleveland, was appointed assistant geologist for the first geological survey of Ohio, which began in the summer of 1837. A graduate of West Point, he practiced law for a while in Cleveland, but made his greatest contribution through his thorough survey work, not only in Ohio, but also around the Lake Superior region and in Wisconsin. The Ohio geological survey was never completed on account of the economic depression of 1837 and the consequent failure of the Ohio General Assembly to appropriate sufficient funds. While covering the southeastern portion of Ohio, however, Whittlesey prepared maps of the earthworks and mounds found there, as did several others engaged in this project in other sections of the state.[51] None of Whittlesey's maps were included with the survey's official reports, although one by John Locke, an assistant state geologist with Whittlesey, appeared in the report.[52] For many years following his work with the geological survey, Whittlesey and Joseph Sullivant, the youngest son of Lucas Sullivant, the founder of Franklinton along the west bank of the Scioto River, continued to survey and to map many of the earthworks found in the state with full anticipation that a published volume would be the result. Both, however, relinquished their own work, some twenty plans in all, to Squier and Davis to be included in the first volume of the *Smithsonian Contributions to Knowledge*. Whittlesey's plan of Marietta is, without question, a marvelous study.

The earthworks themselves covered approximately 345 square acres and a great part of the original settlement of Marietta was laid out over their surface. The walls of the larger square were probably 6 feet high and, at their base, 20 to 30 feet wide; those of the smaller were not as large. Within the larger square were 4 elevated squares. The largest, marked "A", was 188 feet long by 132 feet wide. Graded ways, 25 feet wide and 60 feet long, ran from the ground to the top. In 1837, when this plan was made, the square near the eastern angle of the larger square was mostly gone. The dimensions of these works, diagrammed in "Sections" on Whittlesey's map, fully illustrate the extent of the Adena and Hopewell works found at Marietta. The old stockade shown to the right of the "graded way" was Campus Martius, the fort built along the Muskingum by the first settlers for the Ohio Company.

To be sure, the directors of the Ohio Company, who laid out Marietta in 1788, respected the ancient architects of these works and provided for their preservation. One of their very first orders of business after reaching the mouth of the Muskingum was to provide that portions of the earthworks be set aside for public use. Within the larger enclosure, the elevated square marked "A" became *Quadranou* and the smaller truncated mound, "B", was reserved as *Capitolium.*[53] Several years later, the proprietors of the Company leased these squares to private individuals, but provided first that the land was to be landscaped as prescribed by the Committee to View the Public Squares and that the "Elevated Squares with the Ascents leading to the same be immediately put into grass and hereafter occupied in no other way."[54] As late as 1796, the Proprietors unanimously agreed that "every prudent measure ought to be adopted to perpetuate the figure and appearance of so majestic a Monument of Antiquity."[55] Eighty-six years later, in 1882, a brick manufacturer in Marietta was elected to the city council. After some maneuvering, he persuaded the city fathers to sell him the remaining walls of the Marietta earthworks from which to make bricks.[56]

From almost the middle of the nineteenth century, men such as Caleb Atwater, Ephraim George Squier, Edwin Hamilton Davis, Henry C. Shetrone, Charles Whittlesey, Gerard Fowke, Stephen Peet, William C. Mills and many more labored as best they could to record, if not preserve, the prehistoric earthworks in Ohio for posterity. To be sure, many of the works had been built by the early inhabitants of the state on the very lands that were necessary for the growth and development of modern civilization and, in many instances, important Ohio cities were built on the graves of those from ages past. The generations that built the agrarian and industrial wealth of this state during the last century did not have the time, nor perhaps the patience or will, to preserve carefully and painstakingly the remains of the ancient cultures with which they themselves had no direct kinship. The scholars and men of letters tried, but they lacked the political or the economic power to prevail. Toward the end of the century, however, support was forthcoming from those who had economic security and the intelligence and interest to care. Several scholars and leading citizens had founded the State Archaeological Association of Ohio in 1875. Enthusiastic about their work, and supported by the state legislature, the members of the Association put together an exhibit of over 5,000 artifacts for the 1876 Centennial Exposition in Philadelphia that was second only to the one displayed by the Smithsonian Institute. Finally in 1885, the state legislature chartered the Ohio Archaeological and Historical Society, which began to work for the care and preservation of the remaining major prehistoric sites in the state. By the time the United States went to war with Spain, Ohio had purchased the Fort Ancient site and two years later, in 1900, it received the deed to the Serpent Mound in Adams County. In 1898, George Frederick Wright of Oberlin College, the author of thirty-three articles dealing with prehistoric Ohio, addressed the thirteenth annual meeting of the Ohio Archaeological and Historical Society. In his speech he outlined the work that was necessary for both the archaeologists and the state legislature to ensure the future preservation of the state's prehistoric works. Most importantly, however, he captured the spirit and the national pride felt by those who had conducted their labors of love at Ohio's ancient monuments throughout the nineteenth century. He dramatically concluded:

> For these relics of prehistoric time in which our State so notably abounds cannot be disregarded without great loss of opportunity. Man does not live by bread alone. There are subtle ties of more than sentimental interest which bind us to the whole human race, and which make the simplest relics of man's past work instructive and impressive. Through these most remarkable relics of the Mound Builders to which our attention has been directed we receive impressions concerning the significance of life and the relative value of the ends which we choose as the mainspring of our activity which we cannot afford to lose. Our children should be more thoroughly instructed as to the character and importance of the prehistoric remains which render their State attractive to historians and men of science the world over. Instead of having their thoughts turned to Egypt and Babylon as the only centers of early civilization worth noticing, their attention may well be directed first to the remarkable inclosures and symbolical monuments of Ohio, which speak so eloquently of a departed and long lost phase of human history enacted within our own borders.[57]

1. David Zeisberger, "David Zeisberger's History of North American Indians," ed. Archer Butler Hulbert and William Nathaniel Schwarze, *Ohio Archaeological and Historical Publications,* XIX (1910), 30–31.

2. Charles Whittlesey, "Description of the Ancient Works in Ohio," *Smithsonian Contributions to Knowledge,* III (Washington, 1850).

3. Richard G. Morgan and James H. Rodabaugh, *Bibliography of Ohio Archaeology* (Columbus: The Ohio State Archaeological and History Society, 1947).

4. Edmund de Schweinitz, *The Life and Times of David Zeisberger, The Western Pioneer and Apostle of the Indians* (Philadelphia, 1870), p. 371.

5. Zeisberger, "History of North American Indians," 30–31.

6. "A Plan of an Old Fort and Intrenchment in the Shawanese country, taken on horse back, by computation only," *Royal American Magazine for January 1775* (Boston, 1775), 29–30.

7. David Jones, *A Journal of Two Visits Made to Some Nations of Indians on the West Side of the River Ohio, in the Years 1772 and 1773 . . . With a Notice of the Author, By Horatio Gates Jones* (New York, 1865), pp. 56–57.

8. Jonathan Heart, "Account of Some Remains of Ancient Works, on the Muskingum, with a Plan of these Works," *Columbian Magazine,* I (May 1787), 425–27.

9. Samuel F. Haven, "General Opinions Respecting the Origins of Population in the New World," *Smithsonian Contributions to Knowledge,* VIII (1856), 27.

10. Ezra Stiles, *American Magazine* (March 1788), 246–47; (April 1788), 291–94; Noah Webster, *American Magazine* (December 1787), 15–19; (January 1788), 87–93; (February 1788), 146–56; (July 1788), 537–41.

11. Winthrop Sargent, "Plan of an Ancient Fortification at Marietta, Ohio," presented by Henry I. Bowditch in American Academy of Arts and Science, *Memoirs,* n.s., V (1853–55), 25–28.

12. *Port Folio,* 3d ser., III (1810), 485–86.

13. Thaddeus Harris, *The Journal of a Tour into the Territory Northwest of the Allegheny Mountains; Made in the Spring of the Year 1803. With a Geographical and Historical Account of the State of Ohio. Illustrated with Original Maps and Views* (Boston 1805), pp. 147–76.

14. "Petition to the Legislature of Massachusetts, October, 1812," cited in the chapter "Origin of the American Antiquarian Society," in Caleb Atwater, "Description of the Antiquities Discovered in the State of Ohio and other Western States," in American Antiquarian Society, *Archaeologia Americana. Transactions and Collections,* I (Worcester, Mass., 1820), 5.

For an excellent discussion of the intellectual activity concerning the mounds, see Robert Silverberg, *Mound Builders of Ancient America, The Archaeology of a Myth* (Greenwich, Connecticut: New York Graphic Society, 1968).

15. Henry Howe, *Historical Collection of Ohio* (Cincinnati: C. J. Krehbiel & Co., 1900), II, 417.

16. Atwater, *Transactions and Collections,* I, 5.

17. Caleb Atwater, *Writings of Caleb Atwater* (Columbus, 1833), p. 5.

18. Atwater, *Transactions and Collections,* I, 183.

19. Henry C. Shetrone, "Explorations of the Hopewell Group of Prehistoric Earthworks," *Ohio Archaeological and Historical Publications,* XXXV (1926), 8–9.

20. Atwater, *Transactions and Collections,* I, 142.

21. *Ibid.,* p. 179

22. Henry Howe, *Recollections of Ohio History,* II, 414–22. If this burial was Hopewell, the flint points found were no doubt spear points rather than arrows. The use of the bow and arrow by Indians was not introduced until the end of the prehistoric period.

23. Atwater, *Transactions and Collections,* I, 180–81.

24. Ephraim George Squier and Edwin Hamilton Davis, *Ancient Monuments of the Mississippi Valley: Comprising the Results of Extensive Original Surveys and Explorations,* Vol. I of *Smithsonian Contributions to Knowledge,* Washington, D.C., 1848, 4.

25. *Ibid.,* pp. 63–64.

26. *Ibid.,* pp. 52–53.

27. *Ibid.,* p. 53.

28. *Ibid.,* pp. 51–52.

29. *Ibid.,* pp. 54–55. While Squier and Davis mentioned that 24 mounds existed on this site, their map indicated only 23. Recent excavations still have not identified the exact number.

30. William C. Mills, "Exploration of the Mound City Group, Ross County, Ohio," *The American Anthropologist,* XXIV (October, November 1922), 397–431; William C. Mills, "Exploration of the Mound City Group," *Ohio Archaeological and Historical Publications,* XXXI (1922), 423–584; G. Richard Pick, "The Rise and Fall of Camp Sherman," pamphlet printed by Craftsman Printing and Kem Printing Service Inc., Chillicothe, 1967.

31. U.S. Department of the Interior, National Park Service, "Mound City Group," 1967, pamphlet.

32. Squier and Davis, *Monuments of the Mississippi Valley,* p. 92.

33. *Ibid.,* pp. 50–51.

34. *Ibid.,* pp. 56–57.

35. Frederick Ward Putnam, "Report of the Curator. To the Trustees of the Peabody Museum of Archaeology and Ethnology," in Peabody Museum, *Eighteenth and Nineteenth Annual Report,* 1884 and 1885 (Cambridge, 1886), bound in Peabody Museum, *Reports,* III (Cambridge, 1887), 401–18.

36. William C. Mills, "The Exploration of the Edwin Harness Mound," *Ohio Archaeological and Historical Society Publications,* XVI (1907), 113–93.

37. Squier and Davis, *Monuments of the Mississippi Valley,* p. 62.

38. Gerard Fowke, *Archaeological History of Ohio; The Mound Builders and Later Indians* (Columbus: F. J. Heer 1902), pp. 206–07.

39. Squier and Davis, *Monuments of the Mississippi Valley,* p. 38.

40. William C. Mills, "Exploration of the Seip Mound," *Ohio Archaeological and Historical Publications,* XVIII (1909), 269–321.

41. Henry C. Shetrone and Emerson F. Greenman, "Exploration of the Seip Group of Prehistoric Earthworks," *Ohio Archaeological and Historical Publications,* XL (1931), 343–509.

42. Squier and Davis, *Monuments of the Mississippi Valley,* pp. 57–58.

43. Fowke, *Archaeological History of Ohio,* p. 206.

44. Squier and Davis, *Monuments of the Mississippi Valley,* p. 11.

45. *Ibid.,* pp. 11–14.

46. Raymond S. Baby, "The Hopewell Culture," *Ohio State Archaeological and Historical Quarterly,* LI (1952), 182–85.

47. Fowke, *Archaeological History of Ohio,* p. 244.

48. Stephen D. Peet, "The Sacred Enclosures of Ohio," *American Antiquarian,* XII (1890), 131–54.

49. Squier and Davis, *Monuments of the Mississippi Valley,* p. 87.

50. Donald McBeth, "Bourneville Mound, Ross County, Ohio," *Ohio Archaeologist,* X (January 1960), 12–13.

51. W. W. Mather, *Second Annual Report on the Geological Survey of the State of Ohio* (Columbus: Samuel Medary, 1838), p. 43.

52. Locke's "Fort of Fort Hill in Highland County, Ohio," is found in Mather, *Second Annual Report,* p. 269.

53. Archer Butler Hulbert, ed., *The Records of the Original Proceedings of the Ohio Company,* Vol. I of the *Ohio Company Series* (Marietta, Ohio, 1917), 51.

54. Hulbert, *The Records of the . . . Ohio Company,* II, 80.

55. Hulbert, *The Records of the . . . Ohio Company,* I, 209.

56. John Patterson MacLean, "Ancient Works at Marietta, Ohio," *Ohio Archaeological and Historical Publications,* XII (1903), 38.

57. "Addresses Before the Ohio State Archaeological Society," *Ohio Archaeological and Historical Publications,* VI (1898), 433–48.

CHAPTER THREE

Tours, Battles and Plans

When military history is written, it is usually national in scope. Each state, however, has its own lists of battlefields, fortifications and heroes that are very much a part of the record of its own past. Ohio, the first state to be carved from the territory north and west of the Ohio River, has an exciting and proud military record. The blood of many Americans was shed in battle when the territory stood at the threshold of the frontier and protected the country's exposed western border. Since the Civil War, not only its human resources, but its rich natural resources and industrial power as well have made Ohio vital to the nation's security. Its military past, in part, has been chronicled over the years through manuscript plans of battles, contemporary drawings, and official plans of fortifications built within the state. All of these, too, make up another part of the history of the mapping of Ohio.

Between 1689 and 1763, four intercolonial wars were fought between England and France for control of North America. With the Peace of Paris in 1763, which ended the Seven Years War in America, the British had won control of the continent. On the eve of that last North American conflict between the two great colonial powers, the French sent Celoron de Blainville, a former commandant of Detroit, along with Father Joseph Pierre de Bonnécamps, a Jesuit priest and mathematician, at the head of an armed expedition down the Ohio River in 1749 to exert French influence over the Ohio Valley. At the confluence of each major stream that emptied into the Ohio, Celoron placed a lead plate upon which was inscribed the date and a passage declaring the land drained by that particular tributary under the direct control of the King of France. In Ohio, plates were buried at the mouths of the Muskingum, the Little Miami and the Great Miami rivers. As a result of this expedition, Bonnécamps, who accompanied the Celoron expedition as its chaplain and geographer, produced the map "Carte D'un Voyage Fait dans La Belle Rivière En La Nouvelle France" (Plate I), which, together with his journal of the famous expedition, appeared in the *Jesuit Relations.*[1] The original of this map is in the Paris archives.

The French expedition left La Chine, later Montreal, on June 15, 1749, and entered the Ohio River on July 29. Two months after Celoron's journey began, the French explorer buried his fourth lead plate (the first in Ohio) at the mouth of the Muskingum River, which Bonnécamps called the Yenangue Konnan. In his own journal, Celoron recorded:

> The 15th of August, 1749, we, Celoron, Knight of the Royal and Military Order of St. Louis, Captain commanding a detachment sent by the orders of Monsieur the Marquis de la Galissoniere, Governor-General of Canada, upon the Beautiful River, otherwise called the River Oyo, accompanied by the principal officers of our detachment, have buried at the foot of a maple tree, which forms a triangle with a red oak and an elm tree, at the entrance of the river, Jenuanguekouan (Muskingum), at the western bank of that river, a leaden plate, and have attached to a tree on the same spot, the arms of the king. In testimony whereof we have drawn up and signed the present official statement, along with Messrs., the officers at our camp, the 15th of August, 1749.[2]

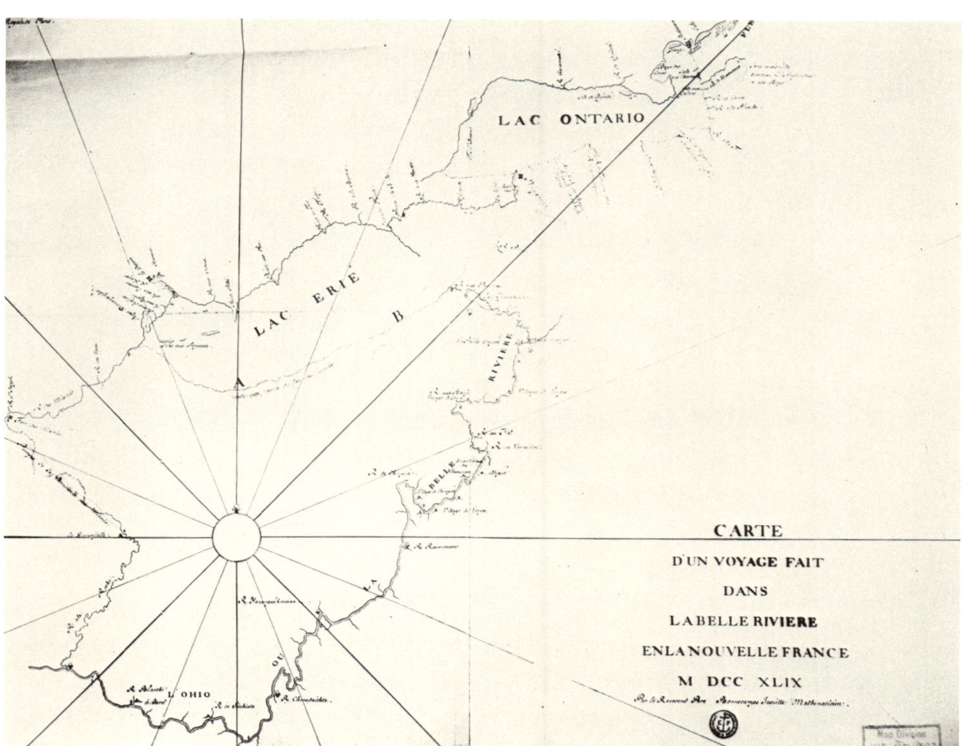

Plate I: Bonnécamps, "Carte D'un Voyage . . . en la Nouvelle France," (1749), accompanied his diary of Celoron de Blainville's expedition. This plate charts the French attempt to expel the British from the Ohio Valley.

This plate was found in 1798 by several boys playing along the banks of the river. Because of the scarcity of lead on the frontier, it was partially melted down for shot before its value was recognized. Caleb Atwater, the Ohio historian, had possession of it for a brief time but sent it to DeWitt Clinton of New York, who in turn gave it to the American Antiquarian Society at Worcester, Massachusetts.

Six days later, on August 21, Celoron reached the Indian village St. Yotoc, or Sinhioto, or Shawnee Town, at the mouth of the Scioto River. At first the Indians were not too friendly, and the French were forced to build a hasty fortification for protection along the southern shore of the Ohio opposite the Scioto. Soon, however, Celoron convinced the Indians that he did not desire to war upon them but merely to warn the British traders located there to leave. Bonnécamps, in his *Journal,* explained that the location of Sinhioto "is quite pleasant,—at least, it is not masked by the mountains, like the other village through which we have passed. The Sinhioto [Scioto] river which bounds it on the West, has given it its name. It is composed of about sixty cabins."[3] On August 27, the French reached the mouth of the *R. Blanche* or White River or Little Miami River. Here was located a small Miami Village of about seven or eight cabins under the leadership of *le Baril* or The Barrel.

Celoron buried his last lead plate at the mouth of the *Rivière à la Roche* or Great Miami River. It was this river that the French expedition ascended to Pickawillany, the Miami village below the junction of Loramie Creek and the Great Miami River, whose chief was La Demoiselle or, as the English knew him, Old Britain. Bonnécamps, the careful observer, remarked that the "*rivière á la Roche* is very well named. Its bottom is but one continuous rock; its waters are extremely shallow."[4] After several days of conferences there with the western Indians, Celoron made his way via *Portage de la Demoiselle aux Miamis to Fort des Miamis,* which was located at the present site of Fort Wayne, Indiana. "The fort of the Miamis was in very bad condition when we reached it; most of the palisades were decayed and fallen into ruin," the Jesuit priest recorded. "Within there were eight houses,—or, to speak more correctly, eight miserable huts, which only the desire of making money could render endurable."[5] From there, Celoron descended the Miami of the Lakes, or the Maumee River to Lake Erie. Bonnécamps observed that "the Miami River [Maumee] caused us no less embarrassment than *Rivière à la Roche* had done. At almost every instant we were stopped by beds of flat stones, over which it was necessary to drag our pirogues by main force. I will say, however, that at intervals were found beautiful reaches of smooth water, but they were few and short. In the last six leagues, the river is broad (and deep), and seems to herald the grandeur of the lake into which it discharges its waters."[6] After a visit to Detroit, Celoron and his men were back in Montreal on November 10. Celoron's expedition reflected France's growing anxiety over the Ohio Valley and the English encroachments there. Within several years, the fourth and final colonial war was fought and England emerged in sole possession of the continent. Bonnécamps's map of Celoron's trip down the Ohio river and through the interior of the state recorded the attempt to hold Ohio for the King of France on the eve of the Seven Years War.

The British were interested in the Ohio Valley because of its abundant wealth in furs, the affluent trade with the Indians and the desire on the part of some colonists to create new British colonies west of the mountains. After 1754, war raged on the frontier; but by 1760, the French had virtually surrendered their holdings in North America to the English. Immediately after the Peace of Paris in 1763, the British rushed to take control of the numerous French posts throughout the vast northwest. England now became concerned about its relationship with its former Indian allies, as well as those who had supported France during the last war. In order to gather information concerning the attitude and disposition of all those Indians living in the vast Great Lakes basin, George Croghan, Deputy Agent for Indian Affairs for the Northern District, sent Thomas Hutchins on a trip in 1762 throughout the region west and north of the Ohio to visit the various Indian villages. Hutchins, who was then an ensign in the Royal American Regiment of Foot, kept a close record of his tour.[7] On April 4, 1762, he left Fort Pitt at the Forks of the Ohio and made his way westward, through the interior of Ohio to Sandusky. From there, he went on to Detroit, Michilimackinac and Green Bay. On July 12 he was at the "Fort at the Miamie"; on September 8 he visited Lower Shawnee Town at the mouth of the Scioto. Hutchins returned to Fort Pitt on September 24, six months after he had left. Upon his return, he reported to Croghan that all was not well with the Indians in the Ohio Country and concluded: "I see it is with the greatest difficulty that the officers can keep them [the Indians] in a good Temper."

The map, "A Tour from Fort Cumberland North Westward round part of the Lake Erie, Huron, and Michigan, including part of the Rivers St. Joseph, the Wabash and Miamis, with a sketch of the road from thence by the Lower Shawnee Town to Fort Pitt, 1762," was made by Hutchins probably shortly after his return. Portions of this map were copied by Guy Johnson, one of the deputy agents for Indian Affairs. (He was a nephew of Sir William Johnson, Superintendent of Indian Affairs in the Northern District.) Johnson entitled his map "A Sketch of the several Indian roads leading from Fort Pitt to Scioto, Lake Erie & Tak.[n] from a Draft made on a Tour throy.[t] Country in 1762 by Mr. Hutchins then in the Indian Department" (Plate II). The deputy agent omitted the upper portions of the original Hutchins map that showed the whole of Michigan and copied only those parts that illustrated western Pennsylvania and Ohio. This map is important to the history of the gathering of geographic data about Ohio because it showed the extent of knowledge gained through personal observations just on the eve of the famous Pontiac Rebellion in 1763.

Following the conclusion of the French and Indian War, the Ottawa chief Pontiac used his Confederacy, an alliance of the various Indian tribes in the old northwest, to attack the string of British posts from Fort Detroit to Fort Pitt in an attempt to expel their new rulers. The British were able to thwart the Indians, and when the Indian alliance was broken, the English military authorities wanted to punish the Indians who had participated in this war. In 1764 two expeditions were sent out from Fort Pitt into the Ohio country to extract declarations of peace from the Indians and to gather up the white captives taken by the Indians during both the French and Indian War and the Pontiac uprising. One expedition, lead by Colonel Henry Bouquet, a Swiss officer in the British army, struck deep into Indian territory and established a small temporary post close to the junction of the Muskingum and the Tuscarawas rivers. Hutchins, who accompanied Bouquet's expedition, recorded the army's march into Ohio on a map entitled "A Survey of the Part of the Indian Country through which Colonel Bouquet Marched in 1764." This map accompanied *An Historical Account of the Expedition Against the Ohio Indians, In the Year MDCCLXIV*

Plate II: Johnson, "A Sketch of the Several Indian Roads," (1762), William L. Clements Library, Ann Arbor, Michigan. This map illustrates Ohio on the eve of Pontiac's War.

Under the Command of Henry Bouquet, Esq., which first appeared in both colonial America and in England in 1765.[8]

A very elaborate edition of this map was later produced and entitled "A Topographical Plan of that part of the Indian Country through which the Army under the Command of Colonel Bouquet marched in the Year 1764 by Thoss: Hutchins' Asst. Engineer"[9] (Plate III). The major portion of this map contained a plan called "A General Map of the Country on the Ohio and Muskingum Shewing the Situation of the Indian-Towns with respect to the Army under the command of Colonel Bouquet by Thoss: Hutchins' Asst. Engineer," which was an elaboration of Hutchins's "A Tour from Fort Cumberland. . . ." The two plans that were contained in the impressive "A Topographical Plan . . ." served in large part as the basis for Hutchins's greatest work, "A new Map of the Western Parts of Virginia, Pennsylvania, Maryland and North Carolina; Comprehending the River Ohio, and all the Rivers, which fall into it; Part of the Mississippi, the whole of the Illinois River, Lake Erie, Parts of the Lake Huron, Michigan, etc. And all the Country bordering on these Lakes and Rivers."[10]

The top portion of the Hutchins map, "March of His Majesty's Troops from Fort Pitt, to the Forks of the Muskingum in 1764," illustrated the location of the sixteen campsites established by Bouquet's force as it made its way from Fort Pitt deep into the Indian territory of the Ohio Country. This rather sizeable force consisted of colonial troops, including the First and Second Corps of Virginia Volunteers, the First and Second Battalions of Pennsylvanians and two companies of Maryland volunteers, as well as the 60th Royal American Regiment, one of the King's own regiments recruited mostly from Protestant Pennsylvanian Germans, and the 42nd Regiment of Foot, or the famous Black Watch, which consisted of regular British forces. It is of interest to follow Bouquet's movements to the Muskingum as recorded by Hutchins.

The first encampment within Ohio along the Conchake Trail was made on Sunday, October 7, 1764. "The camp No. 6," the historian of the expedition related, "lies at the foot of a steep descent, in a rich valley, on a strong ground, three sides thereof surrounded by a hollow, and on the fourth side a small hill, which was occupied by a detached guard."[11] This site is located near the Mt. Zion Church in Middleton Township in Columbiana County. Camp 7, reached the following day, October 8, "eleven miles one quarter and forty nine perches from the last encampment," was situated "by a small run on the side of a hill, commanding the ground about it."[12] This camp, on a small branch of Patterson's Run, was just east of present U.S. Route 30, about one and a half miles north of West Point, Madison Township, Columbiana County.

On October 9, the British column came to a fork in the path in what is now the northeast corner of Wayne Township in Columbiana County. Here stood "several trees painted by the Indians in a hieroglyphic manner, denoting the number of wars in which they were engaged and the particulars of their success in prisoners and scalps. The Camp No. 8 lies on a run, and level piece of ground, with Yellow-creek close on the left, and a rising ground near the rear of the right face. The path after the army left the forks was so brushy and entangled that they were obliged to cut all the way before them, and also to lay several bridges in order to make it passable for the horses, so that this day they proceeded only five miles, three quarters and seventy perches."[13] Camp 8 was located somewhere in the extreme southern part of Columbiana County. The following day, after a march of "seven miles and sixty perches in the whole," Camp No. 9 was established at what is now the intersection of Ohio Route 644 and Augusta Township 257, almost on the line between Hanover Township, Columbiana County and East Township, Carroll County.

Campsite No. 10 was located near the town of Minerva, probably along the banks of Sandy Creek, a branch of the Tuscarawas River. "Keeping the aforesaid creek [Sandy Creek] on their left, they marched through much fine land, watered with small rivers and springs: proceeding likewise through several savannahs or cleared spots, which are by nature extremely beautiful; the second which they passed being, in particular, one continued plain of near two miles, with a fine rising ground forming a semicircle round the right hand side, and a pleasant stream of water at about a quarter of a mile distant on the left. The camp No. 11 has the above-mentioned branch of Muskingum [Tuscarawas] on the left, and is distant ten miles and three quarters from the last encampment."[14] This campsite was about one mile north of Magnolia in the southern portion of Stark County.

On October 13, Bouquet reached the crossing place of the Tuscarawas where Bolivar is now located, and established Camp No. 12. "A little below and above the forks of this river [Tuscarawas] is Tuscarowas, a place exceedingly beautiful by situation, the lands rich on both sides of the river; the country on the north-west side being an entire level plain, upwards of five miles in circumference. From the ruined houses appearing here, the Indians who inhabited the place and are now with the Delawares, are supposed to have had about one and fifty warriors. This camp No. 12 is distant eight miles nineteen perches from the former."[15] It was here, at the "Great Crossing Place of the Muskingum," that Bouquet expected to make his initial contact with the Indians. He delayed in this camp for one day; and when no Indians appeared, he moved "two miles forty perches"

(Overleaf) Plate III: Hutchins, "Plan of Indian Country through which Colonel Bouquet marched," (1764) printed with an account of the expedition, (1765). It clearly shows the British attempt to exert control over the Ohio Valley.

A Topographical
PLAN
of that part of the
INDIAN-COUNTRY
through which the Army
under the Command of
COLONEL BOUQUET
marched in the Year 1764.
by Thos. Hutchins, Asst. Engineer.
March of His Majesty's Troops from Fort Pitt
Scale of Ten Miles
1 2 3 4 5 10
10 Encampt
11 Encampt
12 Enct
Fort Built Oct. 16, 1764
13 Encampt
14 Encampt
15 Encampt
16 Encampt
Muskingum River
Three Legs. Old Town.
New Comers T
Conference House
Old Wyandot T.
White Womans
Forks of Muskingum
Bullets T.
Tom's T.
PART of
Miamis Fort
Mineame River
Sandusky R.
Wyandot Town
Junqueindundeh
Miamis R. 25 Yards wide
6 Yards wide
Small Creek
20 Yards wide
100 Yards wide
10 Yards wide
10 Yards wide
60 Yards wide
10 Yards wide
a Spring
10 Yards wide
Lick T
Lower Shawneese T
Sioto R.
60 Yards wide
The Country from the Miamis Fort to Sandusky is level and Rich.

Forks of Muskingham in 1764.
A GENERAL MAP of the COUNTRY on the OHIO and MUSKINGHAM Shewing the Situation of the INDIAN-TOWNS with respect to the Army under the Command of COLONEL BOUQUET by Thos. Hutchins Ass. Engineer.
Path to the LowerTowns.
7 Encampt
6 Encampt
5 Encamp.
Big Beaver C
4 Encampt
Logs T.
OHIO R
3 Encampt
2 Encampt
1st Encampt
Chartiers C
Ruins of the Saw Mill
F Pitt
Grants Hill
Road from Fort Pitt to Fort Ligonier.
ERIE
Fort Presq Isle,
Fort Le Boeuf,
little Lake
Sugar Creek 10 Yards wide
The Allegheny R
Venango Fort
Allegheny R
Ottawas T
Cayahoga T
Salt Lick T.
Mohoning T
Pematuning
Shaningo
Big Beaver
Kishkuske T.
The Beaver Town at Tuscarawas
Bucks T.
Delaware T.
Muskingum R
Will's T
New T
Ohio R
Fort Pitt
Turtle C.
Bushy Run
Coll Bouquets Field of Battle
Gl. Braddocks Field
Kiskeminetas C
Monongahela R
Fort Ligonier
Yoxhio
Laurel Hill
Great Meadows
Stony Creek
Allegheny Mountains
F. Bedford
Juniata R
Sidling Hill
Fort Cumberland
Potowmack R
Fort Lyttleton
F Loudon
Savage R
Scale of 30 Miles
3 15 30
the Ohio, to Beaver Creek from other Maps, and not from a Survey.
Canhawa R

further down river and established Camp No. 13 on October 15. A fort was built here as indicated on the map, and "some distance from the camp" a conference bower was constructed. An artist's concept of this bower appeared in the lower right-hand corner of the map. On October 17, Bouquet met at the bower with Kiyashuta of the Senecas, Custaloga of the Wolfe clan and Beaver of the Turkey clan of the Delaware, and Keissinaulchtha of the Shawanee. The official accounts of the expedition stated that "the colonel, with most of the regular troops, Virginia volunteers and light horse, marched from the camp to the bower erected for the congress. And soon after the troops were stationed, so as to appear to the best advantage, the Indians arrived, and were conducted to the bower. Being seated, they began, in a short time, to smoke their pipe of calumet, agreeable to their custom. This ceremony being over, their speakers laid down their pipes, and opened their pouches, wherein were their strings and belts of wampum."[16] The Indians promised peace and agreed to return their white captives taken during the many years of war on the western frontier.

Determined to pursue his advantage and not to allow the Indians time to prepare for resistance, Bouquet moved on October 22 from Camp No. 13, "nine miles to camp No. 14, crossing Margaret's creek [Sugar Creek] about fifty feet wide." This campsite was near Winfield, Dover Township, in Tuscarawas County. The following day the army marched "sixteen miles one quarter and seventy seven perches" to Camp No. 15, located near the present village of Chili, Crawford Township, in Coshocton County.

On October 25 the army moved further down river to Camp No. 16 and made its deepest penetration into the Ohio Country. The account related:

> They marched six miles, one half and sixteen perches to camp No. 16, situated within a mile of the Forks of Muskingham; and this place was fixed upon instead of Wakautamike, as the most central and convenient place to receive the prisoners; for the principal Indian towns now lay around them, distant from seven to twenty miles; excepting only the lower Shawanese town situated on Scioto river, which was about eighty miles; so that from this place the army had it in their power to awe all the enemy's settlements and destroy their towns, if they should not punctually fulfill the engagements they had entered into.—Four redoubts were built here opposite to the four angles of the camp, the ground in the front was cleared, a store house for the provisions erected, and likewise a house to receive, and treat of peace with, the Indians, when they should return. Three houses with separate apartments were also raised for the reception of the captives of the respective provinces, and proper officers appointed to take charge of them, with a matron to attend the women and children; so that with the officers mess houses, ovens, etc., this camp had the appearance of a little town in which the greatest order and regularity were observed.[17]

The map did not illustrate the second fort built by Bouquet at the Forks of the Muskingum. Convinced that resistance was futile, it was here that the Indians returned to the Colonel 306 whites who had been captured from the settlements during nearly a decade and a half of constant border warfare. His mission accomplished, Bouquet returned to Fort Pitt on November 18, 1764.

By the fall of 1774, on the eve of the American Revolution, Virginian frontiersmen had penetrated westward, beyond the mountains, and along the southern shore of the Ohio River as far as the site of Louisville, Kentucky. These encroachments by the Virginians into lands considered inviolable by the Shawnee, the most powerful of the Ohio Indians, brought conflict to the Ohio Valley that resulted in Lord Dunmore's War. Lord Dunmore, the last royal governor of Virginia, raised two forces, one directly under his own command and another under that of Colonel Andrew Lewis. Dunmore and his men marched from Fort Pitt to Wheeling and then on to the mouth of the Hocking River. After building Fort Gover, the army proceeded up the Hocking to the famous Pickaway Plains on the Scioto. Lewis, meanwhile, moved from southwestern Virginia, down the New and Kanawha river valleys to Point Pleasant on the Ohio. Here, on October 10, 1774, was fought the Battle of Point Pleasant, which attracted worldwide attention because it represented the classic frontier struggle between the white man and the noble red man deep in the pristine American wilderness.[18] At Camp Charlotte, the small encampment established to the east of the Shawnese villages along the Scioto, Dunmore concluded a satisfactory peace that kept the Ohio Indians neutral for at least the first two years of the Revolutionary War. It was at this peace conference that Logan, the Mingo or Seneca chief, submitted his famous speech that synthesized the plight of the Indian caught by the advance of the white's westward movement.

Plate IV: Renick "Indian Towns on the Pickaway Plains," (1844), prepared for *The American Pioneer* (1844) but published as a separate map. It details the famous Shawnee villages on the upper Scioto River.

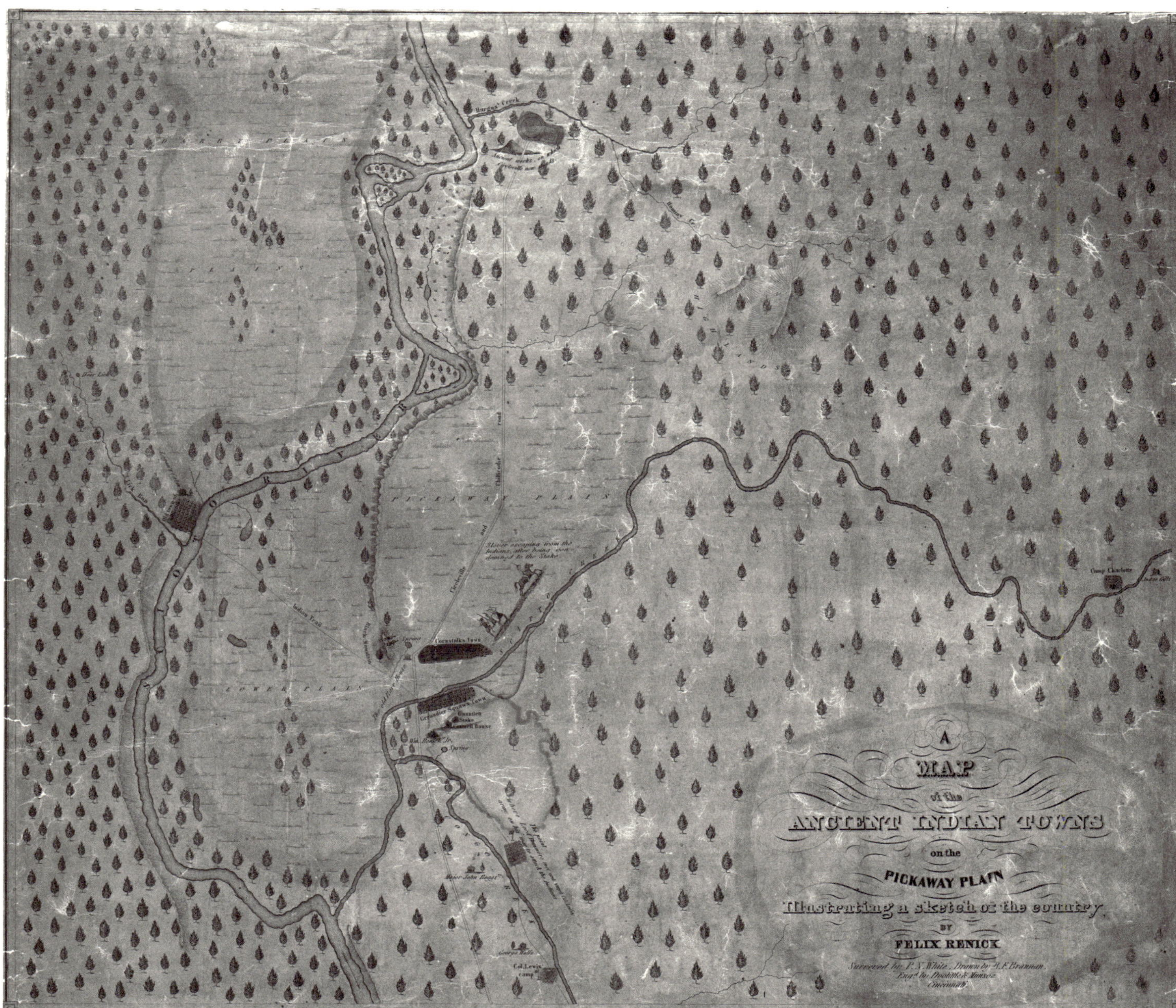

"A Map of the Ancient Indian Towns on the Pickaway Plain Illustrating a Sketch of the Country" (Plate IV), prepared by Felix Renick, depicted the situation of the Shawnese villages on the famous plains in central Ohio shortly after Lord Dunmore's War at the close of the eighteenth century. Renick was born in 1771 in Virginia and moved to Ross County in 1798. He was influential in establishing the beef cattle industry in the Scioto River Valley and was the founder of the Ohio Company for importing English cattle, which introduced the purebred English shorthorn into the state. He was an educated man who enjoyed leading a gentleman farmer's life at his home, High Rock Farm in Ross County. Renick had a profound interest in history and served as the president of the Logan Historical Society, a group founded at Westfall on July 28, 1841, "to collect and preserve for the use of posterity, all that can be collected of early or incidental history, of the early settlement and successive improvements of the Western country, and, as collateral therewith, of all North America."[19] This society was named for Logan, the Mingo chief, who had shown particular friendship to the Americans until several members of his family were killed in 1773. The Society's monthly publication, *The American Pioneer,* which was compiled in Chillicothe but printed in Cincinnati, was rather short lived. Its first edition appeared in January 1842 and its last in October 1843. Renick was a regular contributor and had prepared his map for that publication. The magazine was suspended, however, before his work appeared. Consequently the map was printed as a single sheet probably in 1844, by Munson and Doolittle of Cincinnati.[20] A small revised version of it appeared in Henry Howe's *Historical Collection of Ohio* in 1848.[21]

On his map, Renick illustrated the Shawnee villages as they were located on Pickaway Plains after 1758, believed to be the year that the Indians moved their homes from the mouth of the Scioto. Grenadier Squaw's Town, located on the south bank of Scippo Creek, was named for Nonhelema, the sister of Cornstalk, the Shawnee chief. She was born some time in the 1720s in the southern part of the country. Apparently Nonhelema was a big woman, and a natural and capable leader. The British called her a Grenadier squaw or woman soldier and thus the name of the village. She eventually became a Christian and was known among the whites as Katy.[22] Cornstalk's Town, on the north bank of Scippo Creek, was the village of the famous Shawnee. He had led the Shawnees against Colonel Andrew Lewis at the Battle of Point Pleasant and sued for peace to Lord Dunmore following the action there. He had pledged peace to the whites, and in October 1777, while being held as a hostage by the Virginians, he was murdered along with his son, Elinipsico, at Fort Randolph, which had been constructed at the mouth of the Great Kanawha River. The third village Renick showed was Chillicothe, located four miles below Circleville along the west bank of the Scioto River. After the Shawnees left Ohio, it became Westfall, a white settlement. "Westfall was once, a small white village. It is now a name only—cornfields, pastures, etc.—have taken the place of tavern, blacksmith shop & house. It was first 'Old Chillicothe' 'Shawneetown'—then 'King's Ferry' again 'Westfall,' and last and best, a splendid farm."[23] Apparently Renick knew the exact location of Lewis's camp. In a letter written in 1842, he stated that "it was immediately on one of the first traveling traces used between Chillicothe and the Pickaway Plains. It differed in some respect from Camp Charlotte. There was no appearance of anything having been done toward fortifying it."[24]

Not only did the Renick map locate the important points during Lord Dunmore's War at the Pickaway Plains, but it illustrated the miraculous escape of John Slover from the Shawnee at Cornstalk's Town. Slover, along with Jonathan Zane, was a guide for Colonel William Crawford's expedition against Sandusky in 1782. Following the engagement at Battle Island in what is now Wyandot County, Crawford's forces began a disorganized march back to the Ohio. Along with several other men, Slover was captured on June 9 near the Tuscarawas River in Wayne County. Earlier, in 1759, he had been taken from his home along New River in Virginia and lived for six years with the Miamis and Shawnees in their villages in Ohio. In 1773 he was released to relatives in Pittsburgh.[25] In 1782, however, Slover was taken to the Indian villages along the Scioto and sentenced to death by burning. After experiencing torments, including running the gauntlet, he was able to escape. The explanation which accompanied the map (Plate IV) described the situation:

> The above map, with the appended remarks, was prepared for the "American Pioneer," but that publication having been suspended, the author was advised, by some of his friends, to have it engraved, and offered to the Public. With this advice he now complies; not doubting but the Work will receive all the patronage it may be thought to merit. Be its favor great or small, he will at least have the gratification of knowing that he is handing down to posterity an accurate representation of matters and things of bygone days.
>
> The scope of the map comprises—the ancient Indian Towns of the Pickaway Plains and vicinity—the Encampments of Dunmore and Lewis, in their expedition against those Towns, in 1774—and also a small representation of the more Ancient Works where the Town of Circleville now stands.
>
> Of all places in the Western country, the section represented in the drawing pre-eminently deserves the name of "classic ground." There, in olden times, burned the council fires of the Red Man; here the affairs of the nation, in general council, were discussed, and the important question of peace and war decided. Here the "pent-up agony" of the chiefs and braves burst forth, in those impassioned strains of thought and feeling which compare so favorably with the best eloquence of ancient or modern times. Instance, among many others, the speeches of Logan and Cornstalk, at Dunmore's Treaty—or, more properly speaking, his *truce,* with the Indians at Camp Charlotte. Here was their rendezvous, in times of war; and their place of retreat, in case of defeat. Here, they made their last stand for the salvation of their country. Valor consecrated the spot. The eloquence of their Braves, in council, was equalled by their daring in the field; and their feats in arms evinced a love of country and a contempt for death unsurpassed in history. Antiquity has thrown its halo around the brave of ancient times, but braver men have never lived than those who fell upon the soil we occupy and call our own, in defence of the graves of their fathers and their native land. From these Plains, numerous small bands departed, to commit depredations on the inhabitants of the neighboring States. On these Plains, the allied tribes collected, marched forth, met Gen. Lewis, and fought the memorable and sanguinary battle of Point Pleasant. And, after their defeat, in that battle, to these fields they retreated, collected their forces and took their stand at Old Chillicothe—one of the towns represented in the above map. This was a strong position, in their mode of warfare; and there they prepared themselves, in anticipation of the "Long Knives," to make another desperate struggle for their country and their homes. The second conflict was, however, prevented, by Gov. Dunmore overtaking General Lewis in person, and turning him back, when he was in sight of one of the towns, previous to carrying out the truce above alluded to, at Camp Charlotte. This Treaty had, undoubtedly been previously arranged between Dunmore and the Indian agents, or he never would have ventured to send Lewis back, and remain himself, with the less efficient division of his army, exposed to the mercy of the Indians, who, he knew, were collected, in their full force, at Old

Chillicothe, within seven miles of his encampment. His danger, however, was less imminent, as it was the confirmed opinion of several of the officers who served under him in that expedition, (than whom no men were more capable of judging of matters that took place under their own observation,) that the "treaty," as it was called, at Camp Charlotte, was all a farce—merely intended to cover up the perfidy of Dunmore, in not forming a junction with Lewis. This he had ample time to do, and it would have saved many valuable lives and the waste of much blood. It was further believed, by those officers, (among whom the undersigned grew up,) that, through the perfidy of Dunmore, the Indians, as allies of Great Britain, might boast of spilling the first blood in the Revolutionary War; and, that Lewis is justly entitled to the honor of gaining the first battle fought in that memorable struggle. Time and space will not admit, at present, of going into detail, in giving the circumstances on which their opinions were founded. In justice to them, however, the history of these matters may be presented in some future publication.

Among the circumstances which invest this region with extraordinary interest, is the fact that to those towns were brought so many of the truly unfortunate prisoners who were abducted from the neighboring States. Here they were immolated on the altar of the red man's vengeance, and made to suffer, to the death, all the tortures savage ingenuity could invent, as a sort of expiation for the aggressions of their race. Strange does it seem that human beings, on whom Nature had bestowed such riches of intellect, could be brought, by force of habit, not only to commit, but to delight in committing, such enormous cruelties as they often practiced on many of their helpless victims—acts which had the direct effect of bringing down retaliation, in some form or other, on their own heads. But, that they should contend to the last extremity for so delightful a spot, will not be wondered at, by the most common observer, on a view of the premises. For picturesqueness, fertility of soil, and every other concomitant to make it desirable for human habitation, it is not surpassed by any other locality in the Western country, or, perhaps, in the world. The towns were supplied with good spring water; some of the adjacent bottom lands were susceptible of being made to produce, as Nature had left them, one hundred bushels of Indian corn to the acre, and all other grains and vegetables in proportion.

The Black Mountain represented on the Map, (so called by the natives, but why so named tradition hath not informed us,) is a ridge somewhat in the shape of an inverted boat, elevated from 130 to 150 feet above the bottom prairie immediately in its vicinity, and commands, from its summit, a full view of the high plains and the country around it to a great extent. This facility the natives enjoyed, for they were in the practice, yearly of burning over the country, which kept down the undergrowth, while the larger growth was so sparse as not materially to intercept the view. This elevated ridge answered the Indian some valuable purposes. No enemy could approach, in day time, who could not, from its summit, be described at a great distance; and, by repairing thither, the Red Man could often have a choice of the game in view, and his sagacity seldom failed him in the endeavor to approach it with success. The burning-ground, in the suburbs of Grenadier Squawtown, represented in the Map, was also situated on an elevated spot, which commands a full view of all the other Towns within the drawing, so that when a victim was at the stake and the flames ascending, all of the inhabitants of the other Towns, who could not be present, might, in a great measure, enjoy the scene by sight and imagination. The burning-ground at Old Chillicothe was somewhat similar, being in full view of the burning-ground at Squawtown, The Black Mountain and two or three other small towns in other parts of the Plains.

The Grenadier squaw whose name the above town bore, was a sister to Cornstalk. She was represented as being a woman of great muscular strength; and like her brother, possessed of a superior intellect.

From accounts most to be relied on, it was to Grenadier Squawtown that Slover, who was taken prisoner at Crawford's defeat, was brought to suffer a similar death to that which Crawford, his commander, had undergone a few days before, but from which, through Providential aid, he was relieved and

> enabled to make his escape. The circumstances of his escape have previously published; but as they seen to be inseparably connected with the history of this spot, I hope to be excused for repeating them here. After his capture, on his way thither, he had been very much abused at the different towns he passed through, beaten with clubs, etc. On his arrival here he had a similar punishment to undergo. A council was held over him, and he was doomed to die the death that Crawford had suffered. The day was appointed for the consummation of the horrid deed, and its morning dawned without any unpropitious appearances to mar the anticipated enjoyments of the natives collected from the neighboring towns to witness the scene. At the appointed time, Slover was led forth, stripped naked, tied to the fatal stake and the fire kindled around him. Just as his tormentors were about to commence the torture, it seemed that the Great Spirit looked down, and said, "No! this horrid deed shall not be done!" Immediately the heavens were overcast; the forked lightnings in all directions flew; in mighty peals, the thunder rolled, and seemed to shake the earth to its centre; the rain, in copious torrents, fell, and quenched the threatening flames, before they had done the victim much injury—continuing to a late hour. The natives stood dumb-founded—somewhat fearing that the Great Spirit was not pleased with what they were about to do. But had they been never so much inclined, there was not time left that evening to carry out their usual savage observances. Slover was therefore taken from the stake, and conducted to an empty house, to an upper log of which he was fastened by a buffalo-tug, tied around his neck, and his arms were pinioned behind him by a cord. Two warriors were set over him, as a guard to prevent his escape in the night. Here again Providence seemed to interfere in favor of Slover, by causing a resistless sleep to come over his guard. Until a late hour, the Indians sat up, smoking their pipes and talking to Slover—using all their ingenuity to tantalize him, asking "how he would like to eat fire," etc. At length, one of them lay down, and soon fell asleep. The other continued smoking and talking with Slover some time. After midnight, a deep sleep came upon him. He also lay down, and soon thought of nothing, save in dreams, of the anticipated pleasure to be enjoyed in torturing their prisoner next day. Slover then resolved to make an effort to get loose, and soon extricated one of his hands from the cords. He then tried to unloose the tug around his neck, but without effect. He had not long been thus engaged before one of the Indians got up and smoked his pipe. While he was thus engaged, Slover kept very still, for fear of a discovery. But, the Indian being again overcome with sleep, again lay down. Slover then renewed his exertions, but, for some time, without effect, and he resigned himself to his fate. After resting awhile, however, he resolved to make another and a last effort. He put his hand again to the tug, and, as he related, he slipped it over his head without difficulty. He then got out of the house as quietly as possible, sprang over a fence, into a cornfield. While passing through the field, he came running over a squaw and her children, who were sleeping under a tree. To avoid discovery, he deviated from a straight track, and, rapidly hurried to the upper plain, where, as he had expected, he found a number of Indian horses grazing. Day was then fairly breaking. He untied the cord from the other arm, which, by this time, was very much swelled. Selecting, as he thought, the best horse he could see, he made a bridle of the cord, mounted him, and rode off at full speed, (as represented in the Map.) About 10 o'clock the horse gave out. Slover then had to travel on foot, with all possible speed; and, between musquitoes, nettles, brush, briars, thorns, etc., by the time he got home, he had more the appearance of a mass of raw flesh than an animate being.

It was during the American Revolution that the first American post was constructed in Ohio. Following the murder of Cornstalk at Fort Randolph, Indian hostilities on the western frontier increased. The year 1777 was costly to the Americans throughout the Ohio Valley and the success of the Indians against the western settlements in Virginia and Pennsylvania, indeed, made it "the year of the bloody sevens." By the year's end, however, the American forces had gained several military victories against the British which had cooled the fervor of their Indian allies. To take full advantage of this situation, an expedition against Detroit, the center of British intrigues among the Indians in the west, was considered necessary. The construction of Fort Laurens along the Tuscarawas River, deep in the Ohio Country, was part of that plan and was a major aspect of Ohio's participation in the Revolutionary war.

In May 1778 Brigadier General Lachlan McIntosh, a native of Scotland, was assigned to the post at the Forks of the Ohio with the purpose of leading an expedition against Detroit. Plagued by petty rivalries between Pennsylvania and Virginia, however, Congress could not support such a military thrust. Consequently, a more modest expedition against the Indians in the Ohio Country was attempted. In October 1778, McIntosh moved from Fort Pitt and established Fort McIntosh at the mouth of Beaver Creek. From there, he followed Bouquet's route to the Tuscarawas and along the west bank of that river, just below the mouth of the Sandy Creek where Bolivar now stands, he constructed Fort Laurens, named in honor of Henry Laurens, the president of Congress.

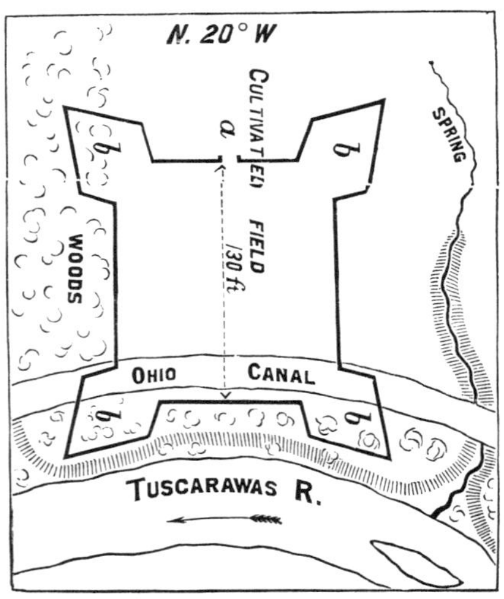

Plate V: (Whittlesey), "Fort Laurens," (1876), drawn from data gathered from a geological survey made in 1850. This post was the only United States fortification built in Ohio during the American Revolution.

No official plan of Fort Laurens exists, but in 1876 a sketch (Plate V) of the fort at the Tuscarawas was compiled from surveys made in 1850 by Charles Whittlesey, engaged with the geological surveys of Ohio, and published in Charles H. Mitchener's *Historic Events in the Tuscarawas and Muskingum Valleys and in other portions of the State of Ohio.*[26]

In a letter to Mitchener, Whittlesey explained that when he surveyed the site, the outlines of the fort were still visible although the eastern portion had been destroyed by the channel of the Ohio-Erie Canal. "The base of the parapet was seven feet broad, its height four and one half feet and the depth of the ditch two and one half feet, with a breadth of eight feet," Whittlesey stated. "It was a regularly laid out work though small, and was probably picketed along the inner edge of the the ditch, connecting the earthwork and stockade." John Cuppy, a native of Pennsylvania who died in Montgomery County, Ohio in 1861, was at Fort Laurens in 1778 under McIntosh's command. He explained:

> Where Fort Laurens was built, was no timber, on a high bank, and a barren back for half a mile or more; and then men had to carry in the timber, four or five to a stick. Made mostly of large, hardwood timber, split, some six inches thick, bullet-proof, planted in trenches three feet deep, solidly packed around, and extending fifteen feet above ground. [He] Thinks the fort, which was on the west bank of the Tuscarawas, enclosed about an acre of ground, and was the longest on the river. No pickets along the river bank, no high, overlooking ground either near Fort McIntosh or Laurens. The gate was on the west side of the fort, no spring; relied upon the river for supply of water. There was one block-house, about 20 feet square, which was directly to the right of the gate, and next to it, and formed a part of the outside in place of picketing: the block-house, about six feet above the ground, the block-house was made a foot wider on the wall side, and made to over-jut, so if Indians came up, the garrison could shoot down through this open jut directly upon an enemy below; and the floor of puncheons on a level with the over-jut and the timbers built up some eight feet, so as completely to protect those within from the enemy without, and port-holes all around about five feet from the floor, and some two or three feet apart, through to which for the garrison to fire in case of an attack, with a rude roof slanting one way, and that within the fort. [He] Thinks there were cannon on the expedition. There were also about 2 cabins built on each end of the fort, not quite together and in a line with the picketing, and helped to form the enclosure, and also had over-jutting and with portholes; but smaller than the block-house; and these were for shelter for provisions and baggage.[27]

Another description of the fort that was similar to the outline of the fort surveyed by Whittlesey stated:

> Fort Laurens was seated on the west side of the Tuscar[awas] [branch of the Muskingum] on the bank half a mile below the mouth of Sandy Creek [the present Bolivar]. It was a stockaded fort, with two gates, one next to the river and one the opposite side — pickets 12 or 14 feet high — Two rows of pickets, one placed on the bank of earth that was thrown from the ditch, which was about 4 feet wide and 3 feet deep — this covered 3 sides of the fort, the river covering were lined with ranges of cabins, built of logs and covered with split clapboards — on each side of the river gate, rows of pickets extended to low water in the river, for the protection of the soldiers in going and returning for water built in November and December 1778, evacuated in 1779, August. It covered about one acre of ground.[28]

Following the close of the American Revolution, the Indians, encouraged by the British who refused to vacate the territory northwest of the Ohio River, continued their warfare along the American frontier. Despite the attempts on the part of Congress to discourage it, many settlers and traders who sought the riches of the new land, crossed the Ohio and settled on lands yet unopened for occupation. In an effort to restrict this immigration flow and to control the territory gained from England,[29] Colonel Josiah Harmar, the commander at Fort McIntosh, sent Captain John Doughty in October 1785 to establish a fort "at or near the mouth of the Muskingum," and "to stockade or pallisade himself for his own security, as he should judge proper."[30] There, Doughty located Fort Harmar, which was the most advanced post on the western frontier. Harmar explained:

> This position at Muskingum will answer the valuable object of removing the intruders from the public land should any hereafter presume to encroach upon them . . . . It will also cover the Continental surveyors, in some measure, in their laying off the seven ranges of townships and convince the Indians that Congress means to protect the legal settlers on the lands ceded to them by treaty. For these reasons I judge it a preferable position to the mouth of the Scioto which I though to be rather too far down the [Ohio] River.[31]

The manuscript sketch of the "Plan of Fort Harmar" (Plate VI) depicted the post as a pentagon which, within its walls, contained approximately ¾ of an acre.[32] The post was so located that it not only commanded the mouth of the Muskingum, but from its walls the soldiers inside had a wide sweep of the Ohio as well. The walls of the fort, as portrayed in the small but detailed sketch of the post made from 100 yards distance (Plate VII), were constructed of large timbers which rose approximately 12 to 14 feet from the ground and were 120 feet in length. The bastions were constructed of 14 foot poles set into the ground and fastened together by strips of timbers that were treenailed, or pegged, into position. The barracks for the soldiers were built along the interior of the walls and divided into 4 separate rooms of 30 feet each with a fireplace in every room. The officers' quarters were 2 stories high with 2 rooms on each floor, and each room contained a fireplace. In the "Sketch of Fort Harmar from the Hill on the Northwest" (Plate VIII) made in 1790 by Joseph Gilman who came to the Marietta settlement in 1789 and was a judge in the General Court of the Northwest Territory between 1796 and 1803, a square tower with a flag atop can be seen rising above the post.[33] Here a sentry was posted to keep a vigil over both the rivers. Directly beneath this sentry station was a guard house. Close to the guard house was the magazine, which was covered with dirt to protect it from enemy artillery or from accidental fire

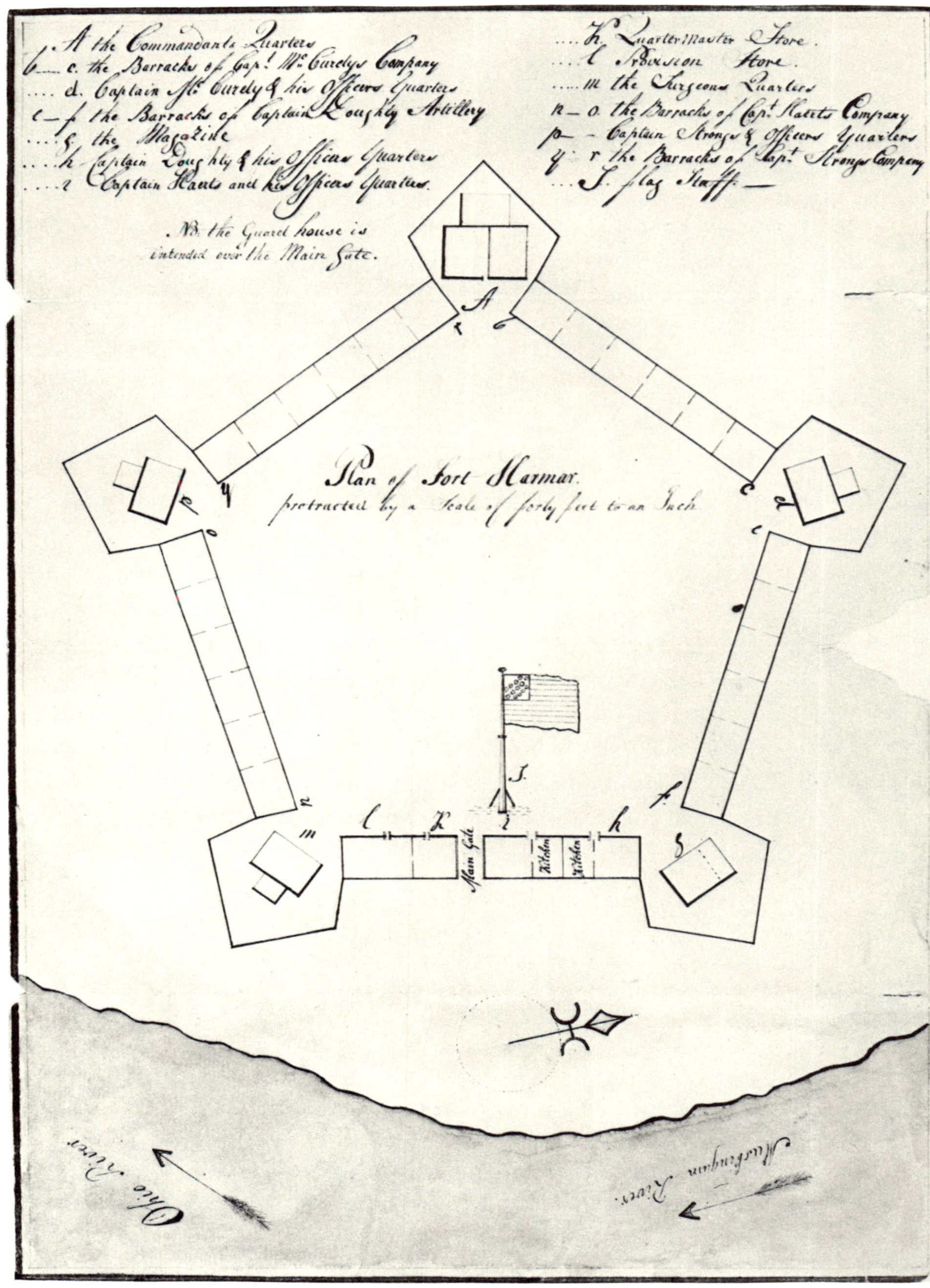

Plate VI: "Plan of Fort Harmar," William L. Clements Library, Ann Arbor, Michigan. Harmar was built to house troops who would remove white settlers from the Northwest Territory before it was surveyed.

from within the post itself. In the center of the fort, but not shown in the plan, was a well which was to furnish the garrison with drinking water in case of siege. Under ordinary circumstances, however, water was obtained from the rivers. Captain Doughty, who planned the fort, also laid out gardens from which produce was obtained in order to feed the small garrison stationed there. The Council House designated as "E" by Gilman in his sketch was the spot where Governor Arthur St. Clair in 1789 negotiated with the Ohio Indians the Treaty of Fort Harmar, which established an Indian territory within Ohio.[34] St. Clair himself was disabled throughout the talks by an attack of gout, and was carried in a large chair from his office in the fort to the council house. The post was named for Colonel Josiah Harmar and was occupied by United States troops until September 1790 when the garrison was ordered to Fort Washington.[35]

Congress directed that another fort be constructed up the Ohio River from the mouth of the Muskingum, in order to protect the surveyors under the direction of Thomas Hutchins as they laid out the townships in

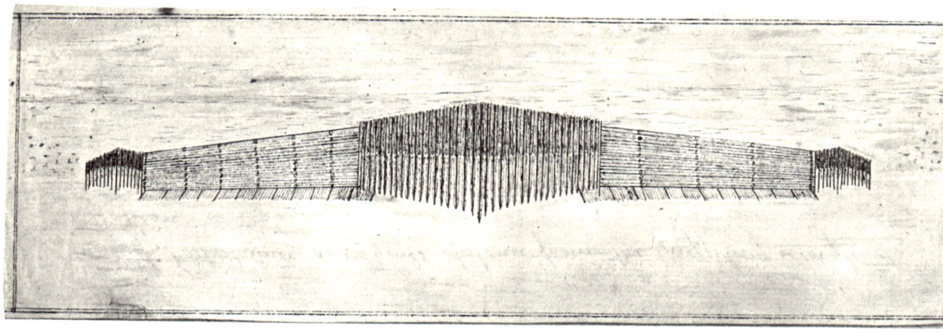

Plate VII: "A View of Fort Harmar," drawn by an unknown artist. The plan is in the William L. Clements Library, Ann Arbor, Michigan.

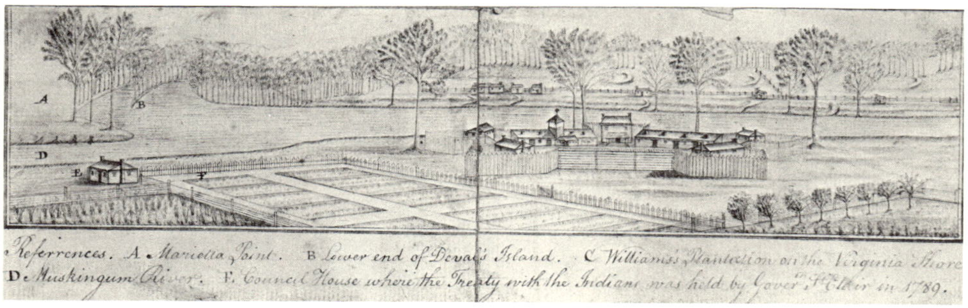

Plate VIII: Gilman, "Sketch of Fort Harmar," Dawes Memorial Library, Marietta College, Marietta, Ohio. It details the fort after the founding of Marietta.

the Seven Ranges, and to keep squatters from Pennsylvania from settling on the west side of the river. Late in 1786 a group of men under the command of Captain John Francis Hamtramck, a very capable officer serving on the western frontier, began to build the post on a high bank on the site of present Steubenville between 50 to 60 yards from the river. Anxious to complete the fort as quickly as possible, Hamtramck explained to Colonel Harmar, his superior officer, that "with 33 men from each company (the rest being unfit to work from the want of shoes), on the 27th of October we commenced our fortification, [and] promised as a reward six gallons of liquor to the first house completed, that is up and the under floor on, and the chimneys built on the 2nd floor." Apparently this inducement worked because Hamtramck reported that in the 2½ days the houses were ready for the roofs, in 6 days the roofs and flooring were being applied, and within 8 days the houses were completed. "The furiousness of my men on this occasion," he continued, "can only be attributed to their great attachment to whiskey."[36]

The fort, which was completed in 1787, was named Fort Steuben for Baron von Steuben, the Prussian officer who helped Washington to form his army during the American Revolution. The "Plan of Fort Steuben" (Plate IX) showed that the post was small but efficiently designed. The walls that connected the four blockhouses were approximately 150 feet in length. Each blockhouse was almost 25 feet square and 11 feet high. The interior of each was divided by a partition which made 2 rooms. The rooms were 7 feet from floor to ceiling, each permitting the quartering of 14 men. The officers' quarters were next to the main gate and contained both bedrooms and parlors. The kitchen was also contained within this complex. The fort had a quartermaster's store, magazine and artificer's shop built along the interior wall that ran perpendicular to the Ohio River and is not shown in this sketch. A guard house was constructed on piers on either side of the sally port, which apparently had a commanding view of the approach from the river. The post also had a "black Hole, [or] place of confinement" for the unruly. Fort Steuben was abandoned soon after Anthony Wayne's victory over the Indians in 1794 and eventually burned down. The settlers, who had built their homes close to the fort for security,

Plate IX: (Hamtramck), "Plan of Fort Steuben," (1787), included in the Lyman Draper Manuscripts. Fort Steuben was built to protect surveyors as they divided the Seven Ranges.

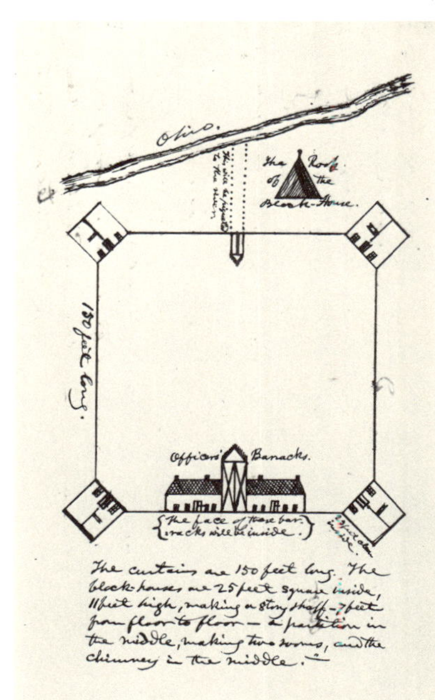

remained, however, and from this nucleus the town of Steubenville emerged.[37]

As a result of the Peace of Paris, which ended the American Revolution, the United States could claim possession over the lands west of the mountains as far as the Mississippi River. The Indians who lived within this region had long been promised permanent residence there by the British, who had controlled that area since 1763. When the United States acquired this territory, however, the Indians, fearful of American encroachment into their lands, looked to the British for protection. Congress, which attempted to restrict settlement north and west of the Ohio, was forced to begin a process which would remove the Indians and make way for settlement of the Ohio Valley.

In 1783 Congress called for all the Indians living within what is now the state of Ohio to withdraw to a line west of the Miami River. In order to carry out this policy, Congress negotiated the Treaty of Fort Stanwix of 1784 with the Iroquois. In this treaty the Indians denounced their ancient claims to the lands in the Ohio Valley. In January 1785 the United States successfully negotiated a treaty with representatives of the Wyandot, Delaware, Chippewa and Ottawa nations at Fort McIntosh, a post built at the mouth of Beaver Creek in Pennsylvania. In the Treaty of Fort McIntosh, the Indian representatives ceded to Congress their lands east, south and west of a line which ran down the Cuyahoga, across the Tuscarawas portage to Fort Laurens, west to Pickawillany, and then to the portage on the Maumee. Anxious to treat with those Indians living further west, Congress called for a third conference. After some discussion concerning the appropriate site, it was decided to hold the meeting near the mouth of the Great Miami River.

Colonel Harmar, then commander of Fort McIntosh, sent Captain Walter Finney, an officer in the Pennsylvania militia, along with a company of men, to construct a post near the mouth of the Miami in October 1785.[38] On October 22 Finney and his party, which included two peace commissioners representing the United States government, reached the Great Miami River. After spending several days to decide upon an appropriate location, Finney selected a spot about one mile above the mouth of the Miami on the north bank of the Ohio and began to fell the trees to make a clearing. General Richard Butler, one of the commissioners, wrote in his journal that on October 25, Captain Finney "began to lay out and clear the ground for four blockhouses and quadrangular work, which was pushed on with great alacrity by himself and officers." By November 8 the walls of the post were in place, and the fort itself was ready to receive the Indians' representatives by January 14, 1786, when the conference began.

Fort Finney (Plate X) according to Major Ebenezer Denny, was built on the

> Best ground for our station about a mile above the mouth, where the boats were brought and everything unloaded. All hands set to work chopping, clearing, etc., and preparing timber for block-houses and pickets; and on the 8th instant had ourselves enclosed. Hoisted the United States flag, and christened the place Fort Finney, in compliment to Captain Finney, the commanding officer. Our work is a square stockade fort; substantial block-houses, two stories, twenty-four by eighteen feet in each angle; curtains one hundred feet of stout pickets, four feet in the ground, and nine feet above; situate one hundred and fifty yards from the river [Miami], on a rising second bank. A building, eighteen by twenty feet, within the east and west curtains, for the accommodation and reception of contractors' stores and

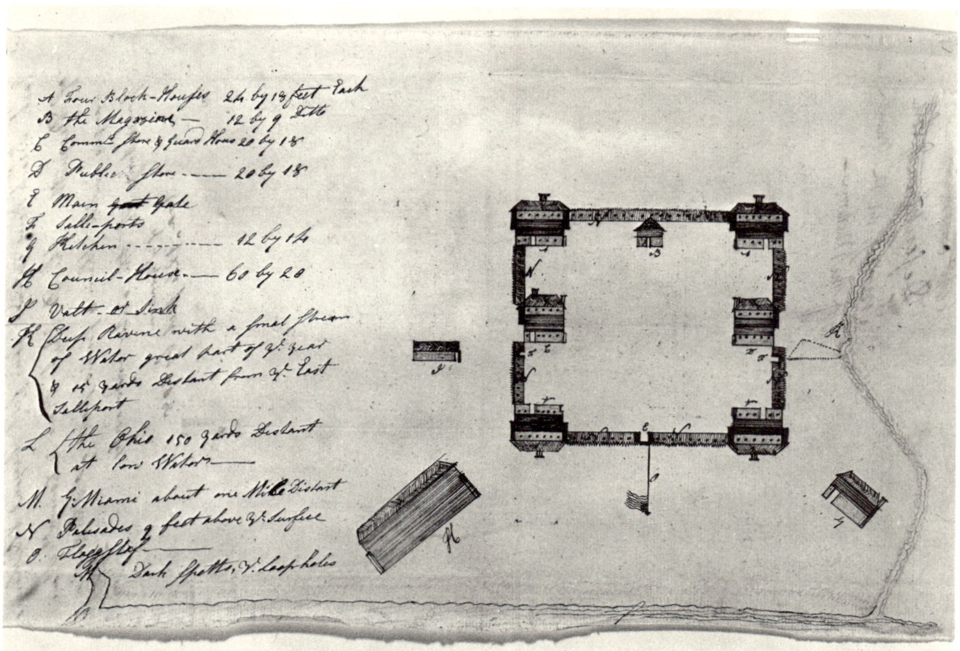

Plate X: "Fort Finney," (1785), was located on the Ohio River near the mouth of the Great Miami River. This plan is in the McHenry Papers, and is used here courtesy of the Indiana Historical Society, Indianapolis, Indiana. Fort Finney was built to house a conference with certain of the Ohio Indians.

> Indians goods; and one small, but strong building, centre of north curtain, for magazine. A council house, twenty by sixty, detached, but within gun shot. Commissioners and their followers pitch their tents within the fort, and erect wooden chimneys. . . . Men employed finishing the block-houses, and clearing off the timber and brush to some distance outside.[39]

It was in the Council House that the American peace commissioners and the spokesmen for the Shawnee Indians and other western nations met between January 14, and February 1, 1786. The Treaty of Fort Finney was concluded and the Indians ceded to the United States their claims to the southern half of Ohio. The majority of the Indians within Ohio, however, refused to accept any of the treaties negotiated with the United States, and provocations between them and the whites continued until open hostilities broke out in 1790.

As the Congress tried to achieve a peaceful settlement with the Indians in the West, there were many in the East who wanted to open the lands north of the Ohio for settlement. One group, made up of Revolutionary War officers, formed the Ohio Company and petitioned Congress for land. The members wanted to satisfy their claims against the government for particpation in the recent war for independence and, at the same time, to speculate in western land. After some negotiations with Congress, the company was sold 1,500,000 acres immediately west of the Seven Ranges. At the mouth of the Muskingum River, opposite Fort Harmar, the Ohio Company set aside 5,760 acres of land for a town site.[40] Shortly after their arrival there on April 7, 1788, the officials of the company began construction of Campus Martius, a post designed to provide housing for some of the new settlers and, at the same time, to offer them security.

Great interest was taken in the East of the Ohio Company's speculative land venture west of the mountains. To help satisfy some of this curiosity, Winthrop Sargent drew "A Plan of Campus Martius at the City of Marietta, Territory of the United States, N.W. of the River Ohio" (Plate XI) for the November issue of *The Columbian Magazine* in 1788. Campus Martius was built on a bank approximately thirty feet above the Muskingum, a little more than 150 yards from it and less than a mile from the Ohio. The

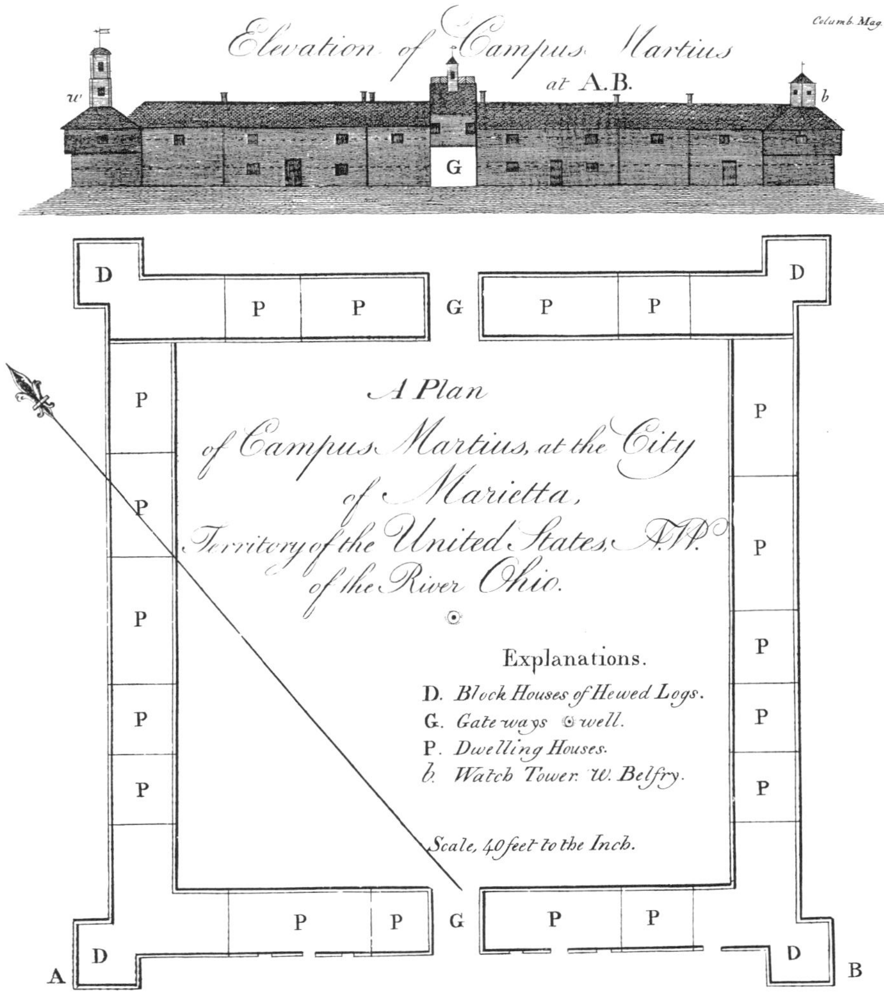

Plate XI: Sargent, "A Plan of Campus Martius," (1788), published in *Columbian Magazine* (November 1788). Campus Martius was a private fortification built by the Ohio Company.

horizontal hewn log walls or curtains of the post actually formed the rear walls of two-story dwellings placed side by side within the fort. Once the settlement was well established and the Indian threat removed, many of these homes were relocated to town lot sites. At the 4 corners were blockhouses 18 feet square. The second floor extended beyond the curtains 18 inches to allow a constant view of the wall below. At the very top of the blockhouses were watchtowers, large enough to hold 4 men each and constructed with thick oak siding in order to make them musketball proof. On top of the blockhouse in the northwest corner (Plate XII) above the watchtower was a small balcony over which was a cupola.[41] This structure housed a bell that had been brought from Boston. The blockhouse beneath this cupola had a reception hall large enough for 300 people. The blockhouse at the southwest corner of the fort was occupied by Governor Arthur St. Clair and used as his office while the blockhouse at the northeast was used by the directors of the Ohio Company. Above the gate in the west wall was a chamber, not shown in the plan, which was intended to protect the main opening and also served as Sargent's office during his term as secretary.

During the Indian wars in Ohio which began in 1790, the defenses at Campus Martius were strengthened. A row of palisades, resting upon stone rails, were implanted from corner to corner. Twenty feet in front of this was a row of pickets. In front of the pickets, a third defensive line was established that consisted of a row of abbatis made from the branches of

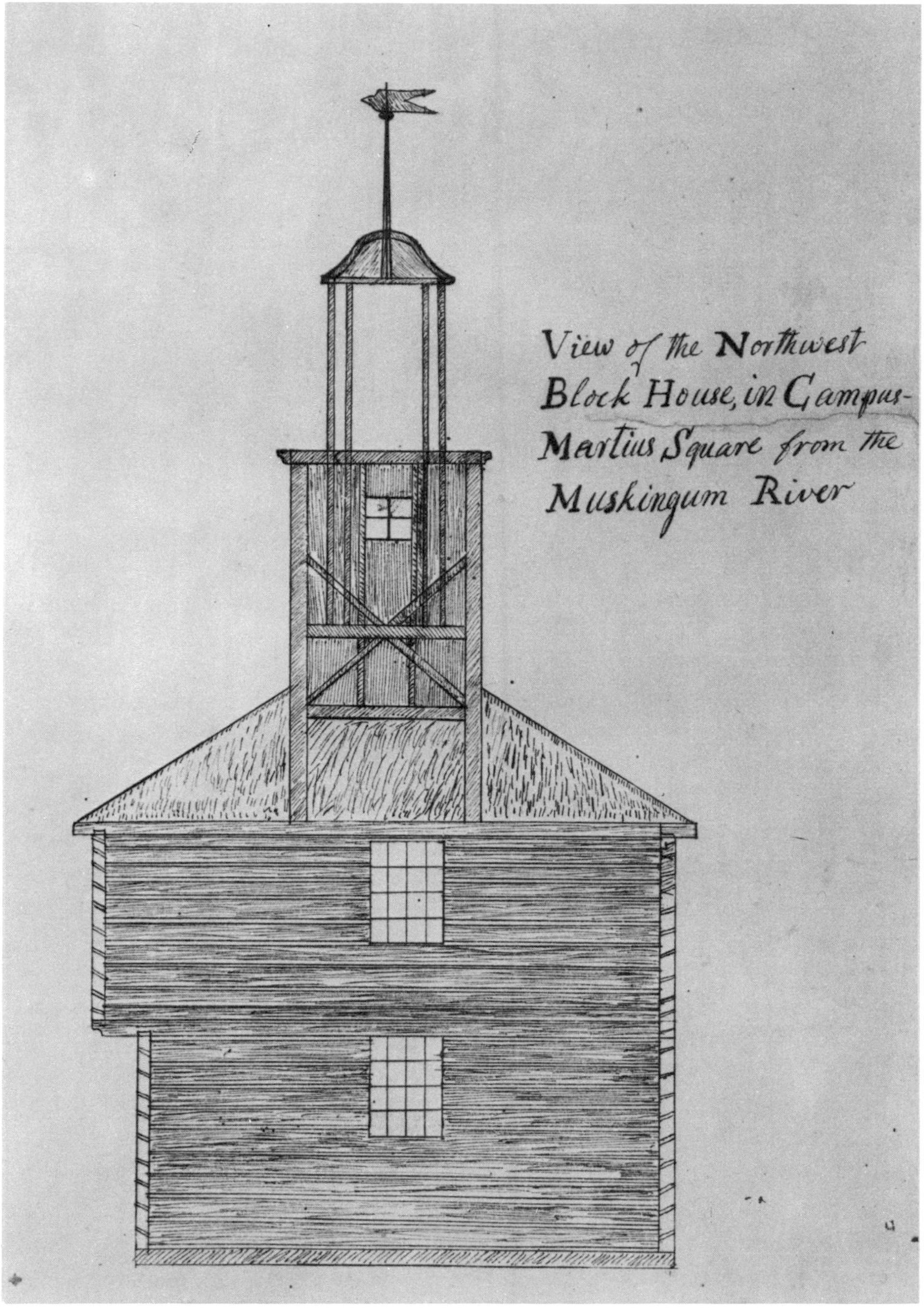

Plate XII: "View of the Northwest Block House," of Campus Martius, Dawes Memorial Library, Marietta College, Marietta, Ohio, (1788).

trees which had been sharpened and planted on a slant in the ground with the points facing any approaching enemy. From this private military post, the directors of the Ohio Company collected information concerning the movement of the Indians, planned the defenses of their lands, and assisted with these defenses throughout their vast holding in whatever way they could.[42]

The Indian wars in Ohio between 1790 and 1794 were precipitated in great part by Congress's attempt to exert its control over the area north of the Ohio. Certain elements among the Indians, however, refused to recognize either treaty negotiated at Fort McIntosh or Fort Finney, and continued to look to the British for military aid in case of a confrontation with the United States. Shortly after Governor Arthur St. Clair assumed control of the Northwest Territory in 1788, he attempted to affirm the principles already agreed to by the Indians in Ohio in the Treaty of Fort Harmar in January 1789. A war faction among the Indians, however, refused to leave their traditional homes in Ohio and began to raid into Virginia and Pennsylvania and to disrupt traffic on the Ohio River. In an effort to meet these hostilities with force, St. Clair, with President George Washington's approval, sent General Harmar from Fort Washington, which had been established at Cincinnati, to attack the Indian villages at the confluence of the St. Mary's and St. Joseph rivers, the present site of Fort Wayne, Indiana, which was the center of Indian power in the northwest. Harmar's expedition was a disaster and as a result, heavier fighting broke out across the western frontier north of the Ohio.

Shortly after the founding of the settlement at Marietta, but before the outbreak of open hostilities in 1790, about 20 houses had been constructed on the point of land between the Ohio Company's Campus Martius and the Ohio River. When the Indian wars broke out, this point of land between the Ohio and the Muskingum was fortified and called Garrison Point.

Colonel William Stacey, a native of Massachusetts who came to the Marietta settlement in 1789 and who had been an officer during the American Revolution, was employed by the Ohio Company to prepare the defensive works at this point. The drawing, "Point Garrison—Marietta 1791" (Plate XIII), showed the type of defensive works Stacey planned.[43] A line of palisades ran from the Muskingum eastward to the approximate location of where First Street is now in Marietta, and then followed that street south to the Ohio. The entire wall enclosed approximately 4 acres. Originally, 3 blockhouses were built—the first by the Ohio, the second at the corner of the pickets and the third by the Muskingum. On the top of each blockhouse was a sentry post constructed of heavy oak planks to protect those inside from musket fire. The second floor of the blockhouse near the Muskingum was used as a schoolroom for the children within the fort, while the upper level of the blockhouse in the northeast corner of the wall was used as a family dwelling and the lower level was the guard house. The blockhouse by the Ohio River was used to house Captain Stacey and his family. A fourth blockhouse was constructed in 1792 along the eastern wall by a detachment of United States troops under the command of Lieutenant Tillinghast from Rhode Island. This was the largest blockhouse in the post, as the second floor was 30 feet square and the lower 26. The upper story was used as a magazine, while the first floor was the jail and, on occasion, served as a court room. Following St. Clair's defeat in 1792, Colonel Ebenezer Sproat, who was in command of the defenses at Garrison Point, declared martial law and the gates were closed at sunset.[44]

Plate XIII: "Point Garrison—Marietta," (1791), Dawes Memorial Library, Marietta College, Marietta, Ohio. Built by Captain Stacey, Point Garrison protected settlers at the mouth of the Muskingum during the Ohio Indian Wars.

During the Indian wars, which began with General Josiah Harmar's raid against the Indian villages at the headwaters of the Maumee River on October 17, 1790, and lasted until the negotiation of the Treaty of Greenville on August 3, 1795, when the Indians accepted the peace terms laid down by General Anthony Wayne, the Ohio Company fortified 7 sites within its settlement. These posts were, in addition to Campus Martius, fortifications at Garrison Point, Big Bottom, Wolf Creek Mills, Fort Frye at Waterford, Farmers' Castle and Upper Settlement at Belpre.

In the spring of 1789, about 20 miles north of Marietta, 19 settlers established a grist- and sawmill along the Muskingum at the mouth of Wolf Creek. Shortly thereafter, the settlements of Waterford and Plainfield sprang up near this site. Following Harmar's defeat in 1790, the Indians attacked a small settlement at Big Bottom, just north of the Wolf Creek settlements. The people in this general area gathered to decide what action should be taken to defend themselves against the Indians. They elected Captain William Gray, a Revolutionary War officer and one of the original settlers at Marietta, their leader; he, with the help of others, decided to defend the settlement rather than return down river to Campus Martius. On a bend in the Muskingum, about half a mile from the present town of Beverly, they constructed Fort Frye (Plate XIV).[45] It was decided that the fort should be triangular rather than square because the defenders believed that they had little time in which to construct 4 walls. The south wall was only a few paces from the Muskingum River and approximately 200 feet long; the two remaining walls were of different lengths. At each angle was constructed a two-story blockhouse nearly 20 feet square. Along the interior of the two longest walls were built either two-story dwellings or barracks for the men. Near the northern blockhouse was a large gate and along the river wall, a smaller opening called the water gate, which gave those inside access to the river. Thirty feet from the fort's walls was placed a row of pickets about 12 feet high, behind which firewood was stored. Inside the post was a well and a blacksmith's shop. By the end of February 1791, Fort Frye was completed except for the hanging of the main gate, which was put into place during the first part of March.[46]

Settlements radiated from Marietta not only north along the Muskingum but also south down the Ohio. During the spring of 1789, 40

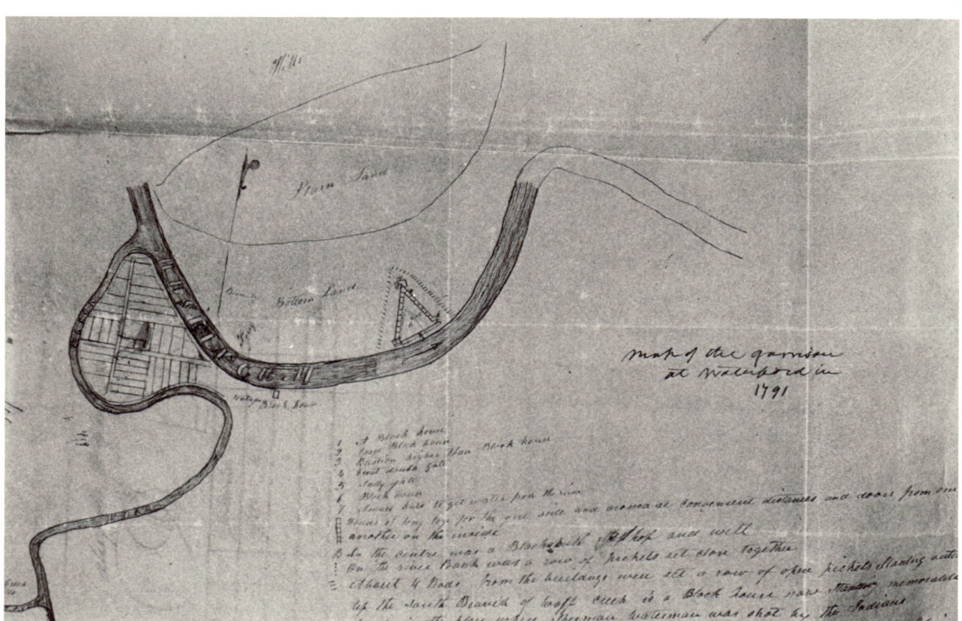

Plate XIV: "Fort Frye," (1789), Dawes Memorial Library, Marietta College, Marietta, Ohio. This post in the Ohio Company lands, protected the early settlers on the Muskingum River.

families settled on the bottom land on the north side of the Ohio opposite the mouth of the Little Kanawha River and formed a small community called Belpre. Here, too, on learning of the Indian attack at Big Bottom, the farmers, instead of going up river to Marietta for protection, gathered together for mutual defense against their red enemy.

Shortly before the Indian wars began, Colonel Nathaniel Cushing, who came to Belpre from Massachusetts in 1789, and Colonel Ebenezer Battelle, a bookseller who came west in 1788, both of whom were Revolutionary War soldiers, had already begun to supervise the construction of 2 large log houses and to erect one of the walls on a high bluff directly opposite Backus Island, or what was later called Blennerhassett Island. The site they selected for their stronghold was excellent because directly behind their structure was a swamp, which more or less protected their rear from attack, while the Ohio River protected their front. During the war there were approximately 220 persons living within the fortification at one time, and because most of the people who had come there for protection were farmers, it became known as Farmers' Castle.

The diagram of Farmers' Castle (Plate XV) illustrates the fortification very closely to what it must have been in 1791.[47] In all, 13 buildings were constructed, each approximately 20 feet square on the first level and 22 feet on the second. Contained in the upper room of each house was a large cask of water to be used only in the event of fire. One task of the officer of the day was to inspect each cask to make sure that its water supply was full. The 13 buildings were enclosed by a wall 10 feet high made of sturdy oak timbers. As a further precaution, to dissuade an attack from the river side, pickets were placed at an angle so that they were leaning out over the bank. This design also allowed the breezes to blow from the river through the

Plate XV: "Farmers' Castle," (1791), Dawes Memorial Library, Marietta College, Marietta, Ohio. This fortification housed settlers at Belpre and was part of the defenses of the Ohio Company.

fort during the hot and humid summer days. Gates 10 feet wide were placed in both the east and west walls to permit the movement of farm animals and teams in and out of the post, and at least one smaller opening was on the river side. At each corner on the land side watchtowers were constructed atop the blockhouses, which allowed sentries to watch for the approach of the enemy. Inside the garrison, in about the middle of the wide avenue that passed between the 2 rows of blockhouses, a cannon was kept. This unique weapon was actually a small swivel gun that had been encased in wood, hooped with iron bands and painted in hopes that it would give the appearance of a larger 6-pounder. The gun was placed on a socketed mount so that it was capable of firing in any direction if needed. Apparently it was quite useful to the defenders inside the post because not only was it fired every morning to overawe the Indians, but it was the focal point of the fort's social gatherings. Major Nathan Goodale commanded the post until his capture by the Indians on March 1, 1793; he later died while in captivity. Cushing followed Goodale as commander of Farmers' Castle until the end of hostilities.[48]

The same year that the lands at the Muskingum were first settled, other immigrants reached the parcel of land between the Little Miami and the Great Miami rivers, purchased from Congress by John Cleves Symmes. Opposite the mouth of Kentucky's Licking River, on the north bank of the Ohio, settlers clustered around the small community of Losantiville, later renamed Cincinnati by Governor St. Clair. Soon it became the center of activity in the Symmes purchase, and because of its location at the junction of two waterways that provided communications to both the hinterlands of Kentucky and the Ohio Country, it was necessary to make the presence of the United States military felt at that point, far advanced on the western frontier.

In August 1789, General Harmar at Fort Harmar sent a small force to Cincinnati to select a site for a post there. On August 16, Doughty and a detachment of soldiers reached the Miami country and began to search for an appropriate spot for a fort. After three days, a location was found midway between the Little Miami and the Great Miami rivers, about 80 feet above the low water mark on the north bank of the Ohio River and opposite the Licking River in Kentucky. Although Doughty admitted that this site was not the best possible, he stated that it was close to the settlement of Losantiville and afforded easy communications and transportation facilities with the eastern regions of the territory. Harmar told Secretary of War Henry Knox that it was "high and healthy, abounding with never failing springs etc., and the most proper position he [Doughty] could find for the purpose."[49]

The fort itself (Plate XVI) was designed by Doughty, who had laid out Fort Harmar. It has been described as being the "acme of perfection in wooden fortifications" and provided for its inhabitants all the possible comforts of any frontier post, including plastered walls, glassed windows, and planked flooring.[50] There is little doubt that between 1789 when the fort was constructed and 1804 when it was razed, it remained the "strongest, [and] most important of the frontier posts in the Northwest Territory."[51]

The actual construction of Fort Washington was supervised by Captain William Ferguson of Pennsylvania, an artilleryman who was killed during St. Clair's campaigning in 1791, and by Lieutenant John Pratt, a Revolutionary War veteran who served as the post's regimental quarter-

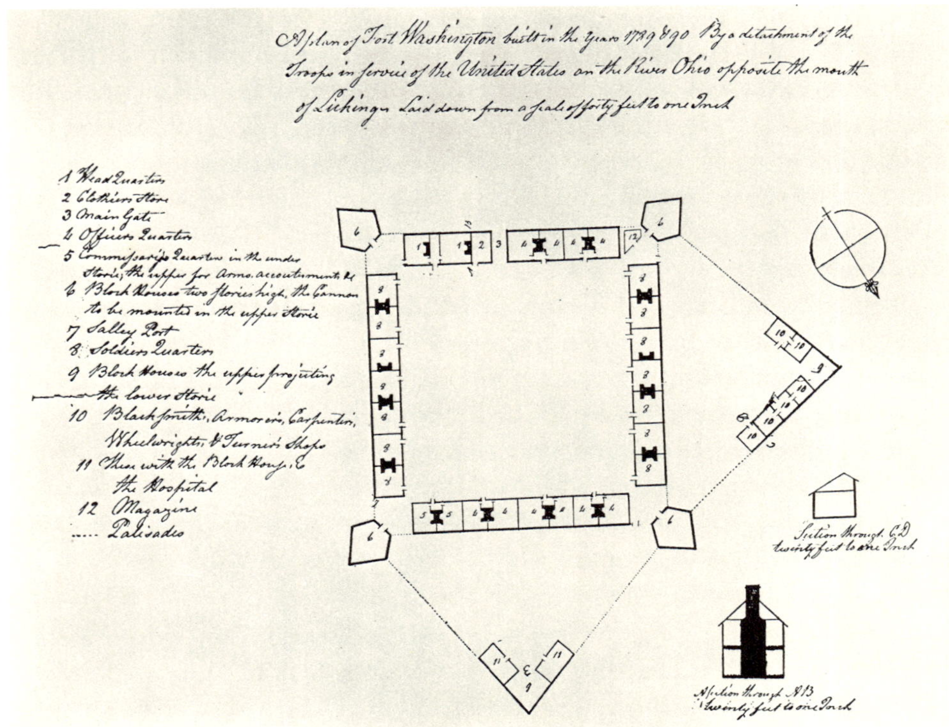

Plate XVI: "Fort Washington," (1790), National Archives, Washington, D.C. This post was designed by Major James Doughty and was the headquarters for the United States Army during the Ohio Indian Wars.

master. Apparently Harmar was quite enthusiastic about this post; he reported to Secretary Knox that "this will be one of the most solid, substantial wooden fortresses, when finished, of any in the Western Territory. It is built of hewn timber, a perfect square, two stories high, with four blockhouses at the angles. . . . On account of its superior excellence, I have thought proper to honor it with the name of *Fort Washington.* The public ought to be benefited by the sale of these buildings whenever we evacuate them, although they will cost them but little." Harmar further explained:

> About forty or fifty Kentucky boats have begun, and will complete it [the fort]. Limestone is the grand mart of Kentucky; whenever boats arrive there they are scarcely of any value to the owners; they are frequently set adrift in order to make room for the arrival of others. I have contracted for the above number for the moderate price of from one to two dollars each; thus much for the plank work. All other expenses (wagon hire, nails, and some glass excepted), are to be charged to the labor of the troops. The lime we have burned ourselves, and the stone is at hand.[52]

Each blockhouse was approximately 20 feet square, 2 stories high and situated so that it commanded the face of the walls as well as the land in view. The walls constituted in part, the rear walls of barracks, which faced the center parade ground. Between the ends of the barracks and the blockhouses palisades were constructed. During the Indian Wars, another blockhouse was built to the west, outside the main wall, and connected by palisades to the northwest and southwest blockhouses. A well dug within the post by John Robert Shaw in 1791 supplied fresh water to the troops. In 1804, the garrison of soldiers stationed there was moved to Newport, Kentucky, and the post was sold at an auction.[53]

Just as settlements had radiated northward from Marietta along the Muskingum, so did new communities spring up throughout the Miami River Valley. In 1790 John Dunlap, who migrated from Colerain, Ireland, to North America and was one of Symmes's chief surveyors, established a small settlement on the east bank of the Great Miami River 17 miles north of Cincinnati. The community was named Colerain, but it was popularly known as Dunlap's Station, for its founder. Many small settlements scattered throughout the wilderness were given formal names, but it was the custom on the frontier to refer to them by names of their founders.[54] Dunlap's Station "consisted simply of their little cabins clustered together upon a space of about an acre, built to face each other and, with a singular want of forethought their roofs so placed as to slope outward, and the eaves so low that it is said the dogs were accustomed to jump from the stumps without to the top of them, and so get into the enclosure."[55] When the Indian Wars broke out, a palisade was constructed to connect the various buildings in order to present the appearance of a fortification. The "Plan of a Settlement call'd Dunlap's Station" (Plate XVII), drawn by Lieutenant Bartholomew Shaumburgh illustrated the type of defensive works that was hastily erected by many isolated settlements on the Ohio frontier during the Indian Wars.[56] Shaumburgh was commander of Fort Jefferson from April 1792 until June 1794. In an effort to reinforce its defenses, 15 militiamen were sent from Fort Washington to Dunlap's Station. During the winter of 1790, an additional 15 men under the command of Lieutenant Jacob Kingsbury, a native of Connecticut who had served in the American Revolution, were sent to Dunlap's Station from the post on the Ohio. On January 7, 1791, just five days after the attack at Big Bottom in the Muskingum Valley, Dunlap's Station was assaulted by about 300 Indians and remained under siege for several days until a relief column from Fort Washington arrived. During that time the defenders ran short of lead balls for their rifles and had to melt their pewterware to make musket balls in order to replenish their ammunition supply. Afraid of being isolated and overrun by Indians following Governor Arthur St. Clair's defeat in 1792, the inhabitants left Dunlap's Station for greater security in the settlements along the Ohio. The settlement, however, was reoccupied shortly thereafter, and remained a fortified station throughout the remainder of the war.[57]

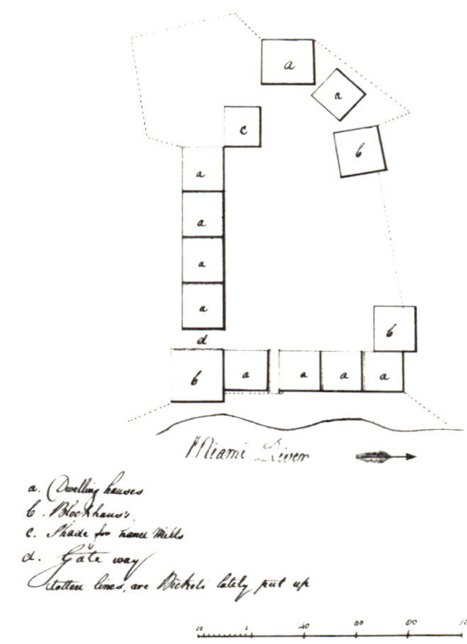

Plate XVII: Shaumburgh, "Dunlap's Station," (1791), William L. Clements Library, Ann Arbor, Michigan. This settlement in Symmes's Purchase, isolated on the frontier, was fortified by its inhabitants, who defended themselves between 1790 and 1794.

Symmes had planned to establish small settlements throughout his purchase in order to drive land prices up at a rapid rate. Because of its location up the Miami Valley, he was particularly concerned that Dunlap's Station survive during the war to illustrate to those interested in western lands that despite hostile conditions settlements in his purchase could prosper. Symmes himself commented that "Colerain [Dunlap's Station] has always been considered the best barrier to all the settlements, and when that place became re-peopled the inhabitants of the other stations became more reconciled to stay." Apparently General Harmar, too, shared Symmes's view because he had planned an elaborate fortification for that site. The "Plan and Elevation of a Stockade Work, with Block-House Bastions, designed for the defense of the settlement of Colerain, on the Great Miami River, 1791" (Plate XVIII) was included among his papers.[58] This fort was never constructed, but its diamond configuration illustrated an interesting change from the more common rectangular forts found on the American frontier.

Despite these efforts of self-defense made by the various settlements, it was necessary for the American government to defend its northwestern frontier. Therefore, a second expedition was sent against the villages in northwestern Ohio along the headwaters of the Maumee River, the center of Indian power in the northwest. This time, Governor Arthur St. Clair himself led a force of about 3,000 men. The strategy followed by St. Clair included the construction of a string of forts from Fort Washington

Plate XVIII: "Colerain," (1791), William L. Clements Library, Ann Arbor, Michigan. Harmar planned to make this settlement part of the government's defensive works against the Indians but it was never built.

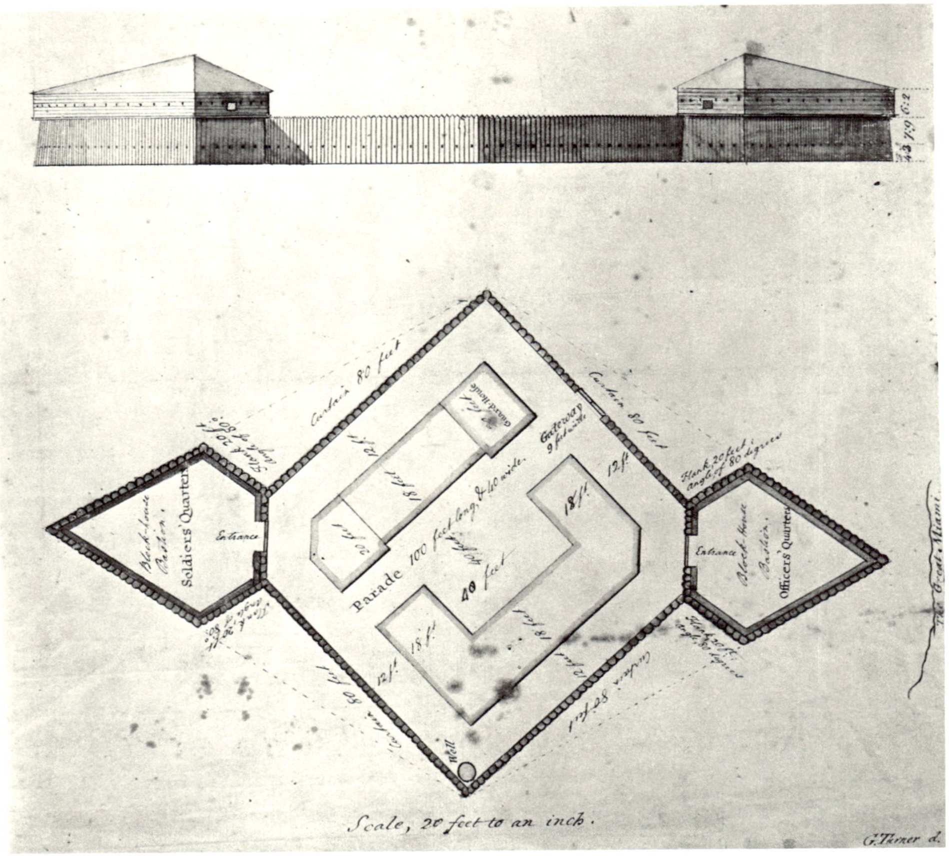

northward toward the Indian settlements. These forts, it was assumed, would keep open the lines of communication to enable an orderly advance upon those villages, while at the same time, providing protection for the outlying settlements and keeping the Indians in check.

The first link to be forged in this defensive chain was Fort Hamilton, located on the east bank of the Great Miami River on the present site of Hamilton in Butler County. In August, St. Clair had sent Major John Stites Gano in command of a survey party to select a suitable route for the army to follow to the crossing place on the Great Miami River and to search for a defensive position there. Gano, who was one of the original settlers of Columbia, the first settlement in Symmes's Purchase, was also a topographical engineer. He surveyed two sites on the east side of the river. On August 30, 1791, St. Clair directed Major John Francis Hamtramck to choose one of the sites selected by Gano, and Major Ferguson was "to mark out the ground for a small fort, to be enclosed with pickets."[59] Captain John Armstrong, a Revolutionary War veteran who later became the Treasurer of the Northwest Territory, supervised the construction of the post and was its commanding officer until the spring of 1793.

Construction of this important post on the Ohio frontier was no easy task. St. Clair had suggested to Hamtramck that he should

> employ all the men that can possibly be spared for that purpose, in cutting down pickets, pointing them, and carrying to the place where they are to be put up, agreeably to such directions as Major Ferguson may give. The work will require about twelve hundred pickets. It is my idea that the best way to get it soon finished is to give the men an easy task for the day, and when they have performed it, that they should be dismissed. I suppose three men will very easily cut down, butt, and point five pickets, and that seven more (six of whom with handspikes) will carry them to the ground. If the working parties, then, are divided into squads of ten men, in every one of which there should be three good axe men, and furnished with poles of exact length the

> pickets are to be cut to, and a sufficient number of officers and non-commissioned officers to oversee them, the business would be very soon completed, and I take the liberty to recommend this method.[60]

Apparently this suggestion, if indeed put into practice, was not successful. The General had hoped that the fort would be completed by September 25, but he admitted to Secretary of War Henry Knox, "I have been too sanguine in my calculation of the time necessary for erecting the fort" and noted that the walls were not yet up.[61] Work on the fort progressed slowly. Armstrong was constantly harassed with supply and construction material shortages, insufficient labor, illness among his men, threat from Indian attack, and desertion.[62] Although the post was used during the remainder of St. Clair's campaign, it was not until February 1792 that Armstrong reported to General James Wilkinson, in command at Fort Washington, that "the garrison is now in a perfect state of defense."[63] Armstrong, as any good commander on the frontier responsible for the safety of his command, continued to strengthen his defenses and, in fact, in mid-1793, after General Anthony Wayne assumed command of the western army, the facilities at Fort Hamilton were greatly expanded to ready the post as a vast supply depot for his advance deep into the Indian heartland. Stables and storehouses were constructed, and palisades were added to the entire fortification.

Conditions were harsh at the post. Armstrong did everything within his power "to preserve every man's hair a month longer,"[64] but desertions so threatened the morale of the army that severe disciplinary measures were necessary. Execution for desertion on the frontier was common under Harmar's, St. Clair's and Wayne's commands.[65] General James Wilkinson, concerned with the safety of the entire command at Fort Hamilton and the morale of the western army itself, instructed Armstrong that "should any man desert you, the scouts are to take the track, pursue, overtake, and make prisoners of them, and for every one so apprehended and brought back you may engage them twenty dollars. If the deserter is discovered making for the enemy it will be well for the scout to shoot him and bring his head to you, for which allow forty dollars. One head lopped off in this way and set upon a pole on the parade might do lasting good in the way of deterring others."[66]

As the "Plan of Fort Hamilton" (Plate XIX) illustrated, it was well suited for the defense of the frontier and easily defended against direct assault. Situated 25 miles from Fort Washington, it was accessible for resupply via barge or raft on the Great Miami River. Its perimeter was nearly 1,000 feet. The walls were made from specially selected trees, about 4,000 in number, that were from 9 to 12 inches in diameter and extremely straight and cut into 20-foot lengths. The logs were butted together and securely buried in a ditch approximately 4 feet deep. Near the top, a piece of timber called a ribbon was fastened by driving a wooden peg through it into the logs to give added strength to the wall. One-half the logs were used for the outer wall. The remaining 2,000 were formed in a similar manner and positioned behind the joints of the first wall to give greater protection to those inside and, at the same time, to strengthen the palisade itself. Around the entire wall, a ditch was dug to carry off rain water. Because of the location of the fort on the first bank, the east wall was higher than the west to prevent anyone from spying into the garrison from atop the river's second bank, which was higher than the first and located a short distance to the east. Four blockhouses were built, and the 3 positioned in the east wall

Plate XIX: "Fort Hamilton," (1791), in *A History and Biographical Cyclopedia of Butler County, Ohio with Illustrations and Sketches of its Representative Men and Pioneers* (1882). The first post built in a chain of forts planned by General Harmar to lead to the Indian settlements on the Maumee.

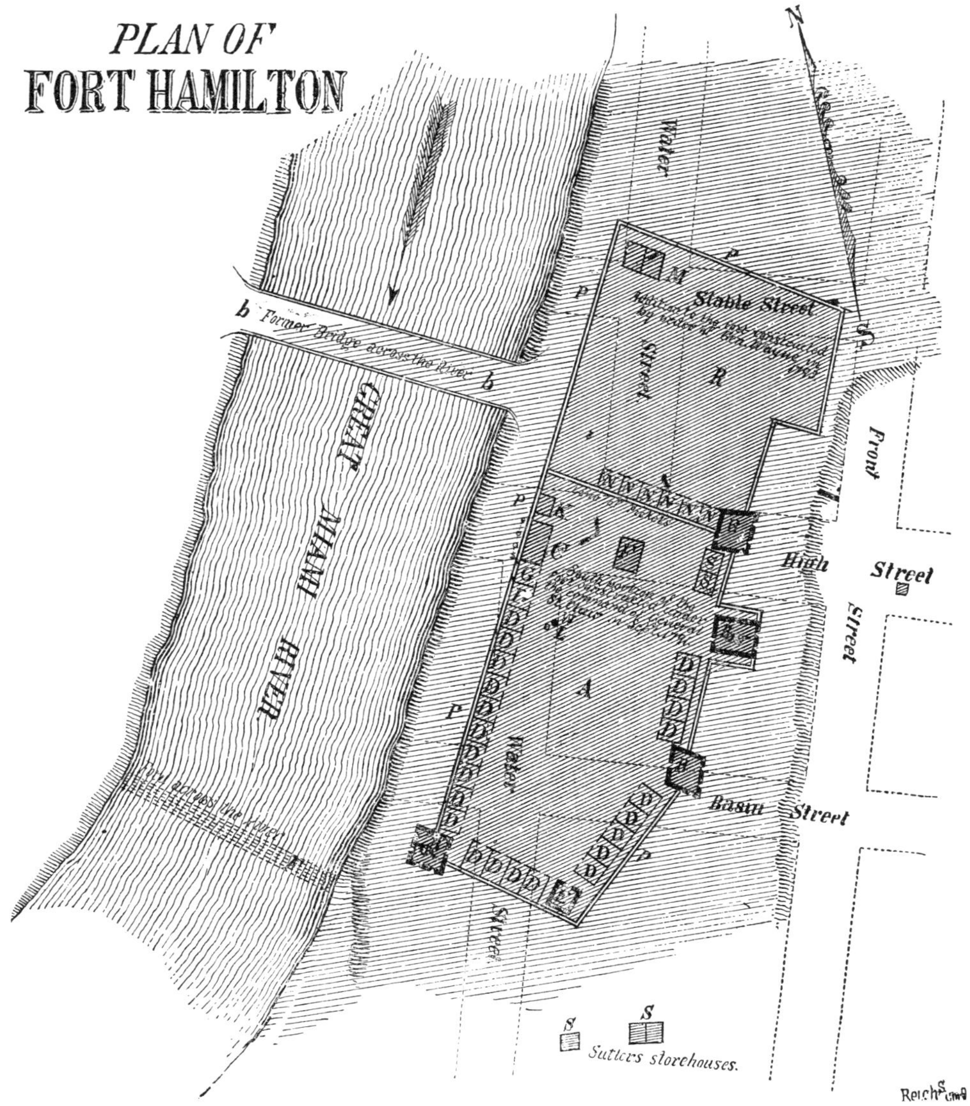

A—The southern portion of the fort, constructed under the co mmand of General St. Clair in September, 1791.
B B B B-Four block houses, on two of which cannon were placed.
C-Quarters of commandng officer, commonly called General Wilkinson's house.
D D D D, &c.—Barracks for officers and men.
E—Magazine.
F—Building called officers' mess-room.
G—Gate for entrance into the fort.
I—Hall or open space between the commandant's quarters and the kitchen.
K—Kitchen.
L—Cellar and cistern.
M—Artificers' shops.
S S—Two storehouses.
R—Northern addition to the fort, constructed by order of General Wayne in 1793.
N N N—Stables.
W—The well, with a wheel to draw with.
P P P P P—Line of pickets.
b b—Former bridge.
f—Ford across the river.
s s—Sutler's storehouses.

were of a sufficient height that, once armed with artillery, they commanded the second bank. The fourth blockhouse protected the west wall and commanded the ford which was a little distance down river.

Inside the fort were located the officers' quarters and barracks for nearly 100 men. Next to the main gate in the west wall was the commanding officer's quarters, which became known as General Wilkinson's house. It was a two-storied, weatherboarded frame building 20 feet high, with a large stone chimney in the center. On its west wall was a covered porch that provided shade for the area below, and a walk-on porch for the second floor. Apparently Armstrong and Wilkinson were fairly good friends. When, at the end of June 1792, a visit to Fort Hamilton by Wilkinson was anticipated, either because of this friendship or to impress his senior officer, Armstrong made great preparations. A cellar containing a cistern with a 400-gallon capacity was constructed beneath the commanding officer's quarters. This cistern cooled the cellar, served as a fish pond and kept the General's wine at proper temperature.[67] This building presented the commanding officer with comforts not familiar on the western frontier. When the fort was abandoned in 1796 and its buildings and property disposed of through a public sale, Wilkinson's house was purchased by William McClellan and used as a tavern for a number of years. It was razed in 1813.

Other buildings inside the fort served the garrison. On the opposite side of the gate was a 40-foot framed weatherboarded structure set upon 3-foot-high supporting blocks that served as the officers' mess. After the fort was abandoned, it was the Court of Common Pleas for Butler County for a brief period of time after the county was organized. The magazine in the southeast corner of the post survived the fort. It was made from heavy hewed logs fitted tightly together to form a building about 15 feet long. The hipped roof was shingled and was decorated with a large wooden ball on top. Between 1803 and 1810 it was converted into a jail. Later it served as a house of worship, a school house and a private dwelling. In 1902 the powder magazine was threatened with destruction, but was saved as an historic relic to remind future generations of the time when Hamilton was the country's western frontier.[68] It no longer stands today.

St. Clair's army left Fort Hamilton during the night of October 3 and early in the morning of October 4, and began its slow march toward the Indian villages. The General was most anxious to establish another intermediate post, especially if he was to fight the Indians on their own grounds further to the north. On October 13, he selected an appropriate site for the second fort in his defensive chain in what is known as the Hills of Judea, located in Neave Township in Darke County. Denny recorded that 200 men under the supervision of Major Ferguson were busied immediately in building the new post. Again, as was the experience with Fort Hamilton, construction was dreadfully slow. The weather was hot and rainy, and slowed operations considerably. Of a more serious nature was the short supply of material, as only 80 axes were available.

The plan "South perspective of Fort Jefferson"[69] (Plate XX) showed the fort as quite an imposing structure.[70] Major Ebenezer Denny described it as a "square work, curtains thirty-five yards; regular bastions at each angle. The whole laid [built] with logs laid horizontally; the face of the curtain forming the rear of the barracks."[71] It was constructed on a gently rising knoll but not on the highest point in the immediate area. Because of this, on occasions, the enemy was able to gain the high ground behind the fort and fire directly into the garrison enclosed within the walls below. One such incident cost the life of Captain Mumford, the post's commanding officer, who was killed in April 1792. After this, a blockhouse was constructed on the hill behind the fort.

Since the post was to serve as a major supply depot for the advancing army, it was necessary to provide it with ample security. The blockhouse to

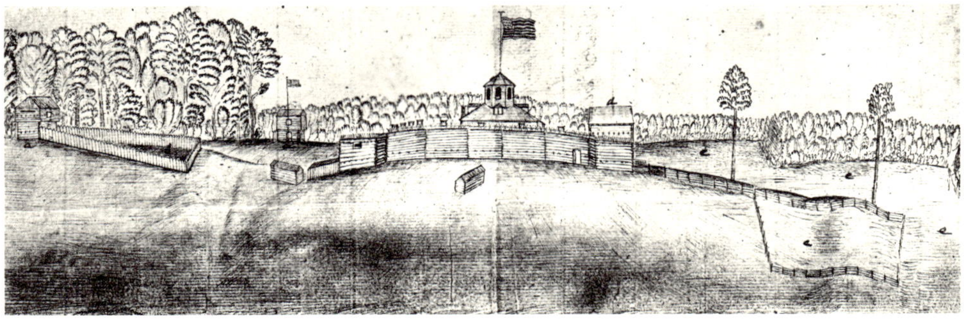

Plate XX: "Fort Jefferson," (1791), in McHenry Papers, courtesy of the Indiana Historical Society, Indianapolis, Indiana. Fort Jefferson was to serve as a major base of supplies for the United States Army's operation in western Ohio.

the left edge of the plan was built to protect the large herds of cattle and horses kept in the adjoining corral. The two small rectangular buildings outside the walls were used to store forage for the livestock, and a large enclosed garden was cultivated to provide fresh food for the garrison. Two cannons were housed at the post, one in the northeast blockhouse and the second in the southwest. When General Anthony Wayne assumed command of the western army, Fort Jefferson became an important point of deposit for supplies that supported Wayne's campaign against the Indians. In the fall of 1796, after the peace treaty with the Indians signed at Greenville, Fort Jefferson was abandoned and burned by its garrison.

On November 9, 1791, St. Clair wrote from Fort Washington to Secretary of War Henry Knox, "Yesterday afternoon the remains of the army under my command got back to this place, and I have now the painful task to give you an account of as warm and as unfortunate an action as almost any that has been fought, in which every corps was engaged and worsted."[72] Several days earlier, on November 3, St. Clair and his army were encamped along the bank of a small branch of the Wabash River where the village of Fort Recovery, Mercer County, is now located. Early in the morning of November 4 the Indians attacked. St. Clair's tired, poorly equipped, and poorly disciplined army, which had been reduced by desertions and illness from approximately 2,600 men to roughly 1,400, was routed by a force of Indians under the command of the famous Miami war chief, Little Turtle. In all, St. Clair's forces suffered 37 officers and 593 privates killed, while 31 officers and 252 enlisted men were wounded.[73] Thus, just as the United States army under Harmar had been defeated the year before, St. Clair's force, too, was badly mauled on the field of battle.

Several reports of this engagement were written, including those by St. Clair himself,[74] Major Ebenezer Denny,[75] and Winthrop Sargent,[76] Secretary of the Northwest Territory. From these written reports, several plans of the battlefield have been constructed. One of the most interesting, however, not included in the papers of Americans who participated in the battle, was "Action between Gen'l St. Clair's Army and that of the American Indians November 4, 1791 collected from his Public Letter,[77] (Plate XXI) from the papers of John Graves Simcoe, who was the Lieutenant Governor of Upper Canada in 1791. This small map accurately recorded the plan of battle and the various units' movements on the field that day. The one aspect of the battle, however, that was not conveyed in this sketch was the sheer panic that gripped St. Clair's army and its disastrous flight to the safety of Fort Jefferson and Fort Washington. Major Denny, in his important *Journal,* chronicled:

> 4th. [November 1791] Camp on a creek twenty yards wide, supposed to be the Pickaway fork of the Omee [Wabash River], ninety-eight miles from Fort Washington. The frequent firing of the sentinels through the night had disturbed the camp, and excited some concern among the officers. The guards had reported the Indians to lie skulking about in considerable numbers. About ten o'clock at night General [Richard] Butler, who commanded the right wing, was desired to send out an intelligent [*sic*] officer and party to make discoveries. Captain Slough, with two subalterns and thirty men, I saw parade at General Butler's tent for this purpose, and heard the General give Captain Slough very particular verbal orders how to proceed. Myself and two or three officers staid with the General until late, when I returned to the commander-in-chief, whose tent was at some distance on the left, and who was unable to be up.
>
> The troops paraded this morning at the usual time, and had been dismissed from the lines but a few minutes, the sun not yet up, when the

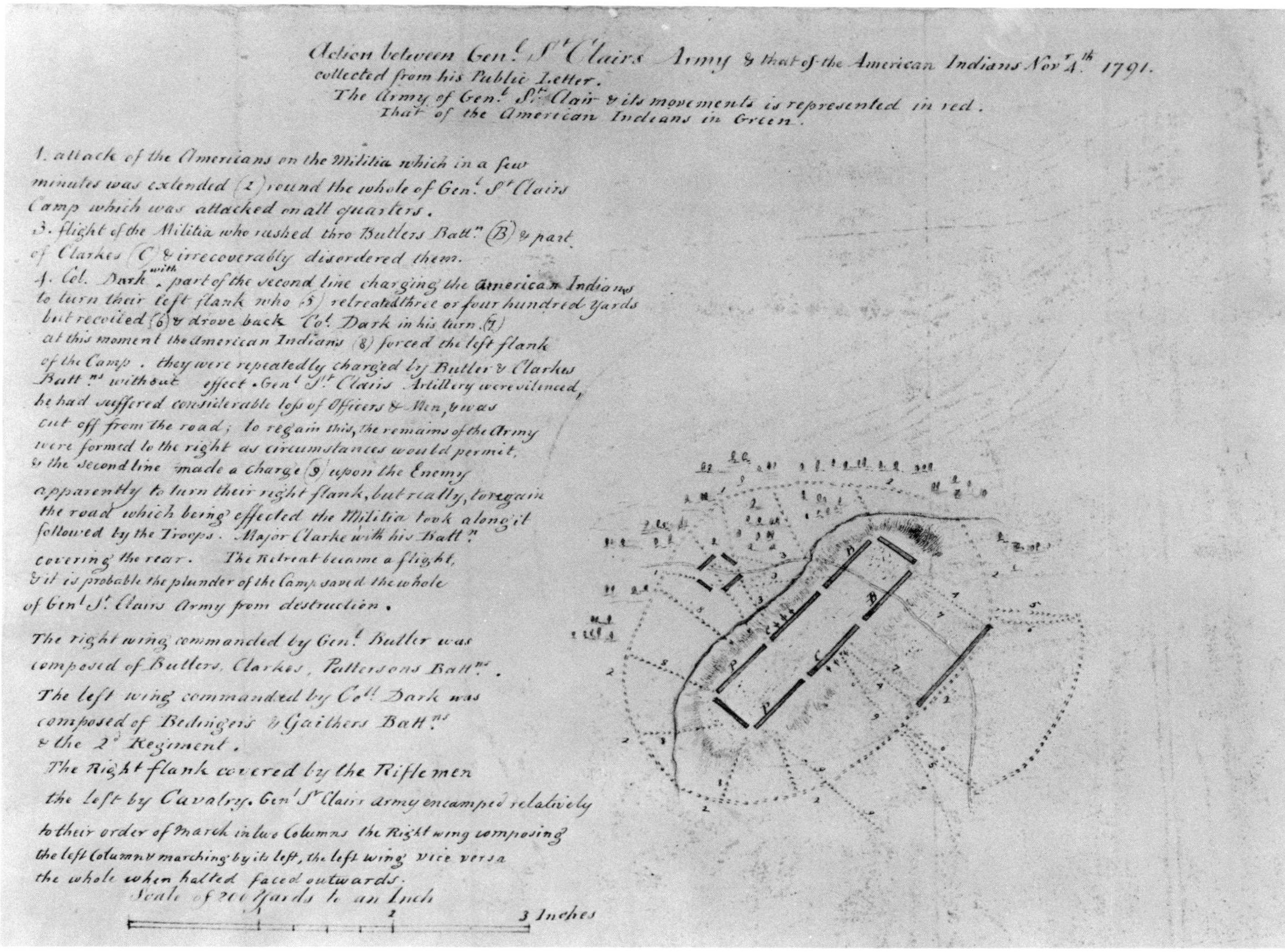

woods in front rung with the yells and fire of the savages. The poor militia, who were but three hundred yards in front, had scarcely time to return a shot—they fled into our camp. The troops were under arms in an instant, and a smart fire from the front line met the enemy. It was but a few minutes, however, until the men were engaged in every quarter. The enemy from the front filed off to the right and left, and completely surrounded the camp, killed and cut off nearly all the guards, and approached close to the lines. They advanced from one tree, log, or stump to another, under cover of the smoke of our fire. The artillery and musketry made a tremendous noise, but did little execution. The Indians seemed to brave everything, and when fairly fixed around us they made no noise other than their fire, which they kept up very constant and which seldom failed to tell, although scarcely heard. Our left flank, probably from the nature of the ground, gave way first; the enemy got possession of that part of the encampment, but it being pretty clear ground, they were too much exposed and were soon repulsed. Was at this time with the General engaged toward the right; he was on foot and led the party himself that drove the enemy and regained our ground on the left. The battalions in the rear charged several times and forced the savages from their shelter, but they always turned with the battalions and fired upon them back; indeed they seemed not to fear anything we could do. They could skip out of reach of the bayonet and return, as they pleased. They were visible only when raised by a charge. The ground was literally covered with the dead. The wounded were taken to the centre, where it was thought most safe, and where a great many who had quit their posts unhurt, had crowded together. The General, with other officers, endeavored to rally these men, and twice they were taken out to the lines. It appeared as if the officers had been singled out; a very great proportion fell, or were wounded

Plate XXI: "St. Clair's Army," (November 4, 1791), in William L. Clements Library, Ann Arbor, Michigan. Details the worst defeat suffered by a United States army at the hands of American Indians.

and obliged to retire from the lines early in the action. General Butler was among the latter, as well as several other of the most experienced officers. The men being thus left with few officers, became fearful, despaired of success, gave up the fight, and to save themselves for the moment, abandoned entirely their duty and ground, and crowded in toward the centre of the field, and no exertions could put them in any order even for defense; perfectly ungovernable. The enemy at length got possession of the artillery, though not until the officers were all killed but one, and he badly wounded, and the men almost all cut off, and not until the pieces were spiked. As our lines were deserted the Indians contracted theirs until their shot centred from all points, and now meeting with little opposition, took more deliberate aim and did great execution. Exposed to a cross fire, men and officers were seen falling in every direction; the distress too of the wounded made the scene such as can scarcely be conceived; a few minutes longer, and a retreat would have been impracticable. The only hope left was, that perhaps the savages would be so taken up with the camp as not to follow. Delay was death; no preparation could be made; numbers of brave men must be left a sacrifice, there was no alternative. It was past nine o'clock, when repeated orders were given to charge toward the road. The action had continued between two and three hours. Both officers and men seemed confounded, incapable of doing anything; they could not move until it was told that a retreat was intended. A few officers put themselves in front, the men followed, the enemy gave way, and perhaps not being aware of the design, we were for a few minutes left undisturbed. The stoutest and most active now took the lead, and those who were foremost in breaking the enemy's line, were soon left behind. At the moment of the retreat, one of the few horses saved had been procured for the General; he was on foot until then; I kept by him, and he delayed to see the rear. The enemy soon discovered the movement and pursued, though not more than four or five miles, and but few so far; they turned to share the spoil. Soon after the firing ceased, I was directed to endeavor to gain the front, and if possible, to cause a short halt that the rear might get up. I had been on horseback from the first alarm, and well mounted; pushed forward, but met with so many difficulties and interruptions from the people, that I was two hours at least laboring to reach the front. With the assistance of two or three officers I caused a short halt, but the men grew impatient and would move on. I got Lieutenants Sedam and Morgan, with half a dozen stout men, to fill up the road and to move slowly, I halted myself until the General came up. By this time the remains of the army had got somewhat compact, but in the most miserable and defenseless state. The wounded who came off left their arms in the field, and one-half of the others threw theirs away on the retreat. The road for miles was covered with firelocks, cartridge boxes and regimentals. How fortunate that the pursuit was discontinued; a single Indian might have followed with safety upon either flank. Such a panic had seized the men, that I believe it would not have been possible to have brought any of them to engage again. In the afternoon Lieutenant Kersey, with a detachment of the first regiment, met us. This regiment, the only complete and best disciplined portion of the army, had been ordered back upon the road on the 31st of October. They were thirty miles from the battle ground when they heard distinctly the firing of the cannon, were hastening forward and had marched about nine miles when met by some of the militia, who informed Major Hamtramck, the commanding officer, that the army was totally destroyed. The Major judged it best to send on a subaltern to obtain some knowledge of things, and to return himself with the regiment to Fort Jefferson, eight miles back, and to secure at all events that post. He had made some arrangements, and as we arrived in the evening, found him preparing again to meet us. Stragglers continued to come in for hours after we reached the fort.

The remnant of the army, with the first regiment, were now at Fort Jefferson, twenty-nine miles from the field of action, without provisions, and the former without having eaten anything for twenty-four hours. A convoy was known to be upon the road, and within a day's march. The General determined to move with the first regiment and all the levies able to march. Those of the wounded and others unable to go on, were lodged as comfort-

> ably as possible within the fort. Accordingly we set out a little after ten and continued our route until within an hour of daylight, then halted and waited for day and until the rear came up. Moved on again about nine o'clock; the morning of the 5th we met the convoy. Stopped a sufficiency to subsist us to Fort Hamilton; sent the remainder on to Jefferson under an additional escort of a captain and sixty men; proceeded, and at the first water halted, partly cooked and eat for the first time since the night preceding the action. At one o'clock moved on, and continued our route until nine at night, when we halted and made fires within fifteen miles of Fort Hamilton. Marched again just before day, the General soon after rode on to the fort. Troops reached in the afternoon.[78]

Following St. Clair's defeat, the Indians were encouraged that, with British aid, they could demand that the Americans not move beyond a line drawn up the Muskingum River to the Cuyahoga, and then to Venango, Pennsylvania, thus keeping the area to the north and the west for their own. By this time, however, the army's second defeat at the hands of the Indians had aroused public opinion, and the United States government in the East realized the seriousness of the situation on its western border. The Washington administration tried to achieve some form of conciliation with its enemy in the Ohio Valley but, in the end, those who understood the circumstances knew that nothing short of a complete military victory over the Indians would bring peace to the Ohio Country.

In order to maintain the advance made into the Miami Valley and to keep open the lines of communication between Fort Washington and Fort Jefferson, the northernmost outpost representing American authority in the west, James Wilkinson, commander of the American fort at Cincinnati, ordered Major John Gano to construct a post halfway between Jefferson and Washington. Thus, during the early spring of 1792, "Fort St. Clair" (Plate XXII)[79] was constructed near what is now Eaton, Preble County. The fort, which took about 3 weeks to construct, was a square structure with large, diamond-shaped blockhouses set in at each corner. Its walls,

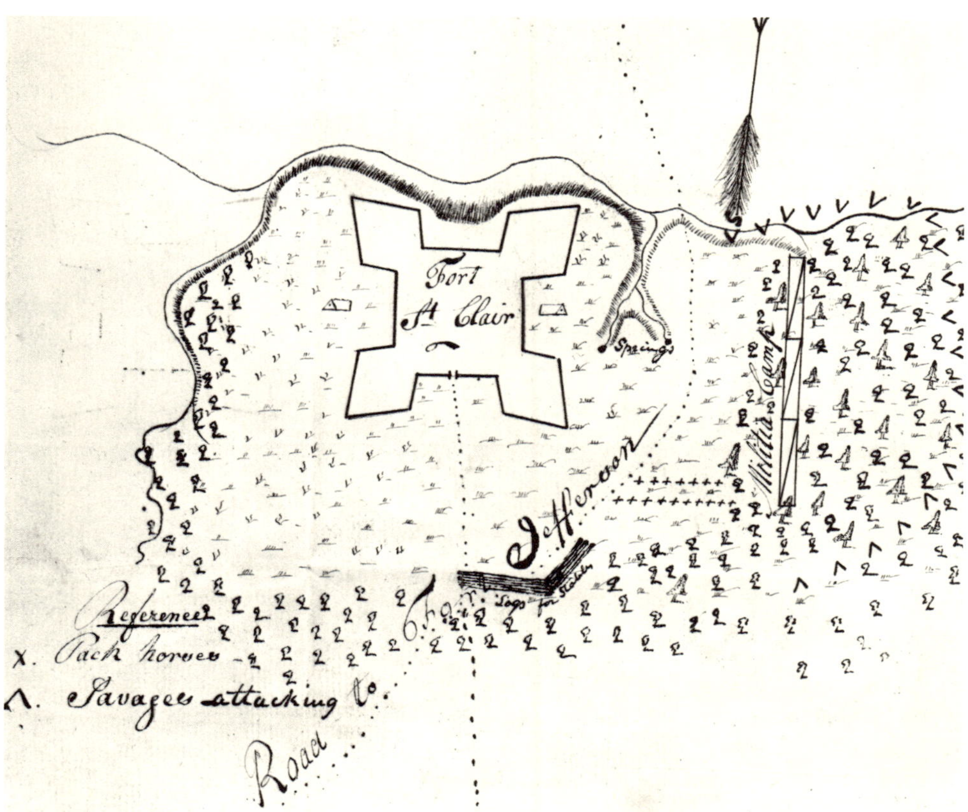

Plate XXII: "Fort St. Clair," (1792), in McHenry Papers, courtesy of the Indiana Historical Society, Indianapolis, Indiana. In order to keep communications open between forts Hamilton and Jefferson, Fort St. Clair was constructed in Preble County.

which were constructed of upright pickets rather than of horizontal logs as was Fort Jefferson, enclosed about a quarter acre. About 20 acres of land was cleared around the fort to allow for good visibility.[80] On March 19, 1792, Wilkinson wrote to Captain John Armstrong, commander of Fort Jefferson, that he had "built upon a square of one hundred and twenty feet a four-sided polygon, with regular bastions. . . . The work is substantial and rather handsome. The area, covered yesterday morning by immense oak, poplars, and beeches, is now clear for parade."[81]

On November 6, 1792, the Miami chief Little Turtle, with a raiding party of about 250 warriors, came upon a supply train, under the command of Captain John Adair of the Kentucky militia, that was returning from Fort Jefferson to Fort St. Clair. The Indians attacked on 3 sides, but left open to the Kentuckians an escape route to the fort. Apparently, the Indians were more interested in the horses in Adair's supply train than in the men. Adair formed his men into 3 divisions and tried to counterattack, but the enemy was too strong. As the Captain explained to General Wilkinson, "About the first appearance of the day, the enemy attacked my camp, within site [*sic*] of this post. The attack was sudden, and the enemy came on with a degree of courage that bespoke them warriors indeed." Because of the pressure from the Indians, Adair's forces retreated to "a kind of stockade, intended for stables; where we then made a stand."[82] This log stable was in a V-shape and stood close to the road to Fort Jefferson. Adair's defensive action saved the lives of his men, but the Indians succeeded in driving off most of the pack animals. During Wayne's campaign, Fort St. Clair served as a valuable supply depot.

On April 13, 1792, General Anthony Wayne assumed command "of the troops in the service of the United States."[83] His mission was to establish a sizeable force in the West with which either to overawe or to defeat the Indians. Such a success would certainly check the growth of British power in the northwest and restore honor to the crippled prestige of the American army after its two previous defeats at the hands of the Indians. Understanding the mistakes made by his two predecessors, Wayne was determined to succeed. Between April 1792, when he was appointed commander of the army, and the spring of 1793, Henry Knox, Secretary of War, and Wayne planned their operations against the Indians. The first year, the general organized his army. At the end of 1793, Wayne began his offensive operations and moved cautiously into the enemy's territory, erecting a chain of 5 forts between Fort Jefferson and the Maumee. Indeed, Wayne knew the seriousness of his mission and explained to Knox in a communique on October 23, 1793, that "the safety of the Western frontiers, the reputation of the Legion—the dignity & interest of the Nation all forbid, a retrograde Manoeuvre or giving up, one inch of ground we now possess—until the Enemy are compeled to sue for peace."[84]

By October 13, 1793, Wayne had moved his army from Hobson's Choice, a base camp he had established near Fort Washington to drill his troops, to a place on what is now Greenville Creek, a branch of the Stillwater Creek, "in order to gain a strong position about six miles in front of Fort Jefferson, so as to keep the Enemy in check (by exciting a jealousy and apprehension for the safety of their own Women & Children) until some favorable circumstance of Opportunity may present to strike with effect!"[85] Here he built Fort Greenville (Plate XXIII) which, Wayne

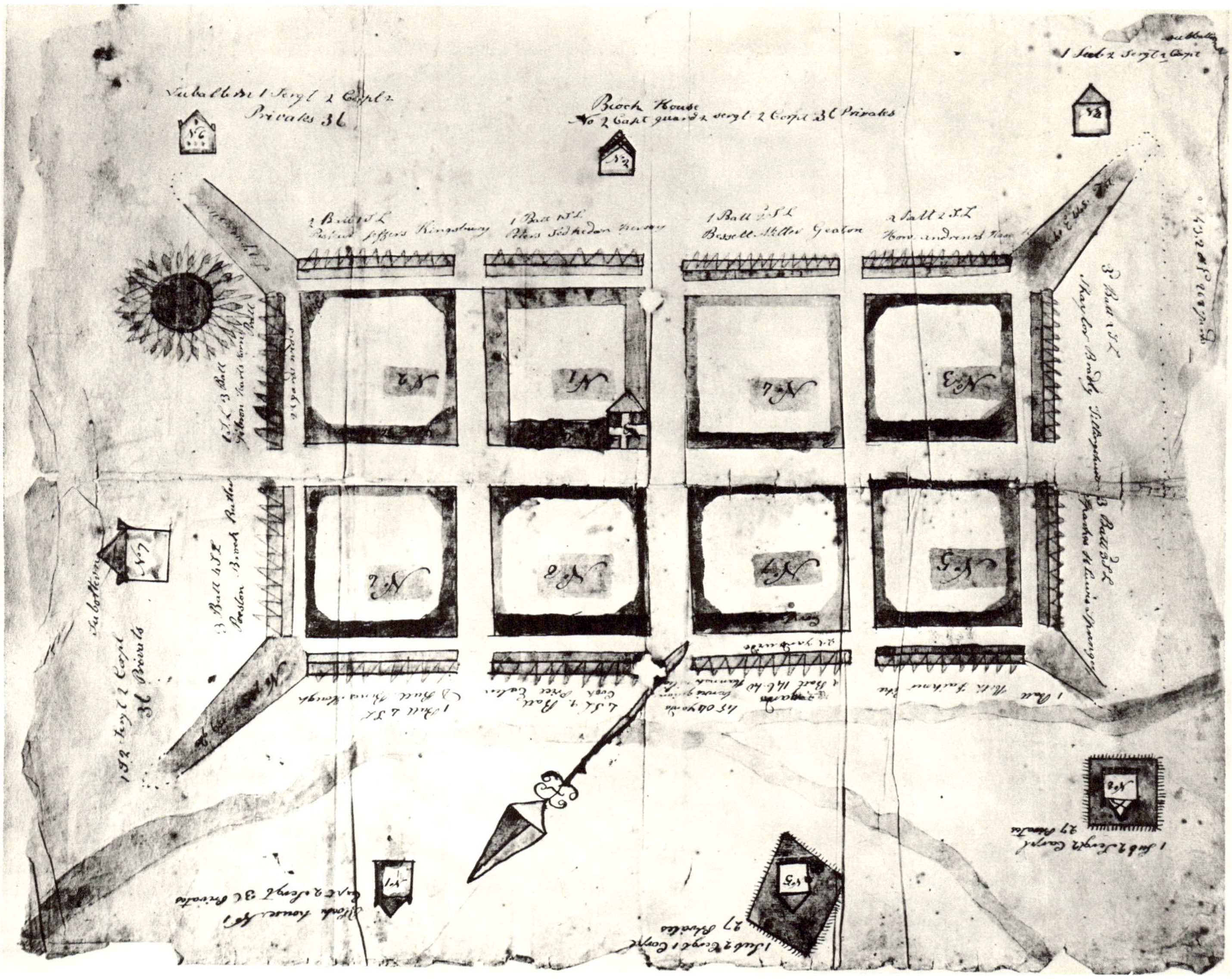

Plate XXIII: "Fort Greenville," (1793), photocopy, in Ohio Historical Society, Columbus. This is the only known contemporary sketch of Wayne's famous headquarters.

explained to Knox, would be "unassailable against the combined force of the Savages. . . ."[86]

The manuscript plan of Fort Greenville, sketched by an unidentified member of Wayne's army, is reproduced from a photostat of the original.[87] Internal evidence illustrated on the plan would indicate that it was drawn probably in late October or early November 1793. For example, Captain John Tillinghurst, the company commander in the 3rd Battalion of the 2nd Sub-Legion, whose command is shown along the redoubts at the west end of the fort, opposite square No. 3, was placed under arrest, along with several other officers, on October 23, 1793.[88] His resignation from the army was accepted by Wayne on November 11, 1793; however, on the plan of the fort, it is illustrated that Tillinghurst was still held a command.[89] Brigadier General Thomas Posey, whose living area is illustrated as having been within square No. 3, resigned from the army on February 28, 1794. However, according to the plan, he was still with the Legion. Also the ends of the elongated extensions at the four corners of the fort are not finished, but show plans for completion by dotted lines. Wayne told the Secretary of War on November 15, 1793 that "The exterior chain of Redoubts will be completed tomorrow, the Huts of the Soldiery are nearly cover'd in, so as to render the camp secure and the troops comfortable. . . ."[90]

At the same time, it should be noted that the areas on the map on the outside of the squares depict troop bivouac locations and identify each unit as well as commanders. On December 10, 1793, Wayne issued orders for "All and every of the officers, non-Commissioned officers and privates now encamped . . . outside of the piquetts, are immediately to strike their tents and to take post with the Battalions or Corps to which they respectively belong."[91] Consequently, evidence suggests that the plan was made before 1794.

Fort Greenville was a large post. Estimates vary as to its size but Captain Daniel Bradley, who was an officer in the 3rd Battalion, 2nd Sub-Legion, recorded that it was an "oblong Square about Six Hundred Yards long and three (hundred yards) across."[92] By July 28, 1794, it was large enough to house over 2,000 regular troops in addition to other supportive personnel. Around its perimeter were the standard defensive works: a ditch, picquetts, redoubt and banquetts. Picquetts, ten feet in length, were set upright in the ground. In front of the wall was a ditch at least two feet in depth, from which the dirt was thrown up a banquett which Wayne instructed should be raised "round the inside of the Stockade, from eighteen Inches to Two feet for the Troops to Stand on, in Order to fire through the loop holes, which are to be cuty [sic] high enough to prevent the Enemy from benefiting from them, should they have the Temerity to press forward to the Picquetts. . . ."[93]

Within the general compound, the battalions were housed. On November 6, 1793, Wayne recorded that the ground was "now Marked out and Assigned to each Wing and Corps for the purpose of Hutting: The Utmost Uniformity must be observed: The Hutts of the Non Commission'd Officers and Privates, are to be fourteen feet in the Clear, and Six to a Company. . . ."[94] Officers, by Wayne's order, were not to be housed until the troops were in their quarters and the Commander-in-Chief himself did not attend to his own billeting until November 13.

Wayne was anxious about the security of the encampment as well as during his army's movement. He detailed areas of responsibility to his two wing commanders, Generals Posey and James Wilkinson, who were to ensure proper fortification and housing. In turn, work assignments were made to each battalion in order that each unit shared equally in the task of construction and security. At times, units did not perform to Wayne's expectations but the blustery commander employed his influence, exerted his privilege of power and promised a gill of whiskey to his men to make sure the work progressed.

Each large numbered square within the walls, aside from housing the troops was assigned to a specific support function. Wayne's headquarters was in square No. 1; Posey was billeted in No. 3 and Wilkinson occupied No. 2. Square No. 7 contained the Contractors, No. 6 the Hospital and No. 4 the Pay Master and Adjutant General. Square No. 8 contained the Laboratory where the powder and shot was prepared for use and the Magazine which stored the ordnance. The area, or square, in the center of the fort formed by the avenues between squares 7, 8, 4 and 1 was reserved as a market place for country produce.[95] Other buildings were constructed in the fort, including store houses, a bakery and a pen for cattle. The wide avenues between the squares, which the plan shows as 29 yards wide, were used as the assembly area for the battalions but were not totally cleared of tree stumps or debris until January 1794. Outside the main fort stood blockhouses, which, no doubt, served as listening posts and a first line of defense. At the same time, they also provided a position from which rifle fire could rake the fort's walls in the event that a direct assault should carry the enemy past the forward defenses. The blockhouses across Mud and Greenville creeks, numbers eight and five, were pallisaded. Apparently Wayne placed great stock in these outer defensive works because on November 13, 1793, he ordered that these posts were to be defended "at every expence of Blood, until Supported by the Legion."[96]

Fort Greenville remained the headquarters for Wayne's army throughout his campaign and it was the site for the famous signing of the Greenville Treaty, which resulted in the opening of over half of the state of Ohio to white settlement. In August 1792, Wayne had pledged to the Secretary of War that he would have an army at the Miami villages by July 1793. On July 28 of that year, Wayne and his army left Fort Greenville to seek out the enemy and to keep that promise.

Apparently Wayne was very impressed with the military abilities of his enemy under certain circumstances. He explained, "I consider the Indian—an enemy—formidable only when he has a choice of time & ground:—in the fall of the year he's strong ferocious & full of spirits—corn is in plenty & Venison and other game every where to be met with,—in the spring he is half starved Weak and dispirited at this Season enterprise upon him—strike at him when least expected (hitherto it has been attempted in the fall) he is timid disconcerted & no longer formidable:"[97] When both Harmar and St. Clair had met the Indians, it was the latter who had chosen the time and place of battle. The new commander of the army, however, was determined not to be surprised, and constructed an elaborate order of march for his forces.

From Fort Greenville, Wayne advanced across a wide front that would not permit easy ambush. Also, he resolved that "whilst the Legion was Advancing & employed in Erecting small intermediate Forts at Convenient distances between Fort Jefferson & the Point intended for establishing a strong & permanent post—I wou'd make it an invariable rule to hault early each day & to secure my camp before evening with small temporary breast works & abbatis such as to cover the troops—so as [to] enable them to repel every kind of attack of the Savages—& to secure the baggage & stores with part of the army shou'd an opening offer to strike at the Enemy. . . ."[98] A diagram of his encampments appeared in the July 1843 issue of *The American Pioneer.*[99] "General Wayne's Daily Encampment" (Plate XXIV) was sent to the magazine published by the Logan Historical Society by John Van Cleve of Dayton, Ohio, whose father, Benjamin Van Cleve, had worked for a supplier for St. Clair's army at Fort Washington and had been at Greenville with Wayne. The editor of *The American Pioneer,* J. S. Williams, sent the diagram to George Will of Adelphia for his comments and authentication. Will, apparently, had been through Wayne's campaign and was familiar with the army's daily routine. In his letter, Will detailed:

> The drawing is a tolerably good representation of general Wayne's encampment at Greenville during our stay there, from September, 1793, until after the treaty of 1795; but we could not always find ground suitable on our daily march to form our encampment just in that form; when the ground and other circumstances admitted we generally formed our camp on that plan. On a march we generally halted early in the afternoon, say two or three o'clock; the quarter-masters of the several sub-legions, with the quarter-master-general, surveyor, and engineer, went ahead with the front guard, selected the ground, laid off the encampment, and marked the

bounds of each sub-legion, so that when the army arrived, the troops proceeded to pitch their tents. After this was done each company had to commence fortifying twenty feet in front of the company. This was done by cutting down timber, trimming off the limbs, and putting up from two to four logs high, according to the timber; so that in one hour from the commencement a complete breast-work was formed around the whole encampment. There were no gates—a few light logs were put up something in the form of bars. The picket guards consisted sometimes of a captain, one subaltern, one sergeant, two corporals, and from thirty-six to forty-eight privates; some of them [picket guards] only of a subaltern, one sergeant, one corporal, and from twenty-one to thirty-six privates. These were generally detailed every morning and were numbered from one to eight, or as many as necessary, (on the campaign we generally had eight,) so that when the army halted in the evening, each guard took its station at the place selected by the

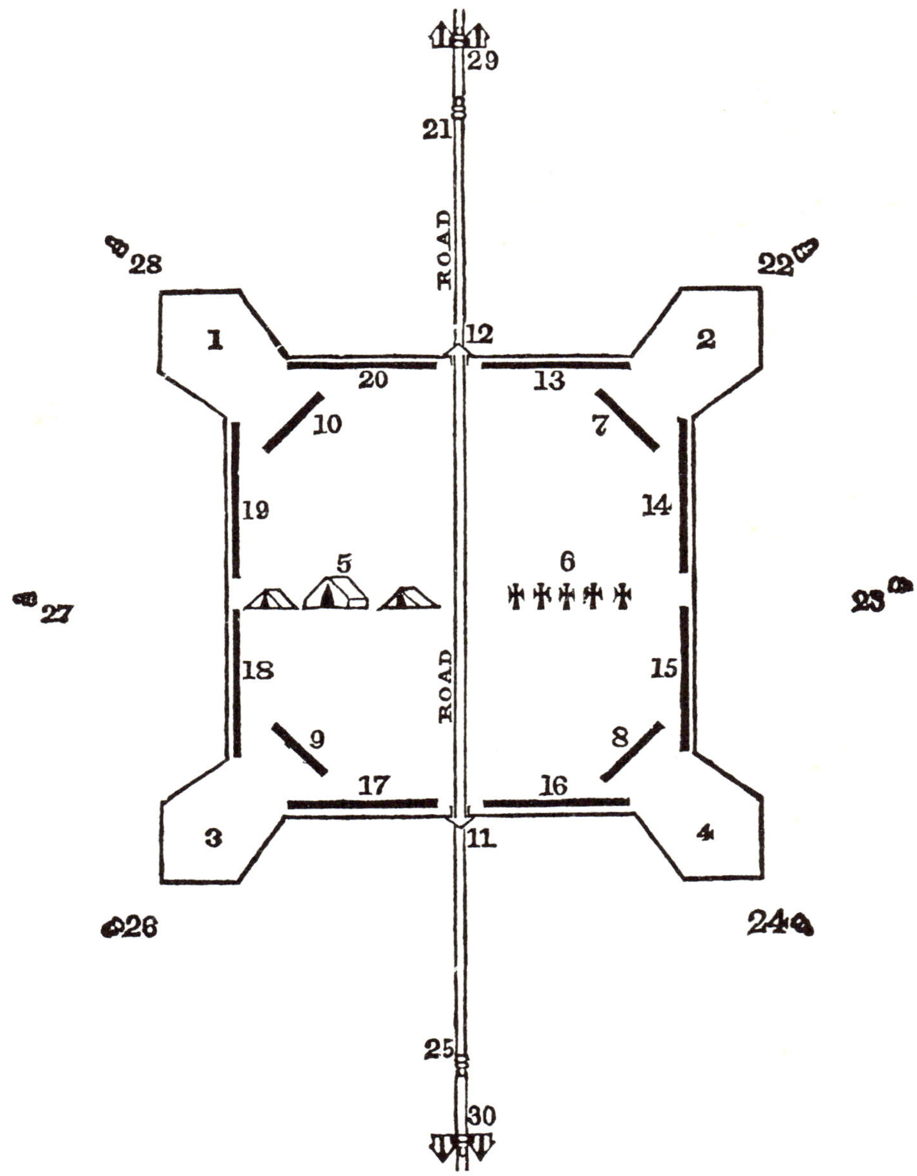

REFERENCE.

1. Lieutenant Massie's bastion.
2. Lieutenant Pope's bastion
3. Captain Porter's bastion.
4. Captain Ford's bastion.
5. Head-quarters.
6. Park of artillery.
7. Second troop of dragoons.
8. First troop of dragoons.
9. Fourth troop of dragoons.
10. Third troop of dragoons.
11. Rear gateway.
12. Front gateway.
13 and 14. Third sub-legion.
15 and 16. First sub-legion.
17 and 18. Second sub-legion.
19 and 20. Fourth sub-legion.
21, 22, 23, 24, 25, 26, 27, and 28. Picquet guards.
29. Advance guard.
30. Rear guard.

Plate XXIV: "Wayne's Daily Encampment," (1794), in *American Pioneer,* II (1843).

> quarter-master, say three hundred yards from the encampment, divided his men into three reliefs, first, second, and third; the first relief he sent out with his sergeant or corporal, placed his sentinels about one hundred yards apart, so that in half an hour from the time the army halted, a complete chain of sentinels was formed around the whole encampment. While the first relief were out, which was two hours, the second and third were employed in cutting down timber and making breast-works to defend themselves in case of an attack. After the first relief had been on sentinel two hours, the second relief went on and stood their two hours, then the third relief, and then the first, and so on until morning. The front guard generally took their station half a mile in advance, the rear guard half a mile in the rear of the army; this was done for the purpose of keeping the enemy from making a sudden attack on the main body of the army. The front and rear guards never consisted of less than a whole company, sometimes more. It was intended that, should the enemy make an attack in front, this guard should maintain its position until the main army could have time to form and come to their assistance. This was the case on the 20th August, 1794, but the front guard were overpowered and obliged to retreat.[100]

Wayne's move toward the Miami villages appeared slow to some; but, in fact, while he built an army strong enough to defeat the Indians, Washington was trying in every way to effect a peace in the Northwest through negotiations. At the end of December 1793, Wayne sent a detachment of men under the command of Captain Alexander Gibson of Virginia to the site of St. Clair's defeat near the Wabash River. Here the men buried the remains of the Governor's defeated legion and constructed Fort Recovery, a post that symbolized the American thrust back into the enemy's territory.

By mid-summer 1794, Wayne's army began to move northward again. After cutting a route of march almost 75 acres wide through the wilderness, the army reached the St. Mary's River on August 1, 1794. Lieutenant William Clark, who later explored the vast Louisiana Territory and beyond to the Pacific, kept an interesting diary of Wayne's advance. He recorded: "We immurged Suddenly, into an open extensive and bountiful Plaine or Pararie which affords an ellegent Seenery, handsomely intersperced with Small Copse of Trees & abounding in every Specie of the Greatest Variety of Herbage. . . . We soon passed through it—(abt. one mile in weadth) and crossed the River [St. Mary's] where our Encampment was formed, but a further Examination, of the Contiguous ground, induced his Excellency [Wayne] to re-cross the River & incamp on its Banks."[101] Here Wayne directed that a fort should be constructed on the south bank of the St. Mary's to serve as another link in his chain of forts. Apparently the General was anxious to be on his way, for he indicated that he wanted it built within two days.[102] The work, however, was slow. Extremely hot and humid weather, coupled with the site's location in swampy ground, prohibited rapid work. Clark recorded that "the works of the Garrison goes on with Wonderous Slowness in So much that the Commander in Chief request of the Comd's officersof Wings & Corps &c to lend all possible assistance and give every encouragement to the forwarding of it & all give their helping hand but time Waited not for us, & night came on e're the work was yet done."[103] It was while here, overseeing the construction of this fort, that Wayne was nearly killed. On August 4, while resting in the tent of his aide, Captain Henry DeButts, a large beech tree fell on both the tent and the General. He sustained internal injuries but refused to relinquish command of the army. The following day he was in the saddle at the head of his men.[104] The post was completed shortly thereafter, and Wayne reported that "another fort was also erected on the

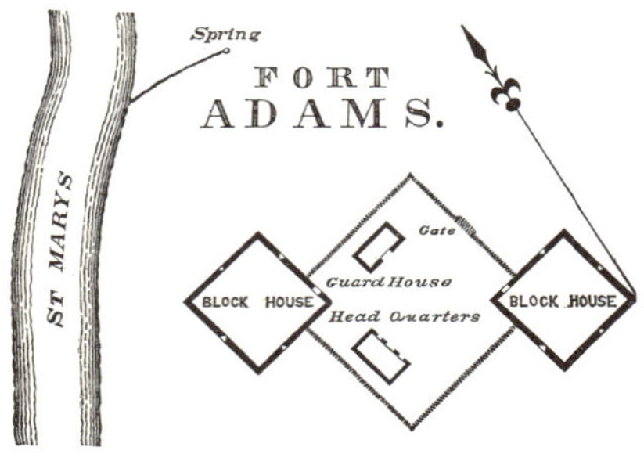

Plate XXV: "Fort Adams," (1794), *American Pioneer*, II (1843). This small post was part of Wayne's plan to protect his line of communication during his northward advance in western Ohio.

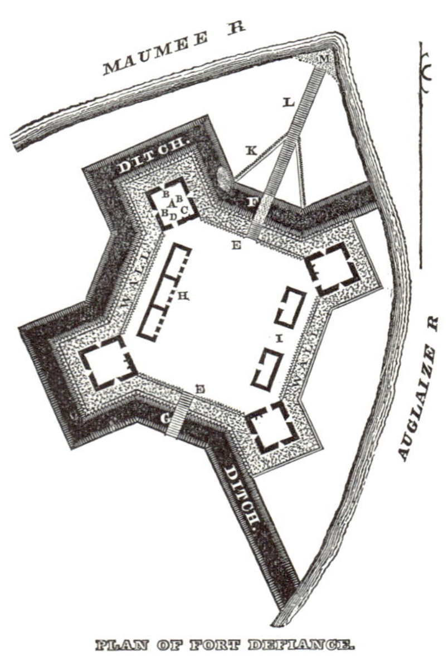

Plate XXVI: "Fort Defiance," (1794), *American Pioneer*, II (1843). Built by Wayne at the junction of the Auglaize and Maumee rivers to show United States presence deep in Indian country.

bank of the St. Mary's, twenty-four miles advanced of [Fort] Recovery, which was named Adams, and endowed with provisions and a proper garrison."[105] Fort Adams (Plate XXV),[106] sometimes called Fort Randolph, was a small post and, as far as the troops were concerned, not luxurious. It was garrisoned by approximately 100 men, who were mostly sick or too disabled to march, and commanded by Lieutenant James Underhill of Vermont. "I can't pass the situation of Lieutenant Underhill unnoticed," Clark said. "The officer was left to the command of Fort Randolph [Adams], owing to his being indisposed; and his command consisting of no more than forty invalids. Thus was he left to finish and defend this miserable hold, in the midst of the enemy's country; without the smallest probability of being reinforced or aided in the completion of the works."[107] Despite Clark's pessimism, which he incidentally displayed through the entire campaign, the practice of leaving the sick behind at isolated posts for garrison duty was not unique. Within a few days, many of those left behind were back at full duty again. Fort Adams did serve as an important supply depot for the army in the West.

By August 8, Wayne's army reached the junction of the Auglaize and Maumee rivers. Here he constructed Fort Defiance[108] (Plate XXVI), which represented the appearance of American military power on the Indian's traditional line of communication between the Great Lakes and the Miami Villages. Van Cleve described the fort as follows:

> At each angle of the fort was a block-house. The one next the Maumee is marked A, having port-holes B, on the three exterior sides, and door D and chimney C on the side facing to the interior. There was a line of pickets on each side of the fort, connecting the block-houses by their nearest angles. Outside of the pickets and around the block-houses was a glacis (a gradual slope), a wall of earth eight feet thick, sloping upwards and outwards from the feet of the pickets, supported by a log wall on the side of the ditch and by facines (wooden reinforcing), a wall of faggots [*sic*] [branches bound for a defensive wall], on the side next the Auglaize. The ditch, fifteen feet wide and eight deep, surrounded the whole work except on the side toward the Auglaize; and diagonal pickets, eleven feet long and one foot apart, were secured to the log wall and projected over the ditch. E and E were gateways. F was a bank of earth, four feet wide, left for a passage across the ditch. G was a falling gate or drawbridge, which was raised and lowered by pullies [*sic*], across the ditch, covering it or leaving it uncovered at pleasure. The officers' quarters were at H, and the storehouses at I. At K, two lines of pickets converged towards L, which was a ditch eight feet deep, by which water was procured from the river without exposing the carrier to the enemy. M was a small sand-bar at the point.

Following the Battle of Fallen Timbers on August 20, the General hurriedly strengthened his post at the confluence of the Auglaize and Maumee rivers, fearful of renewed warfare despite the fact that his army had routed the Indians that day.[109] On September 1, the men of the garrison began to make the buildings within the fort's walls bombproof and "the work now on hand is a glascis [*sic*] with fascines, and a ditch twelve feet wide and eight feet deep; the block-houses are to be made bombproof."[110] This work was completed within several days and on September 20, Wayne reported:

> Permit me now to give you a short detail of our late Manoeuvres &c. In my letter of the 14th. Ultimo [August] I informed you that we had built a strong stockade Fort at the confluence of the Miamis [Maumee] and Au Glaize, which I have since improved into a regular work, surrounded by a good Parapet, sufficient to resist a twenty four pounder, with a fraize [pointed

> stakes] projecting from the berm over the ditch which is fourteen feet wide and eight feet deep—in addition to this, I have directed Major [Thomas] Hunt the commandant to surround the whole to the water edge on each River with a thick Abbatis that were ready cut. & wou'd require but little labor to complete, agreeably to the enclosed Draught—so the *Fort* Defiance, will support that Name even shou'd the British & their Indian Allies eventually attack it nor can it be carried but by regular Approaches aided by a train of heavy Artillery which by recent intelligence are on their way to *Fort Miamis.*[111]

The last fortification Wayne built before the Battle of Fallen Timbers was Fort Deposit (Plate XXVII), along the north bank of the Maumee River, several miles up from the British post of Fort Miami.[112] On August 18, Wayne's army reached Roche de Boeuf or Roche de Bout, a prominent rock which is located at the head of the rapids in the Maumee near Waterville in Lucas County. Wayne reported to the Secretary of War that "we were employed in making a temporary post for the reception of our stores & baggage, & in reconnoitring [*sic*] the position of the enemy who were encamped behind a thick brushy wood and the British Fort."[113] John Boyer, a member of the expedition, wrote that on August 19 "the legion still continued in encampment, and are throwing up works to secure and deposit the heavy baggage of the troops, so that the men may be light for action, provided the enemy have presumption to favor us with an interview, which if they should think proper to do, the troops are in such high spirits that we will make an easy victory of them."[114] The plan of Fort Deposit illustrated the caution that the General used to protect his army against surprise attack. The main camp was protected by howitzers and the entire front shielded by an advanced guard made up of 272 men in small protected outposts. In the lower left corner of the plan, it appeared that Wayne planned a fort within a fort because that area where the baggage, provisions and cattle were kept was especially well fortified and designed to keep the stores safe.

The main objective of the three campaigns—Harmar's, St. Clair's and Wayne's—had been the destruction of the Indian villages at the headwat-

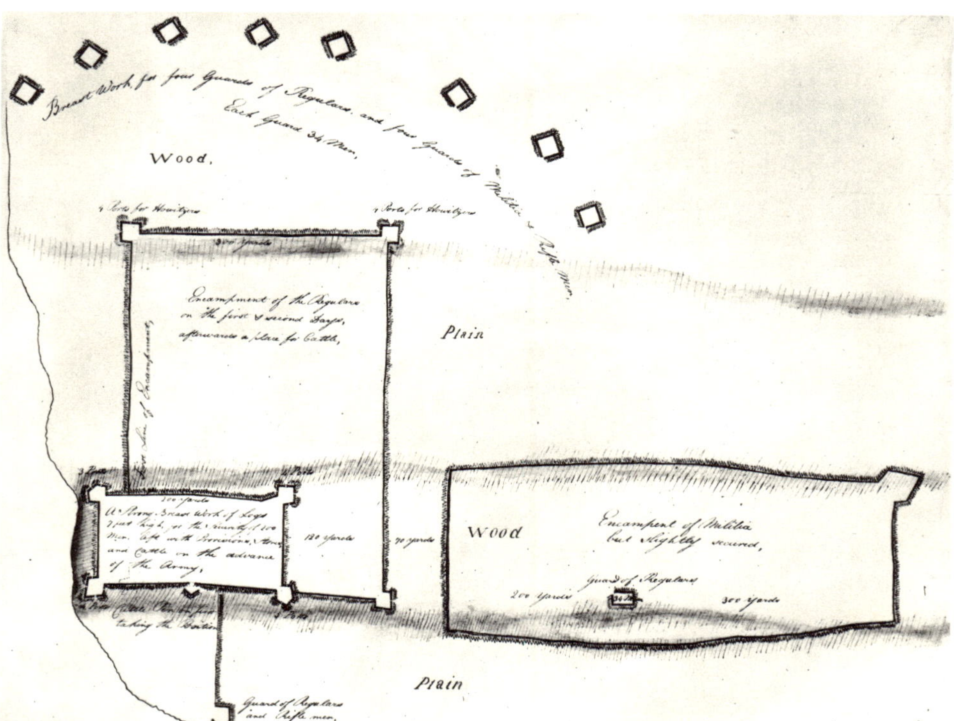

Plate XXVII: "Roche de Bout," (1794), William L. Clements Library, Ann Arbor, Michigan. Built on the eve of the Battle of Fallen Timbers to house stock and baggage.

ers of the Maumee. Wayne had been given orders to terminate the "western Indian war" during 1794. Before Wayne reached the Maumee, however, he was confronted with the problem of a British fortification constructed on American soil on the north bank of the Maumee River near the rapids. The presence of this post demonstrated England's alliance with the Indians throughout the entire four-year war and was intended to give them heart. Accordingly, Secretary Knox wrote to Wayne, "The occurrence of the establishment of a new post within our territory is of a nature not to be embraced in the orders heretofore given you relatively [*sic*] to the respect to be observed to the previously established post of Great Britain. If therefore in the course of your operations against the Indian enemy, it should become necessary to dislodge the part at the rapids of the Miami, you are hereby authorized in the name of the President of the United States to do it, taking care after they shall be in your power to treat them with humanity and politeness and to send them immediately to the nearest British Garrison."[115]

The British Fort Miami, however, was not the first British fortification to be constructed in northwestern Ohio. In 1782 Commodore Alexander Grant, commander of the British forces on the Great Lakes during the American Revolution, ascended the Maumee to the rapids and constructed a small blockhouse there in an effort to check George Rogers Clark's threatened invasion against Detroit. It was on this site that John Graves Simcoe, Lieutenant Governor of Upper Canada, directed Lieutenant Robert Pilkington of the Royal Engineers in 1794 to construct a fortification in order to counter Wayne's military move against the Miami villages. This British post represented the only earthen fortification constructed in Ohio. A ditch was dug which outlined the shape of the structure. The dirt from this ditch was piled up to form ramparts, or the main walls of the fort. Barracks were constructed along these ramparts, much as in an American fort, and covered with logs 12 inches thick. At each corner of the fort was a bastion, armed with cannon. Inside the fort, aside from the soldiers, was an officers' barracks, a log guardhouse, a house for the artillery, a bakehouse, a blacksmith's shop, engineering storehouse and several bombproofs for both the officers and men. At the time of the Battle of Fallen Timbers, Fort Miami probably was protected by four 9-pounders, six 6-pounders, 2 howitzers and 2 swivel guns. It was commanded by Major William Campbell who had a garrison of 120 men from the 24th Regiment and an officer and 10 men from the Royal Artillery.[116] On July 11, 1796, Major John Francis Hamtramck took possession of Fort Miami for the United States. Its first American commander was Captain Andrew Marschalk, who commanded a small garrison of 53 infantrymen and 7 artillerymen. It was abandoned sometime after 1796. At the siege of Fort Meigs, during the War of 1812, the British used the ruins of this old post to quarter some of their troops and store a quantity of supplies.

The "Map showing the present condition of Fort Miami, Ohio," (Plate XXVIII) was made as a result of Toledo Congressman Jacob Romeis's petition to Congress urging that the various fortification sites in Ohio be surveyed by the federal government and marked as historical landmarks. In 1888, when the survey was made under the direction of Colonel O. M. Poe of the Corps of Army Engineers, who had been General William T. Sherman's chief engineer during the Atlanta campaign, the site could have been purchased for $2,500. Apparently the federal government was not interested. The Maumee Valley Monument Association, an organization

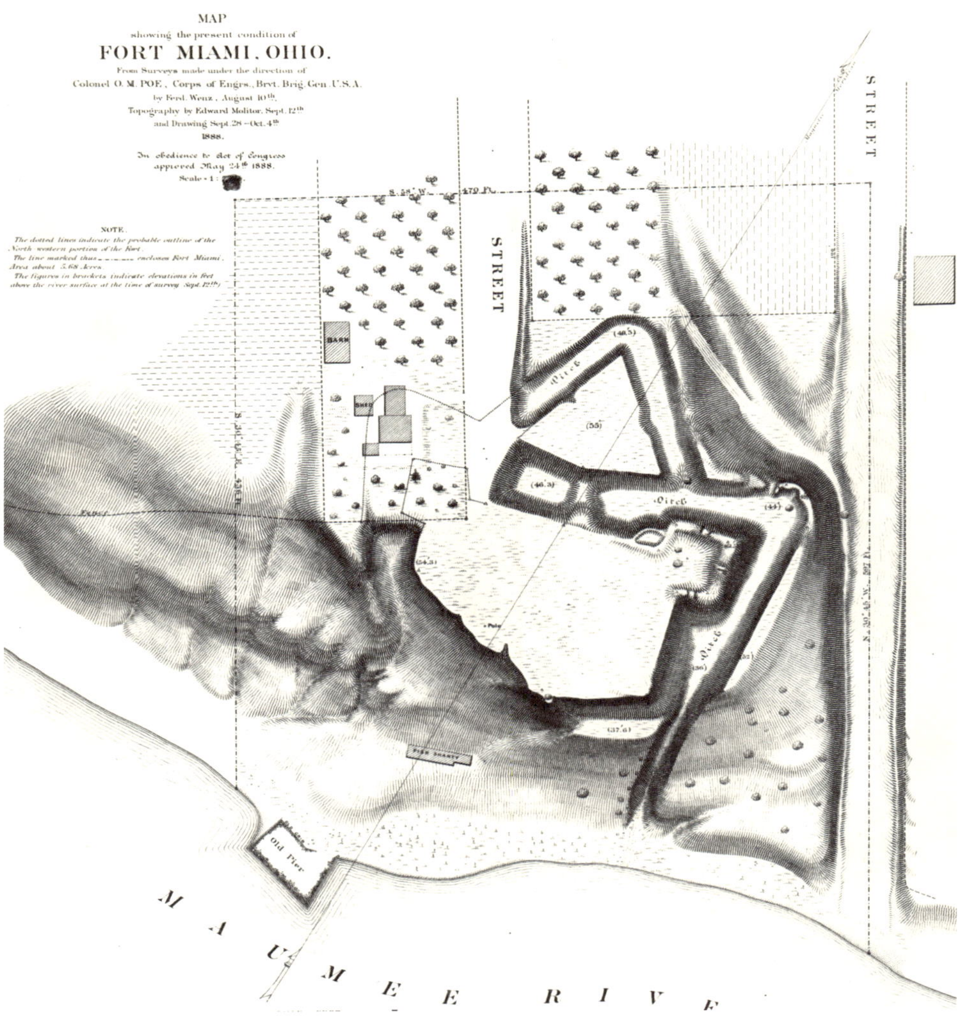

Plate XXVIII: "Fort Miami," (1794), National Archives, Washington, D.C. Record Group 77, pp. 422-4. The only foreign fort built in Ohio.

of concerned citizens interested in the preservation of the historical sites in the Maumee Valley, also lacked the financial resources to acquire the land. Portions of the earthworks of Fort Miami are visible today in a small park in the city of Maumee.

With the conclusion of the Ohio Indian Wars in 1795 and the negotiation of the Treaty of Greenville, a line that separated the white settlements from the Indians' land was established across Ohio from the Cuyahoga to Fort Recovery and on to the Ohio. At approximately the same time, England and the United States concluded Jay's Treaty, in which the British agreed to abandon their posts in Ohio by June 1, 1796. Relative peace fell upon the Ohio frontier. This situation, however, lasted only a few years. The Indians further west, still encouraged by their British allies, began a troublesome harassment of some western trade routes and isolated frontier settlements, and England disrupted American shipping on the high seas. On June 18, 1812, Congress declared war against Great Britain, and the country's second war for independence began.

Ohio was on the exposed northwestern frontier at the time, and it was feared that if England planned to invade the West, the attack would certainly come from Canada through Ohio. On the other hand, if an aggressive war was to be conducted against Great Britain, it was probable that an invasion of Canada would be launched from Ohio's soils. A call for volunteers from Ohio's governor, Return Jonathan Meigs, went across the state sometime before the actual declaration of war. An army was gathered in Dayton and placed under the command of William Hull, governor of the Michigan Territory. When hostilities did break out, Hull and his forces marched to Detroit, crossed the Detroit River and invaded Canada. The situation, however, called for a more determined man than Hull to command the western army. By August 16, Hull had returned to Detroit and surrendered his command there to the British.

Five weeks after Hull's surrender of Detroit, William Henry Harrison, the governor of the Indiana Territory and the hero of Tippecanoe Creek, was placed in command of the army of the northwest on September 17, 1812. His instructions were simple: protect the inhabitants in the Northwest from the Indians and the British army; retake Detroit and invade Canada. That fall Harrison had planned a campaign to retake Detroit and had established a small supply base near the Rapids of the Maumee. The left wing of his army, however, under the command of General James Winchester, was defeated at the Battle of the River Raisin on January 22, and it was necessary for Harrison to take up a defensive position in order to protect the interior. In the meantime, Harrison had assembled almost 1,700 men near the Rapids and it was here that a site was selected for fortification.

On the east bank of the Maumee River, 60 feet above the water, a fairly large fortification was begun on February 2 under the direction of Captain Eleazer Derby Wood, a West Point graduate in 1805 from New York and a member of the Engineer Corps. In his journal, Wood explained how the work on the fort progressed:

> So soon as the lines of the camp were designated, large portions of labor were assigned to each corps in the army, by which means a very laudable emulation was easily excited. Each brigade or regiment commenced the particular portion of work alloted to it with great spirit and vigor. The camp was about 2,500 yards in circumference the whole of which, with the exception of several small intervals left for batteries and blockhouses, was to be picketed with timber fifteen feet long, from two to twelve inches in diameter, and set three feet in the ground. Such were the instructions of the engineer. To complete this picketing, to put eight blockhouses of double timbers, to elevate four large batteries, to build all the storehouses and magazines required to contain the supplies of the army, together with the ordinary fatigues of the camp, was an undertaking of no small magnitude. Besides an immense deal of labor was likewise required in excavating ditches, making abatis and clearing away the wood about the camp; and all this was to be done, too, at a time when the weather was unclement, and the ground so hard frozen that it could be scarcely opened with the mattock and pickaxe. But in the use of the axe, mattock and spade consisted the chief military knowledge of our army; and even that knowledge, however trifling it may be supposed by some, is of the utmost importance in many situations, and in ours was the salvation of the army.[117]

Fort Meigs, as this post was called, was designed for approximately 2,000 troops. Harrison, like Harmar, St. Clair and Wayne, who had conducted extensive military operations on the American western frontier before him, had supply problems and suffered from the lack of a sufficient force. During the spring of 1813, Harrison left the Maumee and tried to raise additional troops from the surrounding states. While he was gone, General Leftwich of Pennsylvania was left in command of the fort. Construction of the fort was not completed, and apparently Leftwich was in no hurry for its completion despite its exposed position. Wood, who had gone to Lower Sandusky to help with the building of the fortifications there, returned to Fort Meigs to find an uncompleted fort. Leftwich, Wood recorded, "pretended that the men could not be made to work, said they were sickly, that the weather was bad, and what was most vexatious indeed, [he] permitted them to burn the picketing timber for fuel, instead of getting it from the woods. After General Harrison left camp, not a single thing towards finishing the works was done, until Captain Wood returned on the 20th from Sandusky."[118] Construction of the post continued until rumors reached the garrison the first part of April that a British force was on its way. Then all efforts were made in earnest to have the fort ready.

An outline of Fort Meigs was shown in the "Plan of Fort Meigs and its Environs" (Plate XXIX) drawn by Captain William Sebree of the Kentucky militia.[119] Sebree was born in 1776 and served for some time in the United States Navy. This most unusual map was discovered among the Captain's belongings by his granddaughter, Mrs. Pensacola Hall Casey of Callaco, Missouri, in 1934 and presented to the Library of Congress.[120] He served at the post along the Maumee during 1813 and his sketch of Fort Meigs was made from notes and drawings that he recorded in his tent while there. Aside from the plan of Fort Meigs, the map itself is a fairly accurate account of the heroic defense of that post made between April 27 and May 9, and again between July 27 and 28, 1813. Sebree himself explained that the purpose of the map was "to preserve from oblivion the useful events of our history, and those occurrances which are characteristic of us, as a nation, must be gratifying to every American." An analysis of this map would necessitate a volume. Standard accounts of the siege are available and should be consulted in order to understand fully the detail of events that Sebree selected for his map. His patriotism and sense of pride in the gallant stand that he and his comrades made in the spring of 1813 are evident. "The typography represents a bold and determined effort for the preservation of liberty against a proud and merciless enemy, allied to the Tomahawk in the hands of the Savage—the transaction is a recent one, the event momentous, and the recollection should be kept alive," he said. "I am sorry the drawing is not more complete; the original sketch and notes were prepared in my tent, subject to many inconveniences with ill health; but I am happy that even this much had been done, to arrest from the destroying hand of TIME; the remembrance of even 'WHICH TRIED THE SOULS OF MEN'." It was no accident that Sebree chose as his motto the reverse side of the great American seal which he displayed in the lower left hand corner of his map. "ANNUIT COEPTIS:" he recorded, "He has favored our undertakings."

Following the second siege of Fort Meigs in July 1813, the British army along with its Indian allies, under the command of Colonel Henry Proctor and the Shawnee chief, Tecumseh, passed through Sandusky Bay and up the Sandusky River to the head of its navigation and the fortifications there. In April 1813, Major Wood, who had laid out Fort Meigs, planned Fort Stephenson at what is now Fremont, Ohio.[121] This fort, along with several others built in the northwestern portion of the state, was to be the second line of defense in case the British were able to break through the fortification along the Maumee. On August 4, despite General William Henry Harrison's original orders to the contrary, Major George Croghan, the nephew of General George Rogers Clark and General William Clark, defended the post against a determined British assault. Croghan's victory there along the west bank of the Sandusky was encouraging news to a nation that had fared rather poorly in its land war against England, and it proved to be one of the major land victories during the entire war.

(Overleaf) Plate XXIX: Sebree, "Fort Meigs," (1813), Library of Congress, Washington, D.C. Sebree's map recounts one of the few important United States land victories during the War of 1812.

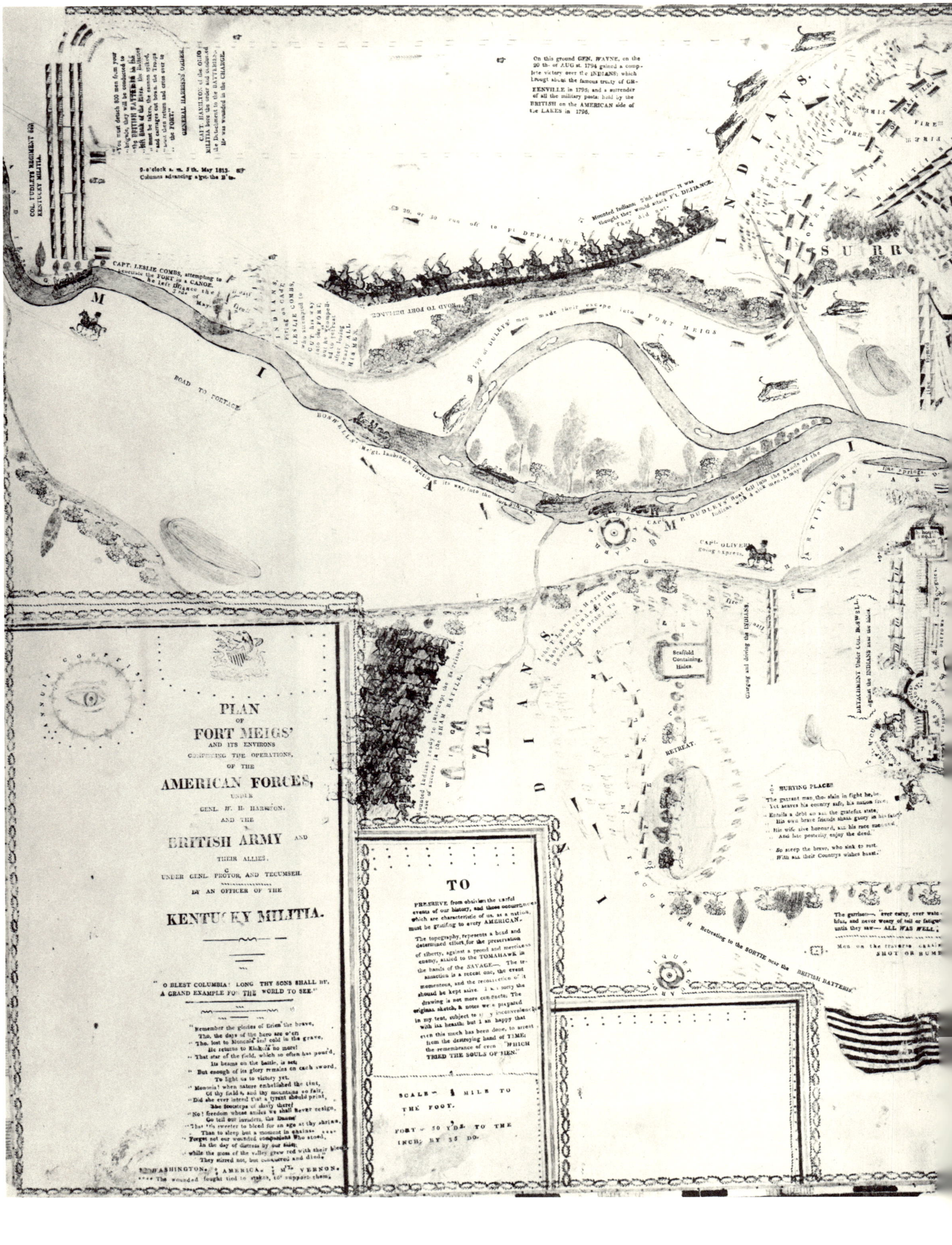

PLAN OF FORT MEIGS' AND ITS ENVIRONS
AMERICAN FORCES, UNDER GENL. W. H. HARRISON, AND THE BRITISH ARMY AND THEIR ALLIES,
BY AN OFFICER OF THE KENTUCKY MILITIA.
" O BLEST COLUMBIA! LONG THY SONS SHALL BE, A GRAND EXAMPLE FOR THE WORLD TO SEE."
On this ground GEN. WAYNE, on the 20 th. of AUG st. 1794 gained a complete victory over the INDIANS; which brougt about the famous treaty of GREENVILLE in 1795; and a surrender of all the military posts held by the BRITISH on the AMERICAN side of the LAKES in 1796.
COL. DUDLEY'S REGIMENT 800 KENTUCKY MILITIA.
GENERAL HARRISON'S ORDER.
9 o'clock a. m. 5 th. May 1813. Columns advancing agst. the B'ts.
CAPT. LESLIE COMBS, attempting to reinforce the FORT in a CANOE.
ROAD TO FORT DEFIANCE
ROAD TO PORTAGE
FORT MEIGS
INDIANS.
Scaffold Containing Hides.
RETREAT.
BURYING PLACE!!
Retreating to the SORTIE near the BRITISH BATTERIES.
Capt. OLIVER going express.
TO
PRESERVE from oblivion the careful events of our history, and those occurrences which are characteristic of us, as a nation, must be gratifing to every AMERICAN.
SCALE— MILE TO THE FOOT.
FORT — 50 YDS. TO THE INCH; BY 35 DO.
WASHINGTON. AMERICA. Mt. VERNON.

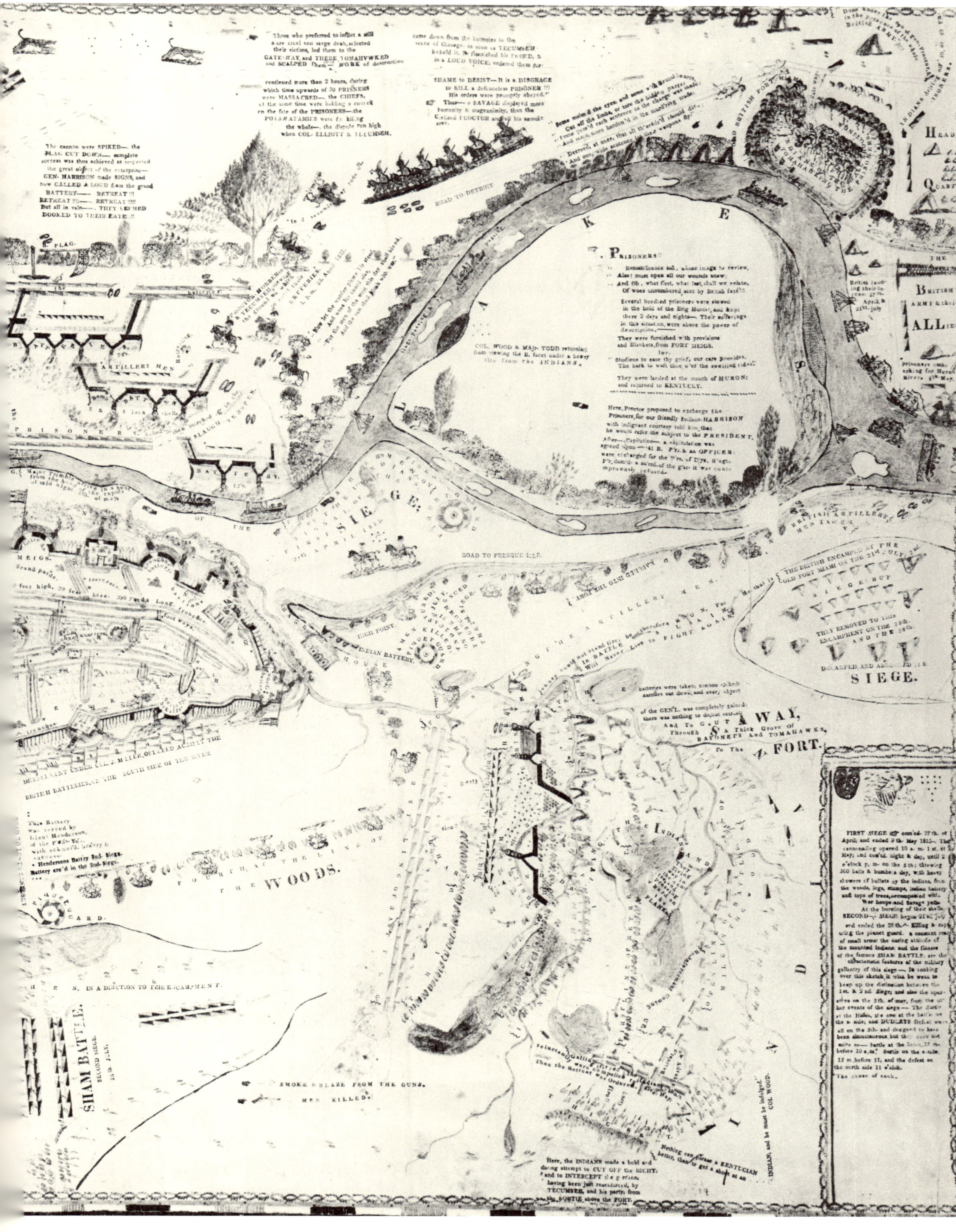

Those who preferred to inflict a still more cruel and sarage death, selected their victims, led them to the GATE-WAY, and THERE TOMAHAWKED and SCALPED Them— WORK of destruction
continued more than 2 hours, during which time upwards of 30 PRISNERS were MASSACRED— the CHIEFS, at the same time were holding a council on the fate of the PRISONERS—the POTAWATAMIES were for killing the whole—. the dispute run high when COL. ELLIOTT & TECUMSEH,
came down from the batteries to the scene of Carnage; as soon as TECUMSEH beheld it, he flourished his SWORD, & in a LOUD VOICE, ordered them for
SHAME to DESIST— It is a DISGRACE to KILL a defenceless PRISONER !!! His orders were promptly obeyed." Thus— a SAVAGE displayed more humanity & magnanimity, than the Civilized PROCTOR and all his associates.
The cannon were SPIKED—. the FLAG CUT DOWN— complete success was thus achieved as respected the great object of the enterprise— GEN. HARRISON made SIGNS, and now CALLED A LOUD from the grand BATTERY— RETREAT !!! RETREAT !!!— RETREAT !!! But all in vain—. THEY SEEMED DOOMED TO THEIR FATE !!!
FLAG.
ROAD TO DETROIT
ARTILLERY MEN
PRISONERS!!
Remembrance sad, whose image to review, Alas! must open all our wounds anew; And Oh, what first, what last, shall we relate, Of woes unnumbered sent by British fate!!!
Several hundred prisoners were stowed in the hold of the Brig Hunter, and kept there 2 days and nights—. Their sufferings in this situation, were above the power of description.
They were furnished with provisions and Blankets, from FORT MEIGS.
Studious to ease thy grief, our care provides, The bark to waft thee o'er the swelling tides.
They were landed at the mouth of HURON; and returned to KENTUCKY.
COL. WOOD & MAJr. TODD returning from viewing the B. fleet under a heavy fire from the INDIANS.
Here, Proctor proposed to exchange the Prisoners, for our friendly Indians. HARRISON with indignant courtesy told him, that he would refer the subject to the PRESIDENT.
HEAD QUARTERS
THE BRITISH ARMY & their ALLIES
SIEGE.
ROAD TO PRESQUE ISLE
MEIGS.
Grand parade.
INDIAN BATTERY.
HIGH POINT.
HOUSE
THE BRITISH ENCAMPED AT THE OLD FORT MIAMI ON THE 21st JULY.
THEY REMOVED TO THIS ENCAMPMENT ON THE 25th AND THE 28th
DECAMPED AND ABANDONED THE SIEGE.
batteries were taken; cannon spiked; carriages cut down, and every object
of the GEN'L. was completely gained; there was nothing to do, but retreat
AWAY, And To CUT Through A Thick Grove Of BAYONETS And TOMAHAWKS, To The FORT.
BRITISH BATTERIES, ON THE SOUTH SIDE OF THE RIVER
This Battery was served by Lieut Henderson, of the Pet'sb'g Vol's. with acknowl'd. bravery & success.
Hendersons Battry 2nd. Siege.
Battery erec'd in the 2nd. Siege.
THE WOODS.
SHAM BATTLE.
SECOND SIEGE.
26th. JULY.
SMOKE & BLAZE FROM THE GUNS.
MEN KILLED.
Here, the INDIANS made a bold and daring attempt to CUT OFF the RIGHT, and to INTERCEPT the garrison, having been just reenforced, by TECUMSEH, and his party, from the SORTIE, above the FORT.
Nothing can please a KENTUCIAN better, than to get a shot at an
INDIAN, and he must be indulged." COL WOOD.
FIRST SIEGE commenced 27th. of April; and ended 9th. May 1813.— The cannonading opened 10 a. m. 1 st. of May; and cont'd. night & day, until 2 o'clock p. m. on the 5th; throwing 500 balls & bumbs a day, with heavy showers of bullets by the indians, from the woods, logs, stumps, indian battery and tops of trees, accompanied with War hoops and Savage yells.

"A Plan of Fort Sandusky [Stephenson], at Lower Sandusky [Fremont]" (Plate XXX)[122] appeared in *Port folio* nearly a year and a half after the battle but before the war with England had been concluded, no doubt, to give heart to the American people. "The defense of fort Sandusky [Stephenson], on the 4th of August 1813, although the work of a few hours, and of a small force, was an achievement brilliant in itself and important in its consequences," *Port folio* said. However diminutive it may appear when compared with many of the military feats of Europe, and even with some that have subsequently occurred on American ground, it is justly entitled to a distinguished place in the annals of our country. "It was among the first events of the present war that gave confidence to our soldiers and compelled the enemy to respect our arms," the article continued. "It furnished, moreover, a memorable instance of what a few bold and determined spirits can perform, when opposed even to more than fourfold their numbers. It is not too much to add, that, under Providence, it was highly instrumental in preserving from the tomahawk and the scalping knife, many of our defenseless frontier inhabitants."

Fort Stephenson, as the post was known, was built on a site previously occupied by a government trading post and enclosed nearly an acre of ground. The timber for the walls of the fort was obtained close by, and it took 25 days to complete the walls. Two blockhouses, the one at the northeast corner and the one in the middle of the north wall, were constructed as part of the original fort. During the summer of 1813, when Croghan took command, he added the blockhouse at the southwest corner and built another next to the one in the north wall. It was from this latter blockhouse that the famous cannon "Good Old Bess," which played a large part in the American victory there, was fired. Croghan also added a ditch 6 feet deep and 9 feet wide around the outside of the fort. The dirt from the trench was then thrown against the walls of the fort, making a gradual slope away from the stockade into the trench, while more dirt was piled to the outside of the ditch, making it appear deeper than it actually was. Included in the fort, in addition to the soldiers' barracks, were a hospital, commissary stores, a well and a magazine.[123]

Croghan, just before the battle, explained, "The enemy are not far distant. I expect an attack. I will defend this post to the last extremity. I have just sent away the women and children, with the sick of the garrison,

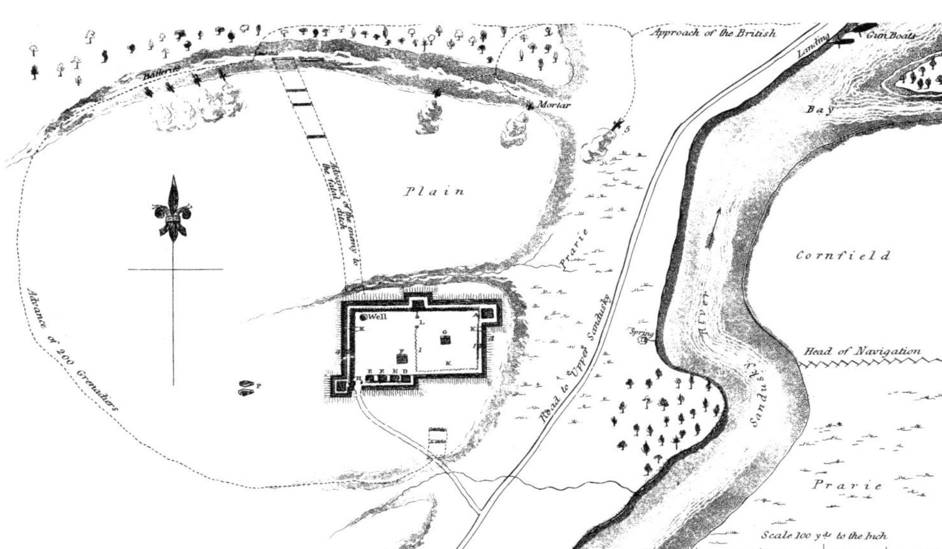

Plate XXX: "Fort Sandusky," (1813), *Port folio,* V (1815). Fort Sandusky, but better known as Fort Stephenson, was the scene of a victory against the British in 1813.

that I may be able to act without incumbrance. Be satisfied. I shall, I hope, do my duty. The example set me by my Revolutionary kindred is before me. Let me die rather than prove unworthy of their name." The official report, written by Croghan himself three days after the battle, best explained the action:

Lower Sandusky, Aug. 5, 1813

Dear Sir:

I have the honor to inform you that the combined force of the enemy, amounting to at least 500 regulars and 700 to 800 Indians, under the immediate command of Gen. Proctor, made its appearance before this place early Sunday evening last, and so soon as the general had made such disposition of his troops as would cut off my retreat, should I be disposed to make one, he sent Col. Elliott, accompanied by Major Chambers, with a flag to demand the surrender of the fort, as he was anxious to spare the effusion of blood, which he probably would not have in his power to do should he be reduced to the necessity of taking the place by storm. My answer to the summons was that I was determined to defend the place to the last extremity, and that no force, however large, should induce me to surrender it.

So soon as the flag had returned, a brisk fire was opened on us from the gunboats in the river, and from a 5½-inch howitzer on shore, which was kept up with little intermission throughout the night. At an early hour next morning, three sixes, which had been placed during the night within 250 yards of the pickets, began playing on us with but little effect. At 4 p.m. discovering the fire from all of his guns was concentrated against the northwestern angle of the fort, I became confident that his object was to make a breach and attempt to storm the works at that point. I therefore ordered as many men as could be employed for the purpose of strengthening that part; which was so effectually secured by means of bags of flour, sand, etc., that the picketing suffered little or no injury. Notwithstanding which, the enemy, about 5 o'clock, having formed into close column, advanced to assault our works at the expected point, at the same time making two feints on the front of Capt. Hunter's lines. The column which advanced against the northwestern angle, consisting of about 350 men, was so completely enveloped in smoke as not to be discovered until it had approached within 15 or 20 paces of the line; but the men, being all at their post and ready to receive it, commenced so heavy and galling a fire as to throw the column a little into confusion. Being quickly rallied, it advanced to the outer works and began to leap into the ditch. Just at that moment a fire of grape was opened from our 6-pounder, which had been previously arranged so as to rake in that direction, which together with the musketry threw them into such confusion that they were compelled to retreat precipitately to the woods. During the assault, which lasted half an hour, an incessant fire was kept up by the enemy's artillery, which consisted of five sixes, and a howitzer; but without effect. My whole loss during the siege was one killed and seven wounded slightly. The loss of the enemy, in killed and wounded and prisoners, must have been 150. One lieutenant-colonel, one lieutenant, and 50 rank and file were found in and about the ditch, dead or wounded. Those of the remainder who were not able to escape were taken off during the night by the Indians. Seventy stands of arms and several brace of pistols have been collected near the works. About three o'clock in the morning the enemy sailed down the river, leaving behind them a boat containing clothing and considerable military stores.

Too much praise can not be bestowed upon the officers, the noncommissioned officers and privates under my command for their gallant and good conduct during the siege.

Yours with respect,

G. Croghan
Major, Seventeenth U.S. Infantry
Commanding Lower Sandusky.[124]

When the Civil War broke out in April 1861, Ohio was ill-prepared to cope with its citizens' patriotic response to President Abraham Lincoln's call for 75,000 troops. Men poured into Columbus from every part of the state, only to find that no provisions for their care had been made. The volunteers were housed in boarding houses, in hotels, and in such public buildings as the state capitol and the penitentiary; some were even placed in the Starling Medical College. Six days after the war began, however, Columbus had erected Camp Jackson, a hastily constructed camp made mostly of board shanties and tents at Goodale Park. Its first commanding officer was Lieutenant Colonel H. Z. Mills, and it was to be the rendezvous point for all troops north of Hamilton County and south of the Western Reserve. The *Ohio State Journal* described the camp as follows:

> The gates of the high picket fence are guarded by sentinels who keep back the baffled and impatient crowd which surges to and fro from morning till night and only gets now and then an eyefull of the inside, by looking through the pallisades. . . . The white tents are pitched in the plain in the centre of the park beneath the yet leafless trees—peaceful above all that martial parade and clamor, and the young grass is crushed and cut in a thousand furrows by the feet and wheels that have passed over it. A large wooden building, hastily erected near the tents, adds nothing to the picturesqueness of the scene but adds immensely to the comfort of the soldiers, for it is full of 'bunks' and is much better defense against the weather than canvas. Wagons are continually coming and going, and the camp is strewn with straw from a stack brought thither for bedding.
>
> . . .
>
> A hastily improvised building near the main entrance to the park is set with long tables and substantial edibles. Great cooking ranges roar with preparation, and the provisional government is marvelously efficient. The officers' quarters are in the building lately occupied by the keeper of the park, and here all the business of the camp is transacted, though the headquarters are, of course, at the Statehouse.[125]

By the end of April, however, Camp Jackson was so overcrowded that it became necessary to make other accomodations.

On May 28, 1861, the *Ohio State Journal* reported that "workmen were engaged in taking down the barracks for the purpose of removing [them] to a new camp to be organized four miles west of the city. It is to be a regular camp. It will contain one hundred acres," the paper continued, and "it is plowed, harrowed and rolled smooth and will make a good place for drilling purposes." This new camp which was also called Camp Jackson until June 20, 1861, when it was renamed Camp Chase, was located about five miles west of Columbus along the National Road. The "Map of Camp Chase" (Plate XXXI) was drawn by W. W. Pollard and is a good representation of what the post looked like about 1863.[126] Pollard was a surveyor and draftsman who had his place of business at 9 East Chapel in Columbus and apparently was hired by the federal government to survey the several federal posts in Columbus. After 1862, when the state was divided into five military districts, Camp Chase became the collection point for recruits from Franklin, Licking, Madison, Champaign, Logan, Union, Delaware, Marion, Morrow and Knox counties. It also was used as a camp for paroled prisoners, Confederate prisoners, and as a muster-out rendezvous point.

A handful of prisoners from the Kanawha Valley in Virginia were kept in confinement in Camp Chase prior to 1862, but the first large influx of Confederate soldiers came following the Union victories at Fort Donelson on the Tennessee River in February 1862. By March 1 over 700 prisoners were housed there and it became necessary to expand the prison facilities. New barracks were constructed and General George B. Wright, Quartermaster General for Ohio who had overall responsibility for prisons in the state, explained that it cost over $5000 to improve the prisoners' quarters alone. This new section of the prison is shown in the southeastern corner of the map and marked 'New Prison.'

The influx of prisoners continued throughout 1862 and 1863 until new additions were again made in 1864. Dr. John H. King, who wrote *Three Hundred Days in a Yankee Prison,* an account of his experiences at Camp Chase while a prisoner of war, explained that the "new houses or shanties were hastily built of rough planks, about eighteen feet wide and eighty feet long with bunks arranged on each side, and hallway through the center about five feet wide." The walls of the prison, King said, were about twenty feet high and well lighted at night. "On a platform near the top," he continued, "sentinels were pacing to and fro and never out of sight

Plate XXXI: Pollard, "Camp Chase," (1861), National Archives, Washington, D.C. Record Group 92, P & R file #106. Camp Chase in Columbus was the major Civil War military post in Central Ohio.

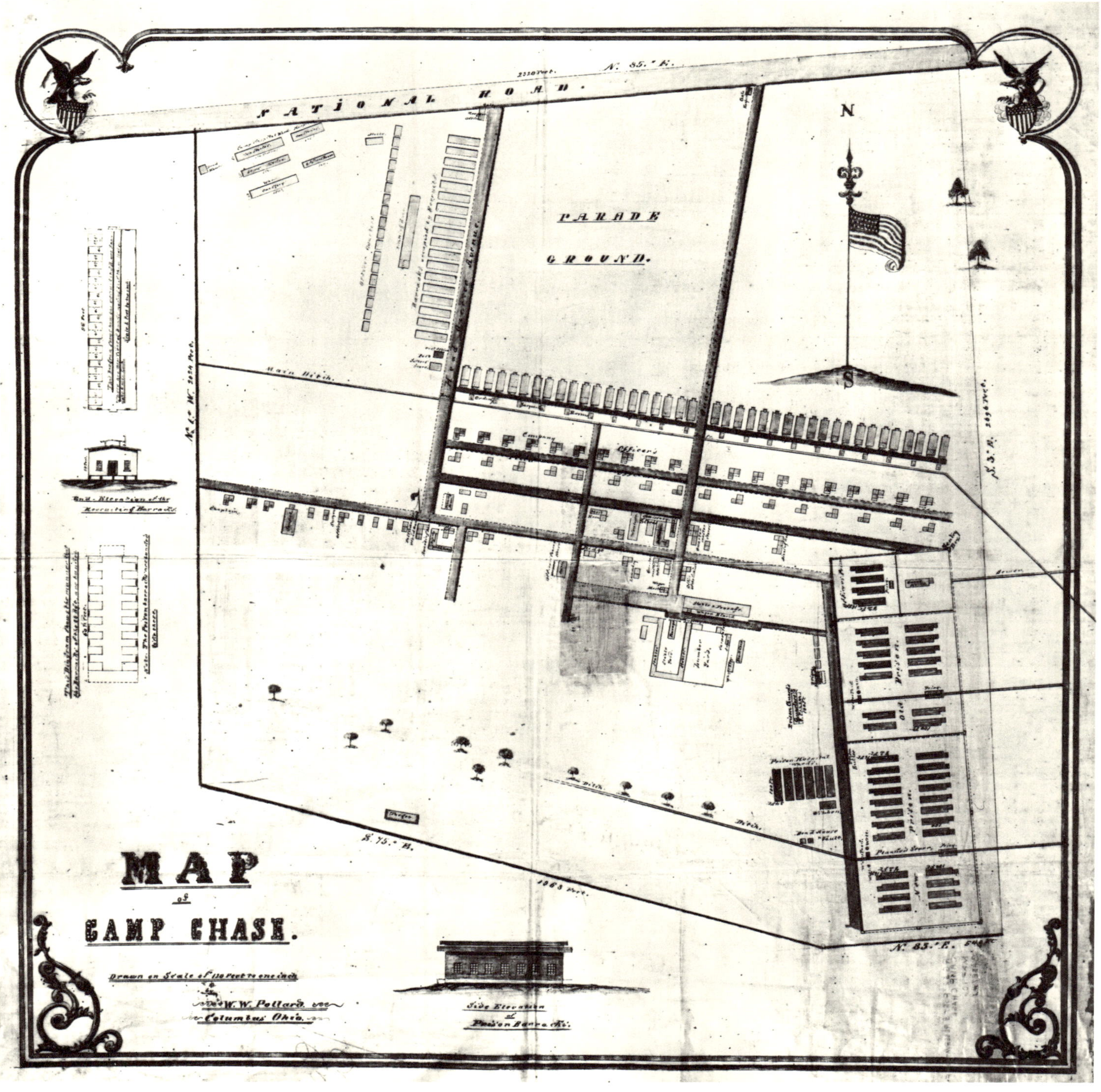

of each other or of the prisoners."[127] Life at Camp Chase for the Confederate soldier was not pleasant, but no prisoner-of-war camp was ever designed with the comfort of the prisoner in mind.

By August 1866 Camp Dennison, the largest federal military camp in Ohio, near Cincinnati, was closed and Camp Chase became the mustering-out center for most of the Ohio troops returning from the war. On June 10, 1865, there were approximately 3,200 Confederate soldiers in the post yet unparoled, but by June 28, all prisoners were on their way home. On July 14 of that year, all the prison property was sold at public auction. Camp Chase itself was continued as a military post until mid-April 1866 when all government property had been distributed to various state institutions; all that remained were the buildings. On May 1 a reporter visited the former busy post and recorded his observations in the *Ohio State Journal:*

> It is no longer a military centre, no more a living thing; the city is deserted, the giant form a skeleton. Hundreds and thousands of armed men paraded as the guardians of the living thing; a single man unarmed keeps watch and ward over the remains of the thing dead waiting for burial. Two years ago you entered the precincts of Camp Chase armed with passes signed and countersigned; were directed by shortspoken orderlies; warned by straight up-and-down sentinels; received with punctilious standoffishness by officials; and came away duly impressed with the military power of the country. Now, you drive up to the gate as you would to that of a cemetery; the guardian presents himself in his shirtsleeves; you tell him your desires; he kicks away a huge stone; opens the gate; cautions you a little, and you enter unchallenged and unheralded to the mighty presence of the great solitude of loneliness. The rows of barracks remain unchanged; the flowers planted by some careful wife of some careless officer are ready to record that 'the hand of woman has been here;' the flagstaff stands without pulley, rope or flag; the chapel with its halfchange in the latter day to the theatre remains a monument of the one, a tell tale of the within. In one, the scaffold on which Hartrup and Oliver were executed stands firm—the grim guardian of the ghostly solitude—and with beam in place and trap half sprung seems waiting for another victim. Everywhere are the marks of the skeleton. The pumpstocks have all been withdrawn from the wells; the windows from the buildings; grass growing on the parade ground. Old shoes tumbled into promiscuous groupings tell which buildings have been last occupied, and the martin boxes give some signs of life. A little fruit tree in the midst of all this loneliness blossoms and puts forth leaves with all the proud defiance of nature, and with a scornful fling with every wave of wind for the works of man perishing on every side.[128]

There were a number of army posts in Columbus during the Civil War besides Camp Chase. Camp Thomas was 5 miles north of the city, while Camp Lew Wallace was north of that; and the United States Arsenal was located where Fort Hayes is now. There were two posts named after Governor David Tod: Camp Tod located near the old state stone quarry, 3 miles west of the town and Tod Barracks, situated on the east side of North High Street just north of the railroad station. All these posts, with the exception of Tod Barracks, were used for a little more than a year during the war.

The *Columbus Gazette* on October 23, 1863, announced that Tod Barracks was being constructed "to be used for the accommodation of the sick, disabled and other soldiers," and that it would have its own brass band for martial music and "the institution is to be complete in every respect, second to none in the Union." The "Map of Tod Barracks" (Plate XXXII) illustrated that this army camp was, indeed, a handsome post. The struc-

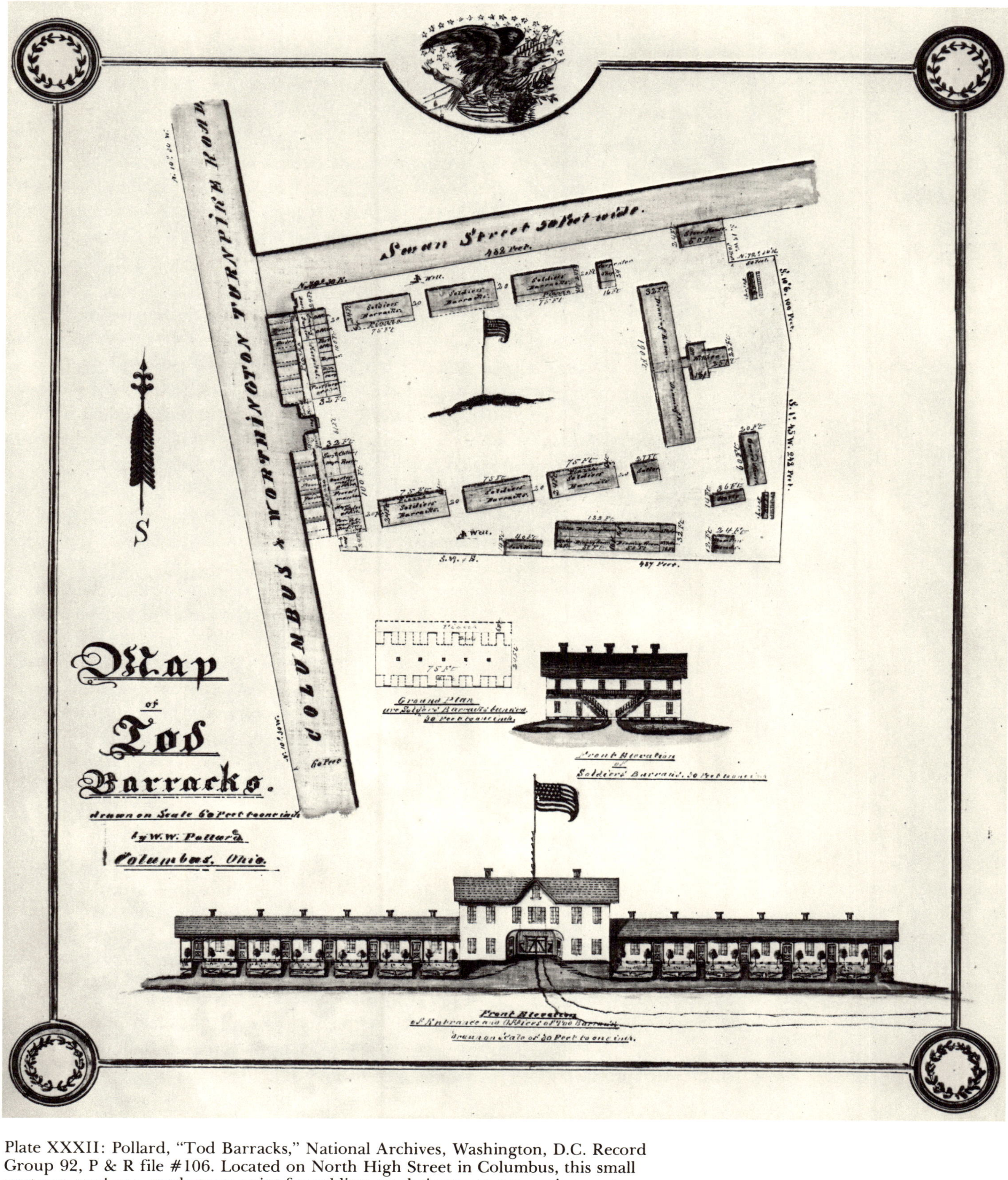

Plate XXXII: Pollard, "Tod Barracks," National Archives, Washington, D.C. Record Group 92, P & R file #106. Located on North High Street in Columbus, this small post was used as a rendezvous point for soldiers on their way to new assignments during the Civil War.

ture that comprised the main entrance to the barracks on North High Street housed all the essential military offices off a long front porch, while the second floor above the main gate was reserved for the officers. The *Gazette* boasted that the enlisted men's barracks would have bunks for the men 2 tiers high and arranged so that they could be taken apart for cleaning and explained that the double piazzas would "be perfect and thoroughly ventilated." A board fence 12 feet high surrounded the entire camp and the parade ground was 425 by 262 feet. In November the *Gazette* said that the Columbus and Indianapolis Railroad had found enough iron rails to extend a trunk line to the new barracks, which would help supply the men with all essentials.[129] The barracks were completed on December 1, 1863.[130]

During the war, Tod Barracks served as a rendezvous point for soldiers; and in February and March 1864, hundreds of men gathered here daily before boarding trains to join their units in the South. It also was a detention point for deserters who had been arrested and were awaiting trials. For a period after the war it served as a mustering-out station for many volunteer units; and, in fact, the last Ohio military unit, the 11th Ohio Cavalry, which had been stationed on the Western Plains, was discharged from the Barracks on July 20 and 21, 1866. The post was officially closed on September 11, 1865, only to be reopened on January 23, 1866. It was closed for the last time on April 1, 1866, and the barracks buildings were sold on April 11, 1867, for $1,400.

Aside from the military posts in Columbus during the war, the federal government built Tripler Military Hospital approximately 3 miles west of the capital city, between the Scioto River and the Columbus, Dayton and Cincinnati Railroad. The "Map of Tripler Military Hospital" (Plate XXXIII)[131] depicted the layout of the grounds and illustrated the elevations of the numerous structures. It was built to provide medical services to those stationed at Camp Chase; and, although there was a hospital there for the prisoners, no doubt the more seriously wounded and ill were attended at Tripler. The hospital was named for Charles S. Tripler, Medical Director at Cincinnati, and it was directed by Dr. J. C. Denise of Dayton throughout its existence. Tripler was a physician of some reputation, although he and Governor William Brough had some disagreement over the administration of the military hospitals in Columbus.[132] In January 1864 the government donated the Tripler Hospital to the state of Ohio for the purpose of establishing a Soldiers' Home. The home remained on that location until it was removed to Dayton in 1867.[133]

Probably the most famous prisoner-of-war camp in Ohio during the Civil War was the one constructed on Johnson's Island in Sandusky Bay. Selected as a prison site in October 1861 by Lieutenant Colonel William Hoffman, half of the island, approximately 300 acres, was leased by the federal government from Leonard B. Johnson for $500 a year. Twenty-eight thousand dollars worth of bonds were sold to construct the prison and Ohioans were to be used as guards. The first prisoners arrived on April 13, 1862, and the last left during September 1865. A total of 6,414 prisoners were kept in the prison; although the living conditions there were never equalled by those in Southern prisoner-of-war camps, 206 Confederates are buried in a beautiful wooded cemetery plot close to the lake on the northeast side of the island.

At times, some prisoners would help while away their time by sketching the prison. The illustration "Prison on Johnson Island. Sandusky Bay

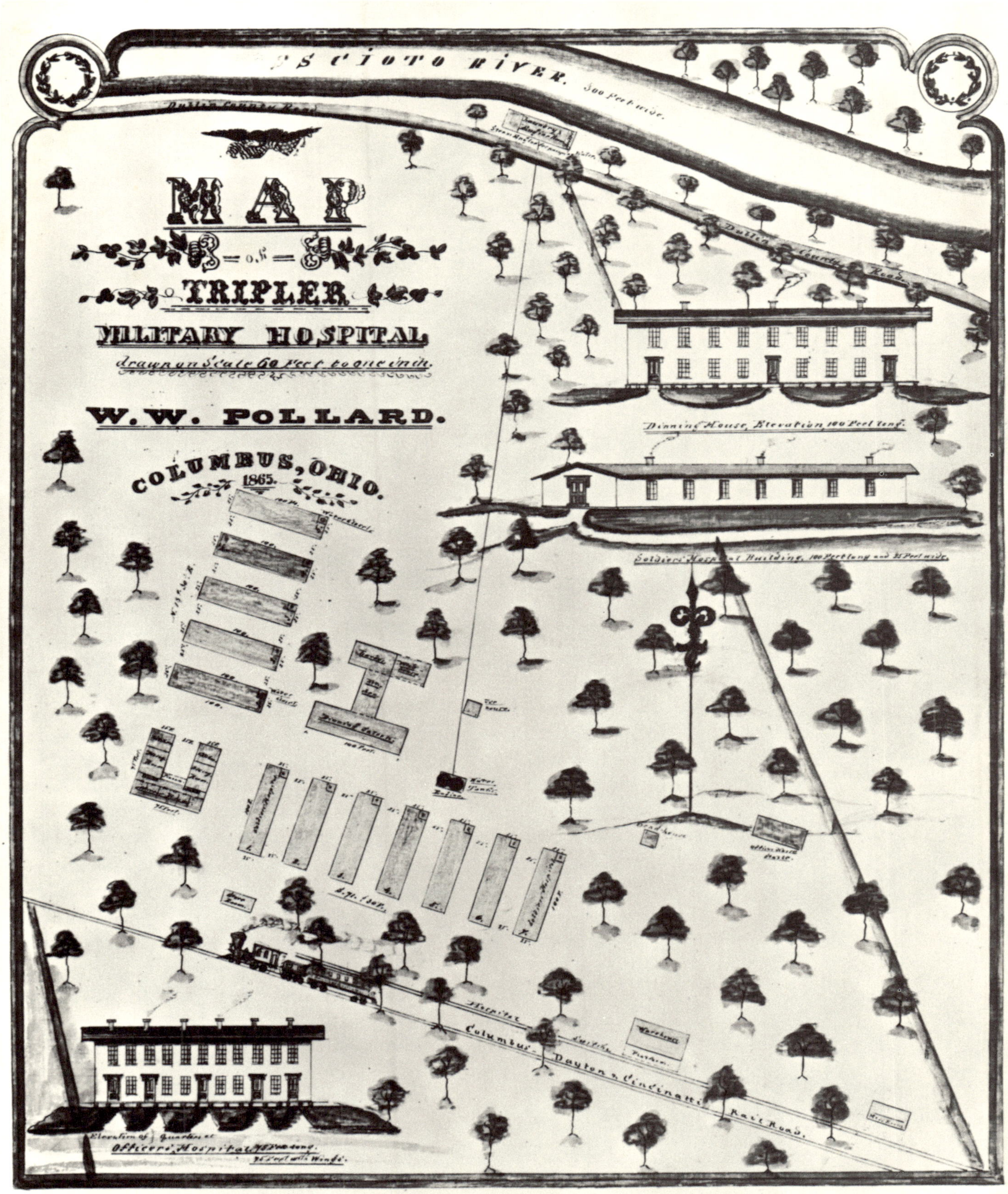

Plate XXXIII: Pollard, "Tripler Hospital," (1865), National Archives, Washington, D.C. Record Group 92, P & R file #106. Established to provide medical facilities for Camp Chase, it later became a Soldiers' Home.

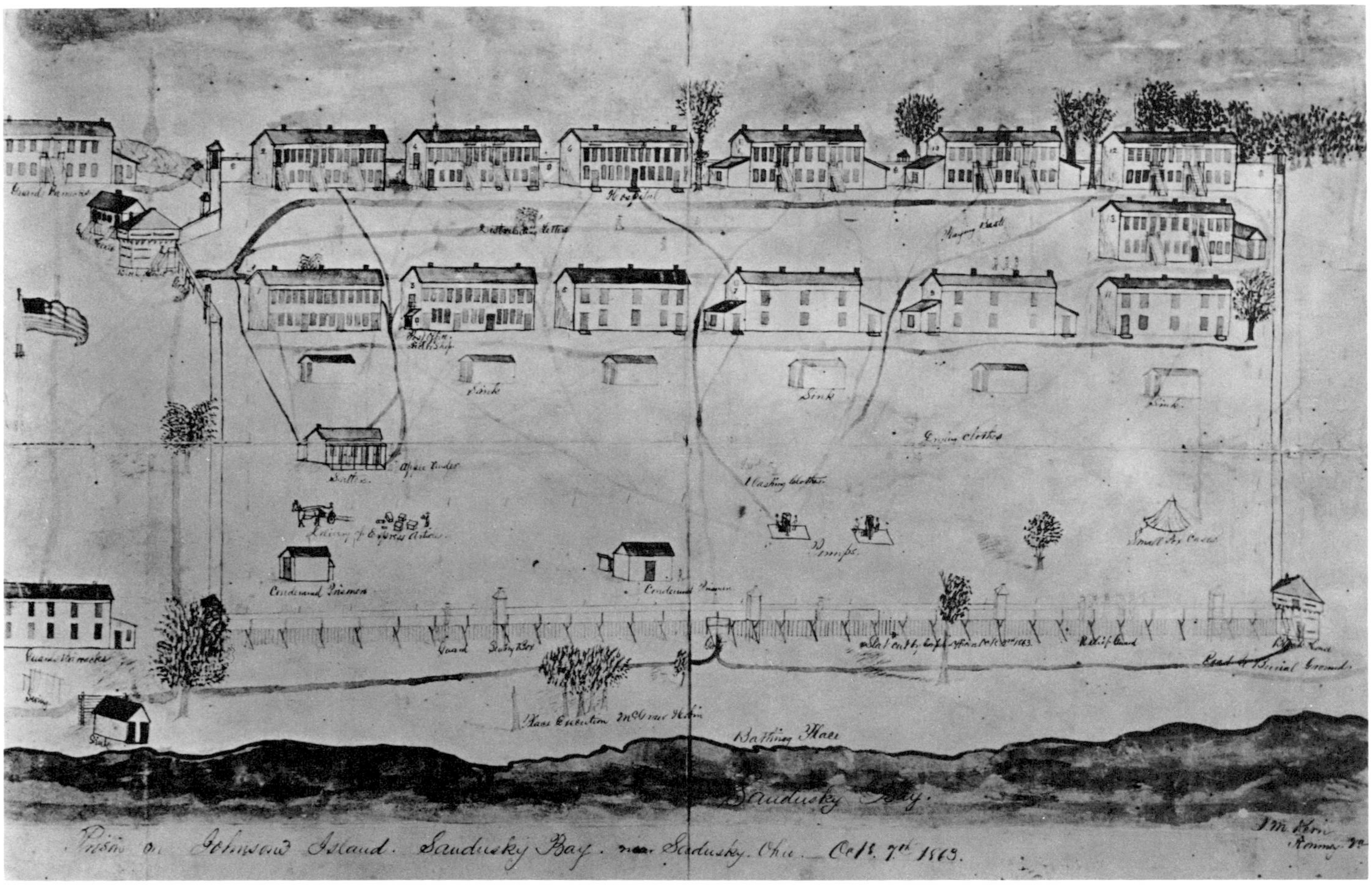

Plate XXXIV: Kern, "Johnson Island," (1965), Southern Historical Collection, University of North Carolina, Chapel Hill, North Carolina. This scene, drawn by one of the prisoners, depicts life in this Confederate prison.

near Sandusky Ohio, Oct. 7th 1863" (Plate XXXIV)[134] penned by Joseph M. Kern who was captured on May 5, 1863 at Wardensville, West Virginia, is such an example. Kern, a native of Romney, Virginia (West Virginia), was serving with Company D, 11th Virginia Cavalry at the time of his capture. He first was housed at Camp Chase in Columbus but later transferred to northern Ohio in June 1863. His quaint sketch not only details the physical plan of the prison, but includes activities of prison life.

The actual prison was contained in sixteen acres of ground. The buildings were constructed from the wood, mostly oak, pine and hickory, found on the island. A 12-foot high wall enclosed the compound and along the inside of the wall, several feet from the top, a wooden walk was built to allow sentries to patrol the perimeter of the prison. At first, 12 two-story barracks were constructed in 2 rows; later, a thirteenth was added at the north end. Each building was 120 feet by 30 feet and constructed 18 inches from the ground. Four of these large barracks were divided into smaller rooms while 9 of them had 2 large rooms on each floor. Between 200 and 300 men were confined in each barrack and they slept in bunks stacked 3 high. There was little ventilation for the men, so each prisoner cut a small hole through the wall near his own bunk. Cooking and eating was done at the end of each barrack building. Kern explained in his diary that "The occupants of each block—(the buildings are so termed) are divided into 2 messes. The kitchens are furnished with a large cooking-stove, and boiler, a coffee boiler, & kettle—The dining room with tables and benches, Plates, Cups, knives & forks. The cooking is done by volunteers, or details as the messes may choose (.)"[135]

Water for the prison was supplied from Sandusky Bay by "pipes & two pumps," Kern explained. However, he continued, "When the wind & tide are out, the pumps are useless and water has to be carried from the outside." Sick prisoners were cared for in a hospital within the compound. "There are also a number of Surgeons in the prison, who attend the sick in Camp and Hospital. The Hospital rations are better than furnished the other prisoners," the Virginian recorded. "The Federal Surgeon visits the hospital every morning. The Hospital is in charge of the Rebel Surgeons. Some few cases of small pox were brought in from Camp Morton, Ind. A tent has been put up in the lower corner of the enclosure."

Kern's sketch illustrated part of the prisoner's everyday activities. The prisoners received mail daily and express boxes periodically; packages from loved ones were examined by federal officers and it was not uncommon for items bound for prisoners to be appropriated by northern guards. Washing and caring for clothes was a continuous chore and for exercise, Kern portrayed prisoners playing "Bash," a form of baseball. Along the wall fronting Sandusky Bay, he indicated an escape attempt on the part of several Confederate officers and, no doubt with a sense of disappointment, wrote, that they "were caught, court martialed and turned back into the prison."

The Confederate prisoners, thrown together by chance, busied themselves to escape boredom and created their own sense of community. Kern recorded that a debate team was organized and discussed issues of moment importance such as "Is our imprisonment on Johnson's Island detrimental to us as individuals?" A musical group, "The Island Minstrels," was formed and gave weekly concerts while the "Rebel Thespians" offered original plays. Kern in his diary also recorded daily routine for prisoners on Johnson's Island which described what the men saw inside the camp. In July 1863, he wrote:

> The daily routine is about the same as at Camp Chase—Roll Call at 8 O'Clock by Yankee sergeants detailed for the purpose—after roll call—policing—About 9 O'Clock the mail arrives—then "Old Pop's" (a Missourian—Mr. Stevenson) "Tailor Shop & Post Office"—is besieged by eager faces—He mounts a stairway & calls the messes letters as they are called—The sutler also stands a siege about the same hour, the attraction being the morning papers—"The Sandusky Register," & Cincinnati Enquirer" At 2 O'Clock the New York Papers are brought in. Passing up the street we see different signs displayed—Tailor Shop—Shoe Repairing—Beer & Cakes—Barber Shop—Pies—Green & Dried Apple Pies—One "Rebel" has opened a "Restaurant" with regular "Bill of Fare"—While passing along you can but note the apple vendors who stationed at regular intervals offer their apples to you—You will also meet the boys with "Ice Cream & Lemonade"—At nearly every "Kitchen" biscuits could be had for 1¢ a piece—"Apple Dumplings & sauce" at 5¢ a piece.

For nearly two decades Ohio was on the country's exposed northwestern frontier, and until 1815 the military activity against the Indians and the British that took place on her soil was vital to the national security and to the country's survival. The maps of the several military expeditions through the territory north and west of the Ohio River, the plans of the battles in western Ohio and the diagrams of the forts built by both the federal government and settlers isolated in the wilderness for defense, help to illustrate the hardships and harsh conditions faced by those who conquered the West. The profiles of the Civil War posts, Tod Barracks, Camp Chase and Tripler Hospital are excellent examples of federal military architecture, and they help to complete the picture of military life during that tragic period in American history. Of course, the plans presented here do not include all of the military installations built in Ohio, because between 1784 and 1945 approximately 159 forts, blockhouses, depots, arsenals and prisoner of war camps were located within the state. This collection, however, and the accompanying text, help to illustrate the state's role in the country's military history, and to depict an exciting segment of the state's history.

1. Pierre Jean [Joseph Pierre?] de Bonnécamps, *Relation du voyage de la Belle rivière fait in 1749, sous les ordres de M. de Celoron. Pierre Jean de Bonnécamps; a Quebec, October 17, 1750,* LXIX of *The Jesuit Relations and Allied Documents. Travels and Explorations of the Jesuit Missionaries in New France,* ed. Reuben Gold Thwaites, 73 vols. (Cleveland, 1900), pp. 150–99. Thwaites stated: "This map (Bonnécamps's "Carte D'un Voyage Fait dans La Belle Rivière En La Nouvelle France") was also preserved, with his memoir, in the above-named archives (Department of the Marine), but cannot now be found; its disappearance seems to have taken place at some time during 1892–94," p. 294. This map is now located in the Bibliotheque Service Hydrographie de la Marine, Bibliotheque Nationale, Paris, France. Bonnecamps's Journal is also reprinted in C. B. Galbreath, *Expedition of Celoron to the Ohio Country in 1749* (Columbus: F. J. Heer Printing Co., 1921), pp. 78–96.

2. Galbreath, *Celoron,* pp. 37–40.

3. *Ibid.,* p. 90. Father Bonnécamps's journal is also printed in Galbreath's edition of *Expedition of Celoron to the Ohio Country in 1749*.

4. *Ibid.,* p. 91.

5. *Ibid.,* p. 92.

6. *Ibid.,* pp. 92–93.

7. A reprint of Hutchins's journal kept on the tour is in Charles A. Hanna, *The Wilderness Trail* (New York, 1911), II, 362–67; see also William L. Jenks, "The Hutchins' Map of Michigan," *Michigan History Magazine,* X (1926), 358–73.

8. Several editions of this work appeared shortly after 1764. A reprint of this account and the map appeared as the first volume of the Ohio Valley Historical Series, published in 1868 by Robert Clarke & Company of Cincinnati. Francis Parkman, the author of the *Conspiracy of Pontiac* and many other volumes, wrote the preface to this edition.

9. This particular map is found in *An Historical Account of the Expedition Against the Ohio Indians, In Year MDCCLXIV Under the Command of Henry Bouquet, Esq.* (Philadelphia: W. Bradford, 1765).

10. See Plate V, Chapter I. The Indian villages and geographical features illustrated on his map of the Bouquet expedition have already been identified.

11. *Historical Account of Bouquet's Expedition Against the Ohio Indians in 1764* (Cincinnati: 1868), p. 47. For an excellent and detailed account of Bouquet's march, see Edward G. Williams, ed., "The Orderly Book of Colonel Henry Bouquet's Expedition Against The Ohio Indians 1764," *The Western Pennsylvania Historical Magazine,* XLII (March 1959), 9–34; (June 1959), 179–200; (September 1959), 283–302.

12. *Historical Account of Bouquet's Expedition,* p. 48.

13. *Ibid.*

14. *Ibid.,* p. 49.

15. *Ibid.*

16. *Ibid.,* p. 52.

17. *Ibid.,* p. 60.

18. Beverley W. Bond, Jr. *The Foundation of Ohio,* Vol. I of Carl Wittke, ed., *A History of the State of Ohio,* 5 vols. (Columbus: 1968), pp. 186–89; Theodore

Roosevelt, *The Winning of the West,* Vol. 5 of *The Works of Theodore Roosevelt* (New York and London: G. P. Putnam's Sons, 1889–96), 222–380.

19. *The American Pioneer,* I (January 1842), 3.

20. For a discussion of the Munson and Doolittle publishing firm of Cincinnati, see Chapter V, p. 000.

21. Henry Howe, *Historical Collection of Ohio,* (Cincinnati: Bradley & Anthony, 1848), pp. 402–03.

22. Louise Phelps Kellogg, " 'Non-hel-e-ma,' Shawnee Princess" in William Albert Galloway, *Old Chillicothe* (Xenia, Ohio: The Buckeye Press, 1934), pp. 283–95.

23. Letter from W. Marshall Anderson to Lyman Draper, May 21, 1876, Draper Mss. 15J92, State Historical Society of Wisconsin. This letter contained a map of Pickaway Plains drawn by Anderson who apparently was familiar with the region when it was inhabited by the Shawnee.

24. Felix Renick, "Mr. Renick's Letter," *The American Pioneer,* II (January 1843), 38.

25. Consul Willshire Butterfield, *An Historical Account of the Expedition against Sandusky under Col. William Crawford in 1782* (Cincinnati, 1873), pp. 126–27.

26. Charles H. Mitchener, *Ohio Annals. Historic Events in the Tuscarawas and Muskingum Valleys and in other portions of the State of Ohio* (Dayton, 1876), pp. 128–29. For the story of Fort Laurens during its short occupancy, see Thomas I. Pieper, "Ohio and the Revolutionary War: Fort Laurens, 1778–1779," M. A. Thesis, Kent State University, 1968.

27. This account of Fort Laurens was from an interview that Lyman Draper had with John Cuppy between August 21 and August 23, 1860, at the latter's home in Wayne Township in Montgomery County, Ohio. Cuppy, at that time was ninety-nine years old. His description of the fort does not correspond with Whittlesey's surveyed sketch. "Recollections of John Cuppy" in Louise Phelps Kellogg, ed., *Frontier Advance on the Upper Ohio: 1778–1779* (Madison: Wisconsin Historical Society, 1916), pp, 158–61.

28. Kellogg, *Frontier Advance on the Upper Ohio,* pp. 183–84, fn. 3.

29. Archer Butler Hulbert, ed., *Ohio in the Time of the Confederation,* Vol. III of the *Ohio Company Series* (Marietta, Ohio, 1918). See Document No. XXVII, "Proclamation of Congress on Western Lands, June 15, 1785" and fn. 65.

30. William H. Smith, ed., *St. Clair Papers* (Cincinnati, 1896), II, 6–7. Consul Willshire Butterfield, *Journal of Capt. Jonathan Heart* (Albany, N.Y., 1885), p. 92.

31. Letter from Josiah Harmar to Henry Knox, October 22, 1785, Butterfield, *Heart,* pp. 92–94.

32. This plan of Fort Harmar is in the map collection in the Map Division, William L. Clements Library, Ann Arbor, Michigan.

33. This sketch was printed in *The American Pioneer* in February 1842, to accompany Samuel Prescott Hildreth's article "Fort Harmar," pp. 25–28.

34. *American State Papers, Indian Affairs,* I, 5–7.

35. The physical description of Fort Harmar was written by Samuel Prescott Hildreth in the article "Fort Harmar," which appeared in the second issue of *The American Pioneer* in February 1842. Later it was reprinted in his *Pioneer History: Being An Account of the First Examinations of the Ohio Valley, and the Early Settlement of the Northwest Territory* (Cincinnati, 1848), pp. 317–24. See also "A Description of Fort Harmar," *Magazine of Western History,* I (November 1884), 26–31.

36. Letter from Captain John F. Hamtramck to Colonel Josiah Harmar, December 15, 1786, *Josiah Harmar Papers,* pp. 265–66, Vols. 1–2, Draper Mss. (Microfilm), Ohio Historical Society.

37. Erkuries Beatty, "Diary of Major Erkuries Beatty," *The Magazine of American History,* I (1877), 175–79; 235–43; 309–15; 432–38. Beatty also sketched Fort Steuben (see note 16, p. 000) and a copy of that plan was published along with his diary.

38. Letter from Josiah Harmar to Henry Knox, October 22, 1785, in Butterfield, *Heart,* pp. 92–94.

39. Ebenezer Denny, *Military Journal of Major Ebenezer Denny* (Philadelphia, 1859), p. 59.

40. See Chapter VI, Plate I for the plan of the Marietta Settlement.

41. This wonderful pen and ink sketch is in the Dawes Memorial Library, Marietta College, Marietta, Ohio.

42. "A Description of Campus Martius", Samuel P. Hildreth, *The American Pioneer,* I (March 1842), 83–89. See also George Jordan Blazier, ed., *Joseph Barker: Recollections of the First Settlement of Ohio* (Marietta, Ohio, 1958), pp. 66–69, 79–80; Denny, *Military Journal,* entry May 28, 1788, pp. 119–20; Hildreth, *Pioneer History,* p. 287; Alfred Mathew, "A Description of Campus Martius," *Magazine of Western History,* I (December 1884), 115–18.

43. This drawing is in the Samuel Prescott Hildreth Collection, Dawes Memorial Library, Marietta College, Marietta, Ohio.

44. Hildreth, *Pioneer History,* pp. 325–29; Blazier, *Joseph Barker,* pp. 71–74.

45. This sketch is in the Samuel Prescott Hildreth Collection, Dawes Memorial Library, Marietta College, Marietta, Ohio.

46. Hildreth, *Pioneer History,* pp. 440–41.

47. This sketch is in the Samuel Prescott Hildreth Collection, Dawes Memorial Library, Marietta College, Marietta, Ohio.

48. Hildreth, *Pioneer History,* pp. 362–64.

49. Letter from Josiah Harmar to Henry Knox, September 12, 1789, in Denny, *Military Journal,* p. 239.

50. Richard C. Knopf, Raymond S. Baby, and Dwight L. Smith, "The Re-Discovery of Fort Washington," *Bulletin of the Historical and Philosophical Society of Ohio,* XI (January 1953), 3–12.

51. *Ibid.,* 40.

52. Letter from Josiah Harmar to Henry Knox, January 14, 1790, Denny, *Military Journal,* 244.

53. For a detailed description of Fort Washington see Robert Ralston Jones, *Fort Washington at Cincinnati, Ohio* (Society of Colonial Wars in the State of Ohio, 1902).

54. James McBride, *Pioneer Biography: Sketches of the Lives of some of the Early Settlers of Butler County, Ohio,* Vol. I (Cincinnati, 1869), 85–86.

55. Henry A. Ford and Kate B. Ford, *History of Hamilton County, Ohio* (Cleveland, 1881), p. 256.

56. This diagram is in the vast collection of the Map Division, William L. Clements Library, Ann Arbor, Michigan.

57. For a complete story of the siege, see McBride's *Pioneer Biography,* I, 17–20; 87–89. Ford and Ford, *History of Hamilton County, Ohio,* p. 62 and pp. 256–57.

58. This plan is among the Josiah Harmar Papers, William L. Clements Library, Ann Arbor, Michigan.

59. Letter from Arthur St. Clair to Major John Francis Hamtramck, August 30, 1791, Smith, *St. Clair Papers,* II, 239.

60. Letter from Arthur St. Clair to Major John Francis Hamtramck, August 30, 1791, *Ibid.,* pp. 239–40.

61. Letter from Arthur St. Clair to Henry Knox, September 18, 1791, *Ibid.,* p. 241.

62. Letters from John Armstrong to James Wilkinson, April 27, 1792; May 15, 1792; July 17, 1792; November 15, 1792, quoted in *A Historical and Biographical Cyclopaedia of Butler County, Ohio with Illustrations and Sketches of its Representative Men and Pioneers* (Cincinnati, 1882), pp. 7–17.

63. Letter from John Armstrong to James Wilkinson, February 5, 1792, *Ibid.,* p. 9.

64. Letter from James Wilkinson to John Armstrong, April 20, 1792, *Ibid.,* p. 10.

65. Letter from Arthur St. Clair to Henry Knox, November 1, 1791, p. 249;

October 23, 1791, entry in Major Ebenezer Denny's diary, in Smith, *St. Clair Papers,* II, 255.

66. Letter from James Wilkinson to John Armstrong, May 11, 1792, quoted in *A Historical . . . of Butler County,* p. 12.

67. Letter from John Armstrong to James Wilkinson, June 1, 1792, *Ibid.,* p. 13.

68. See *Ibid.,* pp. 4–7, 18; William C. Miller, "History of Fort Hamilton," *Ohio Archaeological and Historical Publications,* XIII (1904), 97–111.

69. The plan of Fort Jefferson is in the James McHenry Papers, Indiana Historical Society, Indianapolis, Indiana.

70. See Frazer E. Wilson, *Journal of Capt. Daniel Bradley* (Greenville, Ohio, 1935), pp. 39–42.

71. Entry for October 4, 1791, from the "Diary of Major Ebenezer Denny," in Smith, *St. Clair Papers,* II, 253–54.

72. Letter from Arthur St. Clair to Henry Knox, November 9, 1791, *Ibid.,* p. 262.

73. Frazer Ells Wilson, "St. Clair's Defeat," XI, *The Ohio Archaeological and Historical Society Publications,* 1903, 30–43.

74. Smith, *St. Clair Papers,* II, 262–67; see also the interesting account of the St. Clair tragedy in *A Narrative of the Manner in which the Campaign Against the Indians, in the Year One Thousand Seven Hundred and Ninety-One, was Conducted, Under the Command of Major General St. Clair, Together with his Observations on the Statements of the Secretary of War and the Quarter Master General, Relative Thereto and the Reports of the Committees Appointed to Inquire into the Causes of the failure Thereof: Taken from the Files of the House of Representatives in Congress* (Philadelphia: Printed by Jane Aitken, no. 71 North Third Street, 1812).

75. Denny, *Military Journal,* pp. 164–69, which is accompanied by a foldout drawing of the battlefield prepared by Denny.

76. Benson J. Lossing, *The Pictorial Field-Book of the War of 1812,* New York, 1868, p. 47, fn. 1. The drawing of the battle that Lossing reproduced was a facsimile of Winthrop Sargent's.

77. This map is contained in the Clinton Papers, Map Division, William L. Clements Library, Ann Arbor, Michigan.

78. Denny, *Military Journal,* pp. 164–69.

79. The sketch of Fort St. Clair is in the James McHenry Papers, Indiana Historical Society Library, Indianapolis, Indiana.

80. Ralph B. Ehler, "The Story of Fort St. Clair," *The Ohio Archaeological and Historical Society Publications,* XXXII (1923), 515–19.

81. *A Historical . . . of Butler County,* pp. 9–10.

82. James R. Albach, *Annals of the West: Embracing a Concise Account of Principal Events which have occurred in the Western States and Territories from the Discovery of the Mississippi Valley to the Year Eighteen Hundred and Fifty-Six* (Pittsburgh, 1857), p. 611.

83. Letter from Henry Knox to Anthony Wayne, April 12, 1792, Richard C. Knopf, ed., *Anthony Wayne: A Name In Arms, The Wayne-Knox-Pickering-McHenry Correspondence* (Pittsburgh: Univ. of Pittsburgh Press, 1960), p. 15.

84. Letter from Anthony Wayne to Henry Knox, October 23, 1793, *ibid.,* p. 279.

85. Letter from Anthony Wayne to Henry Knox, October 5, 1793, *ibid.,* pp. 276–277.

86. Letter from Anthony Wayne to Henry Knox, November 15, 1793, *ibid.,* p. 283.

87. The location of the original manuscript has not been determined. In 1940, at least, it was in the possession of Miss Maggie Artz of Hagerstown, Maryland, and identified as a sketch of Fort Greenville by J. W. Foster, who at the time was employed by the Enoch Pratt Free Library in Baltimore. A determined effort by the author to ascertain the whereabouts of the original was futile.

88. Letter from Anthony Wayne to Henry Knox, October 23, 1793, Richard

C. Knopf, (ed.), *Campaign Into The Wilderness,* Vol. II of *The Wayne-Knox-Pickering-McHenry Correspondence,* (Anthony Wayne Parkway Board: Columbus, 1955), p. 139.

89. Richard C. Knopf, *Orderly Books for the United States Legion,* (Anthony Wayne Parkway Board: Columbus, 1954), p. 103.

90. Letter from Anthony Wayne to Henry Knox, November 15, 1793, Knopf, *Campaign,* II, 142.

91. Wayne Papers, Vol. 31, Ohio Historical Society.

92. Xerox copy of the "Journal of Captain Daniel Bradley" entry dated October 22, 1793, Ohio Historical Society. Owned by Mrs. Frances B. Moore, Columbus.

93. Knopf, *Orderly Books,* p. 102

94. *Ibid.*

95. Wayne's *Orderly Book, No. 4,* entry February 12, 1794, Filson Club, Louisville Kentucky.

96. Knopf, *Orderly Book,* p. 104.

97. Letter from Anthony Wayne to Henry Knox, August 24, 1792, *ibid.,* pp. 73–74.

98. Letter from Anthony Wayne to Henry Knox, August 24, 1792, *ibid.,* pp. 74–75.

99. John Boyer, "General Wayne's Daily Encampment," *The American Pioneer* II (July 1843), opposite p. 291.

100. "Mr. Will's Letter," *ibid.,* 291–92.

101. R. C. McGrane, ed., "William Clark's Journal of General Wayne's Campaign," *Mississippi Valley Historical Review,* I (December 1914), 421.

102. *Ibid.,* Entry for August 2, 1794.

103. *Ibid.,* Entry for August 3, 1794, 422.

104. Richard C. Knopf, *The Fort Adams Incident,* (Columbus, 1957).

105. Letter from Anthony Wayne to Henry Knox, August 14, 1794, *American State Papers,* IV, 490.

106. This drawing accompanied "Extracts from B. Van Cleve's Memoranda, *The American Pioneer,* II (September 1843), 293.

107. McGrane, "Clark's Journal," *Mississippi . . . Review,* 422–23. Entry for August 4, 1794.

108. The "Plan of Fort Defiance" was drawn by Benjamin Van Cleve, and published along with a description of the fort in *The American Pioneer,* II (September 1843), 387.

109. Letter from Anthony Wayne to Henry Knox, August 28, 1794, Knopf, *Anthony Wayne,* p. 355.

110. Boyer, "Daily Journal," *The American Pioneer,* p. 351.

111. Letter from Anthony Wayne to Henry Knox, September 20, 1794, Knopf, *Anthony Wayne,* p. 356. No doubt Van Cleve's description of Fort Defiance was accurate; however, Charles E. Slocum, in an article entitled "Some Errors Corrected," *Ohio Archaeological and Historical Publications,* X (1902), 481–83, claimed that he was not totally correct. For instance, Slocum explained that after careful examination of the fort site, it was evident that the ditch Van Cleve said went to the Auglaize River followed the entire wall of the post instead.

112. This plan is included with the Clinton Papers, Map Division, William L. Clements Library, Ann Arbor, Michigan.

113. Letter from Anthony Wayne to Henry Knox, August 28, 1794, Knopf, *Anthony Wayne,* p. 351.

114. Boyer, "Daily Journal," *The American Pioneer,* p. 318.

115. Letter from Henry Knox to Anthony Wayne, June 7, 1794, Knopf, *Anthony Wayne,* p. 337.

116. Howard Peckham, "Fort Miami and the Maumee Communication," *Northwestern Ohio Quarterly Bulletin,* XIV (April 1942), 30–41; Richard C. Knopf, "Fort Miami: The International Background," *Ohio Archaeological and Historical Quarterly,* LXI (April 1952), 146–66; Charles E. Slocum, "Fort Miami and Fort Industry," *Ohio Archaeological and Historical Quarterly,* XII (1904), 120–27. For a

good review of the Ohio Indian Wars, see Bert Anson, *The Miami Indians* (Norman, Oklahoma: University of Oklahoma Press, 1970), pp. 95–138.

117. Robert B. McAfee, *History of the Late War in the Western Country* (Bowling Green, Ohio, 1919), pp. 265–66; George Washington Cullum, *Campaigns of the War of 1812–15, Against Great Britain, Sketched and Criticized* (New York: J. Miller, 1879), p. 372.

118. McAfee, *History of the Late War,* pp. 276–77.

119. This plan is in the Map Division of the Library of Congress, Washington, D.C.

120. Wilfrid Hibbert, "The Recently Discovered Pictorial Map of Fort Meigs and Environs," *Bulletin of the Historical Society of Northwestern Ohio,* VI (October 1934), n. p.

121. McAfee, *History of the Late War,* p. 322. Benson J. Lossing, in *Pictorial Field-Book,* p. 497, n. 1, stated that "Thomas Butler, who had been in Wayne's army," had selected the site for Fort Stephenson. Butler, however, died on September 7, 1805.

122. Included with the articles "Defense of Fort Sandusky" and "Life of Colonel Croghan—For the Port Folio," in *Port folio,* V, 3d Series (March 1815), 209–20. This plan was copied and printed in Lossing, *Pictorial Field-Book,* p. 503, in many editions of Henry Howe's *Historical Sketches of Ohio,* and in *Combination Atlas Map of Sandusky County, Ohio,* (Chicago: Everts, Stewart & Company, 1874.)

123. Lossing, *Pictorial Field-Book,* pp. 497–508.

124. Charles Richard Williams, "George Croghan," *Ohio Archaeological and Historical Publications,* XII (1903), 391–92.

125. *Ohio State Journal,* April 22, 1861.

126. This map is in record Group Number 92, Old Military Records, National Archives, Washington, D.C.

127. John H. King, *Three Hundred Days in a Yankee Prison: Reminiscenses of War Life Captivity Imprisonment at Camp Chase Ohio* (Kennesaw, Georgia, 1959), p. 85; W. H. Knauss, "The Story of Camp Chase," *The Ohio Magazine,* I (September 1906), 233–40.

128. *Ohio State Journal,* May 1, 1866.

129. Columbus *Gazette,* November 13, 1863.

130. Lee, *History of the City of Columbus,* p. 134.

131. This map is in Record Group No. 92, National Archives, Washington, D.C.

132. Whitelaw Reid, *Ohio In the War: Her Statesmen, Her Generals and Soldiers.* Vol. I: *History of the State During the War, and the Lives of Her Generals* (New York, 1868), 194–99.

133. Lee, *The History of the City of Columbus,* p. 163.

134. This sketch is in the Joseph M. Kern Scrapbook, Joseph M. Kern Papers, #2526, Southern Historical Collection, University of North Carolina Library, Chapel Hill, North Carolina.

135. All quotes concerning the prison are from Joseph M. Kern's Diary, Joseph M. Kern Papers, *Ibid.*

CHAPTER FOUR

Surveying the State

It was the surveyors with their lengths of Mr. Gunter's chain, circumferentors and theodolites who first saw much of Ohio. As representatives of speculative land companies, guardians of their own interests, or hired hands of the federal government and individual states, many of these early surveyors plodded through the wilderness to section off lands that were sold unseen in the market place. Their observations of the new land were recorded and, in many cases, used as juicy enticements for those who had the determination to move west. Their vision and wisdom, and in some instances, luck, were used to lay out town plats that became centers of settlement. Men such as Benjamin Tupper, Nathaniel Massie, Thomas Worthington, Lucas Sullivant, Levi Whipple, William W. Morris, Charles Smith, Moses Cleaveland, Absalom Martin, Jonathan Dayton, Augustus Porter, James Kilbourne, Alexander Bourne and many others, surveyors all, some well known but most not, mapped out the state literally inch by inch, and made decisions for locations of settlements and cities that affected Ohio's growth and history for generations.

Most of these surveyors were not cartographers. Their work consisted mainly of rough pen and ink sketches. According to Robert Gibson in his *A Treatise of Practical Surveying,* a popular technical work that in 1808 was in its ninth printing, surveying was the "art which enables us to give a plan, or just representation, of any piece or parcel of land, and to determine the content thereof, in such measure as is agreeable and customary to the country or place where the land is." Surveyors, too, kept field notebooks which were hastily made while in the wilds. These field notebooks described how the surveys were measured, recorded the mathematical processes of the survey, and supplemented the actual plat. Also, in many instances, because the surveyor saw much of the land yet untouched by white society, the field notebooks recorded items such as the types of wild life, Indian trails, roads, and sites which long have been obliterated by modern civilization. The surveys, along with the field notebooks, were usually deposited with some governmental agency. Consequently, many of the original surveys of counties and townships are found in the Land Office in Washington, D.C., the State Auditor's Office in Columbus, or in the eighty-eight county seats in Ohio.

There was much interest in the East in the new lands northwest of the Ohio River, and some of the surveys of the larger tracts of land were enlarged into maps which were printed and had a wide circulation. The first such map was Thomas Hutchins's "Plat of the Seven Ranges of Townships being Part of the Territory of the United States N.W. of the River Ohio which by a late act of Congress are directed to be sold" (Plate I). Hutchins, who had visited the territory as early as 1762, knew more about this region than anyone else at that time. He was appointed Geographer of the United States by Congress and instructed under the Ordinance of 1785 to make the first surveys in Ohio. He, along with eight surveyors, began to survey the land on September 23, 1785; but because of the Indians, who resisted white encroachment upon their lands, their work was halted temporarily. The situation improved and by August of the following year, the first four ranges were surveyed and government supervision of westward expansion had begun.[1]

In the Ordinance of 1785, passed by Congress on May 20, it was determined that a rectangular system of land division, which reflected New England patterns of settlement, would be used to divide the western lands under national control. In order to begin the surveys, the Ordinance pointed out that "the first line, running due north and south . . . shall begin on the river Ohio, at a point that shall be found to be due north from the western termination of a line, which has been run as the southern boundary of the State of Pennsylvania, and the first line, running east and west, shall begin at the same point, and shall extend throughout the whole territory." Once these base lines were established, the land south of the east-west line was divided into townships 6 miles square. Alternate townships were subdivided into sections 1 mile square containing 640 acres each. Each row of townships set off from the east-west base line was known as a Range and, of course, the Ordinance provided for the survey of only the first seven. The sixteenth section of each township was set aside for educational purposes, thus providing for the first federal aid to education. Sections 8, 11, 26 and 29 were originally set aside for Congressional use; but in May 1796 those reserved sections were changed to sections 15, 16, 21 and 22. Actually, few people settled in the Seven Ranges until after 1800, when a more liberal land law was passed by Congress. The entire tract was not completely surveyed until 1802.

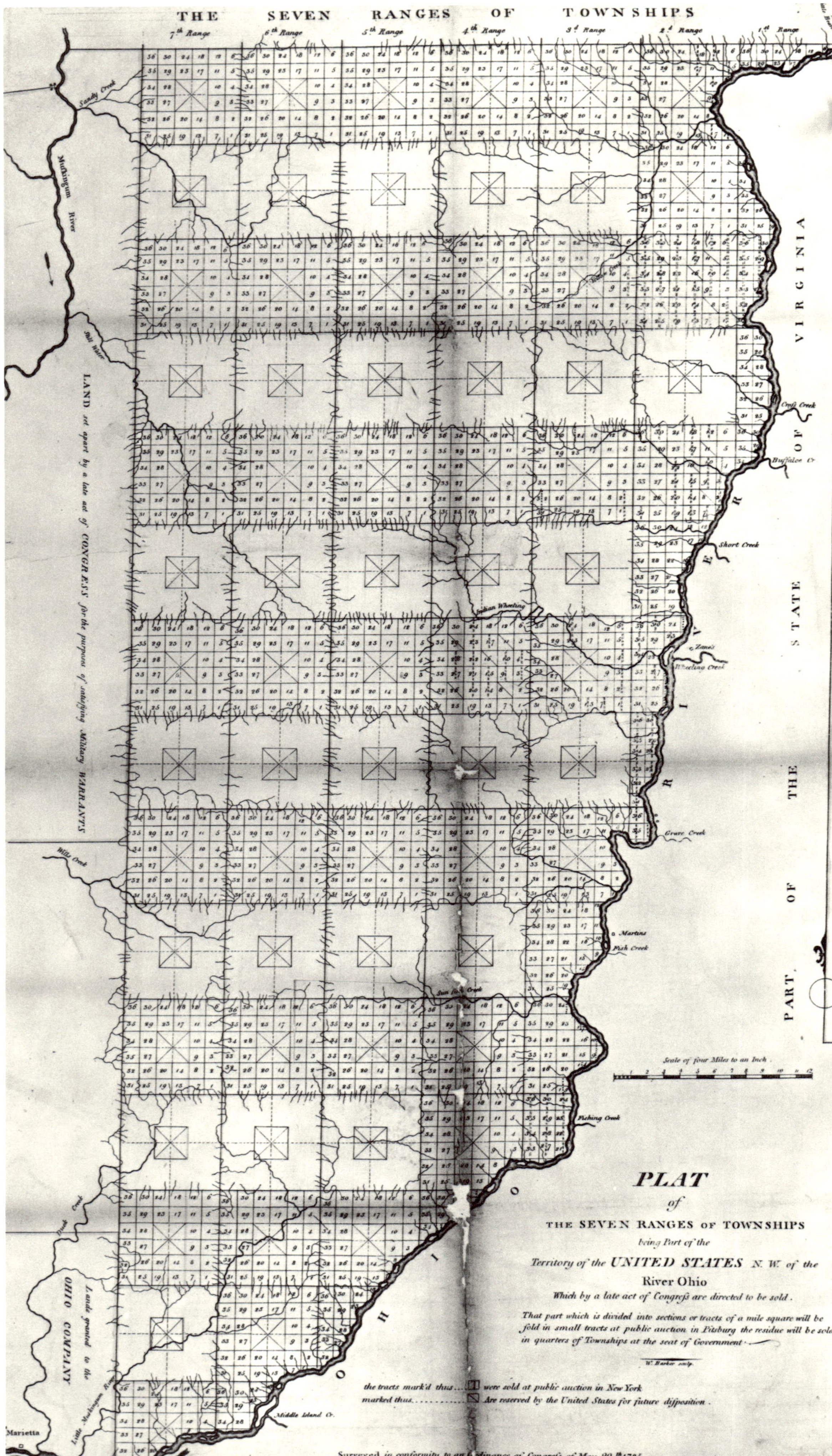
THE SEVEN RANGES OF TOWNSHIPS
7th Range
6th Range
5th Range
4th Range
3d Range
2d Range
1st Range
Sandy Creek
Muskingum River
LAND set apart by a late act of CONGRESS for the purpose of satisfying Military WARRANTS
Wills Creek
Lands granted to the OHIO COMPANY
Marietta
Little Muskingum River
Middle Island Cr.
OHIO
RIVER
PART OF THE STATE OF VIRGINIA
Cross Creek
Buffaloe Cr.
Short Creek
Indian Wheeling
Zanes
Wheeling Creek
Grave Creek
Martins
Fish Creek
Sun fish Creek
Fishing Creek
Scale of four Miles to an Inch.
PLAT
of
THE SEVEN RANGES OF TOWNSHIPS
being Part of the
Territory of the UNITED STATES N. W. of the
River Ohio
Which by a late act of Congress are directed to be sold.
That part which is divided into sections or tracts of a mile square will be sold in small tracts at public auction in Pitsburg the residue will be sold in quarters of Townships at the seat of Government.
Wm. Barker sculp.
the tracts mark'd thus... were sold at public auction in New York
marked thus..... Are reserved by the United States for future disposition.
Surveyed in conformity to an Ordinance of Congress of May 20th 1785.

Hutchins's map first appeared in the 1802 edition of Mathew Carey's *Carey's American Atlas*. It was engraved by William Barker, a map maker who produced many maps for Carey's Philadelphia firm. Barker probably worked from a copy of Hutchins's original surveys after the surveyor's death. The map was made sometime after 1796 because that year Congress changed its reserved sections in the Seven Ranges and also established the United States Military Tract immediately west of the first surveys. Both changes were shown on the map. No doubt many people in the East who were interested in western lands were anxious to examine a detailed map of the first lands opened to settlement by the federal government. Although this map probably was disappointing to many because of its lack of geographic information, at least it illustrated the type of land distribution system adopted by Congress in the Ordinance of 1785.

While Hutchins and his surveyors sectioned off the lands in the Seven Ranges, Benjamin Tupper, one of the members of the original surveying party, gathered information concerning the lands within the region in order to suggest to the Ohio Company suitable lands for its planned settlement. The Ohio Company, formed in 1787 by Rufus Putnam and others as a scheme by which many officers of the Revolutionary Army could redeem their land bounties, had solicited Congress for a 1.5-million-acre tract of land north and west of the Ohio River. After several attempts to convince Congress that a sale of western land to a private company would be profitable to the country, the Ohio Company turned its negotiations over to the brilliant Manasseh Cutler, lawyer, minister, doctor and scientist. After some hard and dramatic negotiations, Congress granted the Ohio Company 1,550,000 acres of land between the Seventh and Sixteenth Ranges on October 27, 1787.

The Ohio Company wasted little time in establishing a settlement on its lands. At the confluence of the Ohio and Muskingum rivers, Marietta, a small stockaded community, was founded in April 1788. Since the Ohio Company was not only concerned with the redemption of veterans' land bounties but also with the making of a profit from land sales for its stockholders, it was necessary to provide those in the East who desired to seek their futures in the West with appropriate information about the lands north of the Ohio River. Cutler, who took the lead in the Company's negotiations with Congress, became its top sales representative. One month after the Ohio Company had signed its contract with Congress, an advertisement appeared in the November 27, 1787, issue of *The Salem* [Massachusetts] *Mercury* offering for sale the pamphlet "A description of the Federal Lands, purchased by the Scioto Companies, confirmed to the United States by sundry Tribes of Indians, in the Treaties of 1784 and 1786, and now ready for settlement," which was written by Cutler himself and published by the Dabney and Cushing printing house of Salem. Six months later, on May 27, 1788, a map entitled "A Map of the Federal Territory from the Western Boundary of Pennsylvania to the Scioto River laid down from the latest information and divided into Townships and fractional parts of Townships agreeably to the Ordinance of the Hon.[b]. Congress passed in May 1785" (Plate II) was also offered for sale by Dabney and Cushing to accompany Cutler's pamphlet.[2]

As a principal stockholder and one of the major architects of the Ohio Company, Cutler, who had suffered much financial loss as a result of the Revolutionary War, was extremely anxious that this venture should be profitable. His pamphlet, which was an effort on his part to encourage immigrants to buy land from the Ohio Company, painted a laudatory

(Opposite page)

Plate I: Hutchins, "Seven Ranges," (after 1796), Carey, *Carey's American Atlas* (1802). The Geographer of the United States, Thomas Hutchins, probably knew more about the Northwest Territory than any other American.

Plate II: Cutler, "Federal Territory," (1785), Ohio Historical Society, Columbus. This map accompanied a promotional pamphlet to help sell lands held by the Ohio Company.

picture of the Ohio Country. Cutler, who visited the company's lands in 1788, almost a year after this tract was published, obtained his information from the reports sent to Congress by Thomas Hutchins while surveying the Seven Ranges, the journal of Captain Harry Gordon who first surveyed the Ohio River, and the personal observations made by Benjamin Tupper, Winthrop Sargent, and other surveyors who had actually visited this area. The authenticity of the entire pamphlet was attested through an endorsement made by Hutchins himself. "*I do Certify* [italics in text], that the facts therein related, respecting the fertility of the soil, productions, and general advantages of settlement, etc.," the famous geographer pledged, "are judicious, just, and true, and correspond with observations made by me during my residence of upward of ten years in that country."[3]

Cutler emphasized the advantages that the new country offered to those immigrants who wanted to work for their future on the Ohio Company's lands. Either by design or because of the lack of adequate information, many of the expressed attributes of the land were exaggerated. Cutler explained that

> the Muskingum is a gentle river, confined by banks so high as to prevent its overflowing. It is 250 yards wide at its confluence with the Ohio, and navigable by large bateaux and barges to the Three Legs, and, by small ones, to the lake at its head. From thence, by a portage of about one mile, a communication is opened to Lake Erie, through the Cuyahoga, wich [*sic*] is a stream of great utility, navigable the whole length, without any obstructions from falls. From Lake Erie, the avenue is well known to the Hudson, in the State of New-York.
>
> The Hockhocking resembles the Muskingum, though somewhat inferior in size. It is navigable for large boats about 70 miles, and for small ones much further. On the banks of this very useful stream are found inexhaustible quarries of freestone, large beds of iron ore, and some rich mines of lead. Coal mines and salt springs are frequent in the neigborhood of this stream, as they are in every part of the western territory. The salt that may be obtained from these springs will afford an inexhaustible store of that necessary article. Beds of white and blue clay, of excellent quality, are likewise found here, suitable for the manufacture of glass, crockery, and other earthenwares. Red bole and many other useful fossils have been observed on the branches of this river.
>
> The Scioto is a larger river than either of the preceding, and opens a more extensive navigation. It is passable for large barges for 200 miles, with a portage of only 4 miles to the Sandusky, a good, navigable stream, that falls into the Lake Erie. Through the Sandusky and Scioto lies the most common pass from Canada to the Ohio and Mississippi, one of the most extensive and useful communications that are to be found in any country.[4]

Of course, Cutler understood that the new land promised advantages beyond its anticipated material wealth. He himself had been educated in the classical traditions of the late eighteenth century. The architect of the Ohio Company believed that the western frontier offered the opportunity for mass public education that would bestow benefits upon those who journeyed west of the mountains in search of a new life. The West was different from the East, Cutler maintained, and he declared:

> In the late ordinance of Congress for disposing of the western lands as far down as the River Scioto, the provision that is made for schools and the endowment of an university looks with a most favorable aspect upon the settlement, and furnishes the presentiment, that, by a proper attention to the subject of education, under these advantages, the field of science may be greatly enlarged, and the acquisition of useful knowledge placed upon a more respectable footing here than in any other part of the world. Besides the opportunity of opening a new and unexplored region for the range of natural history, botany and the medical science, there will be one advantage which no other part of the earth can boast and which probably will never again occur—that, in order to begin *right,* there will be no *wrong* habits to combat, and no inveterate systems to overturn—there is no rubbish to remove, before you can lay the foundation. The first settlement will embosom many men of the most liberal minds—well versed in the world, in business and every useful science. Could the necessary apparatus be procured, and funds immediately established, for founding a university on a liberal plan, that professors might be active in their various researches and employments—even now, in the infancy of the settlement, a proper use might be made of an advantage which will never be repeated.[5]

When Cutler, as a representative of the Ohio Company, negotiated with Congress for western lands, he also represented the Scioto Company and included as part of a larger package that company's request for 3,500,000 acres to help influence the sale. The Scioto Company had been organized by William Duer, Secretary of the Treasury Board under the

Articles of Confederation, and others as a private speculative land company. Duer planned to pay Congress for the tract of land from the Continental Certificates and land bounties he would collect from individuals who purchased small tracts from the company. Furthermore, he believed that he could offer these western lands for sale in Europe and collect Continental Certificates which had worked their way there through Congress's spending during the Revolutionary War. Cutler was able to convince Congress of the correctness of this grant. Although the land was deeded to the Scioto Company, no payment for it was ever made and Duer, who ended his days in debtors' prison, never owned any part of the so-called Scioto Tract.

After Congress agreed to sell land to the Scioto Company, but before that company defaulted on its payments, Duer had actively engaged in the sale of Ohio lands in Europe. He hired Joel Barlow, the poet-diplomatist who wrote *Columbiad,* the first epic poem about the founding of the United States, to represent his company abroad. Barlow solicited the aid of William Playfair, an English schemer with a dubious reputation at best, to help him. Together they formed the *Compagnie du Scioto* to help promote the sale of Ohio lands in France. Barlow had carried a copy of Cutler's promotional brochure to France and had it translated into French to help convince the French that they should migrate to the American western frontier. The French edition was entitled *Description du Sol, des Productions &c. & c. de cette portion des États-Unis, située entre la Pensylvanie, les rivières de l'Ohio & le lac Erie.*

To explain to the French the plan for their settlement along the Ohio River, the *Compagnie du Scioto* published *Prospectus pour l'éstablissement sur les rivières d'Ohio et de Scioto en Amérique.*[6] This brochure mentioned Cutler's work and cited the endorsement that the geographer, Thomas Hutchins, had given it. A French interpretation of the Cutler map accompanied this prospectus. Two distinct copies of this map were made. The first, entitled "Plan Des Achats Des Compagnies De l'Ohio Et Du Scioto" and the second, "Plan Des Achats Des Compagnies De l'Ohio et de la Scioto" (Plate III). The first version was more detailed than the second; it located "*V de Mariana*" at the mouth of the Muskingum and illustrated "*Premiere Ville*" on the north shore of the Ohio, opposite the mouth of the Great Kanawha River.[7] Possibly because of Marie Antoinette's considerable loss of popularity among the French people by 1789, the names of the cities in Ohio on the map were removed, since Mariana, or Marietta, was named for the French Queen. The second version of the map was not very detailed. The map itself was engraved by Antoine-François Tardieu, although the name that appeared was Pierre-François Tardieu. The Tardieu family was famous for its art engraving work in France for over a hundred years, and Antoine-François, a nephew of Pierre-François, signed his uncle's name to many of his earlier works, apparently to draw strength from the famous Tardieu's reputation. Tardieu's map was explained in Barlow's brochure in "*Explication de la Carte Geographique*":

> The divisions that are shown on this map each represent a township five miles square. This system of division was ordered by an ordinance passed by Congress; each of these townships contain 23,040 acres of land; within each township, some of the land was reserved for the church, public schools and Congress.
>
> Congress named the first range of townships to the east, the eighteenth range; the numbering continues from east to west so that the last range is numbered twenty-eight. This numbering system will be kept as long as the

Plate III: Tardieu, "Compagnies De l'Ohio et de La Scioto," (1789). This map was used by representatives of the Scioto Company to lure French immigrants to Ohio.

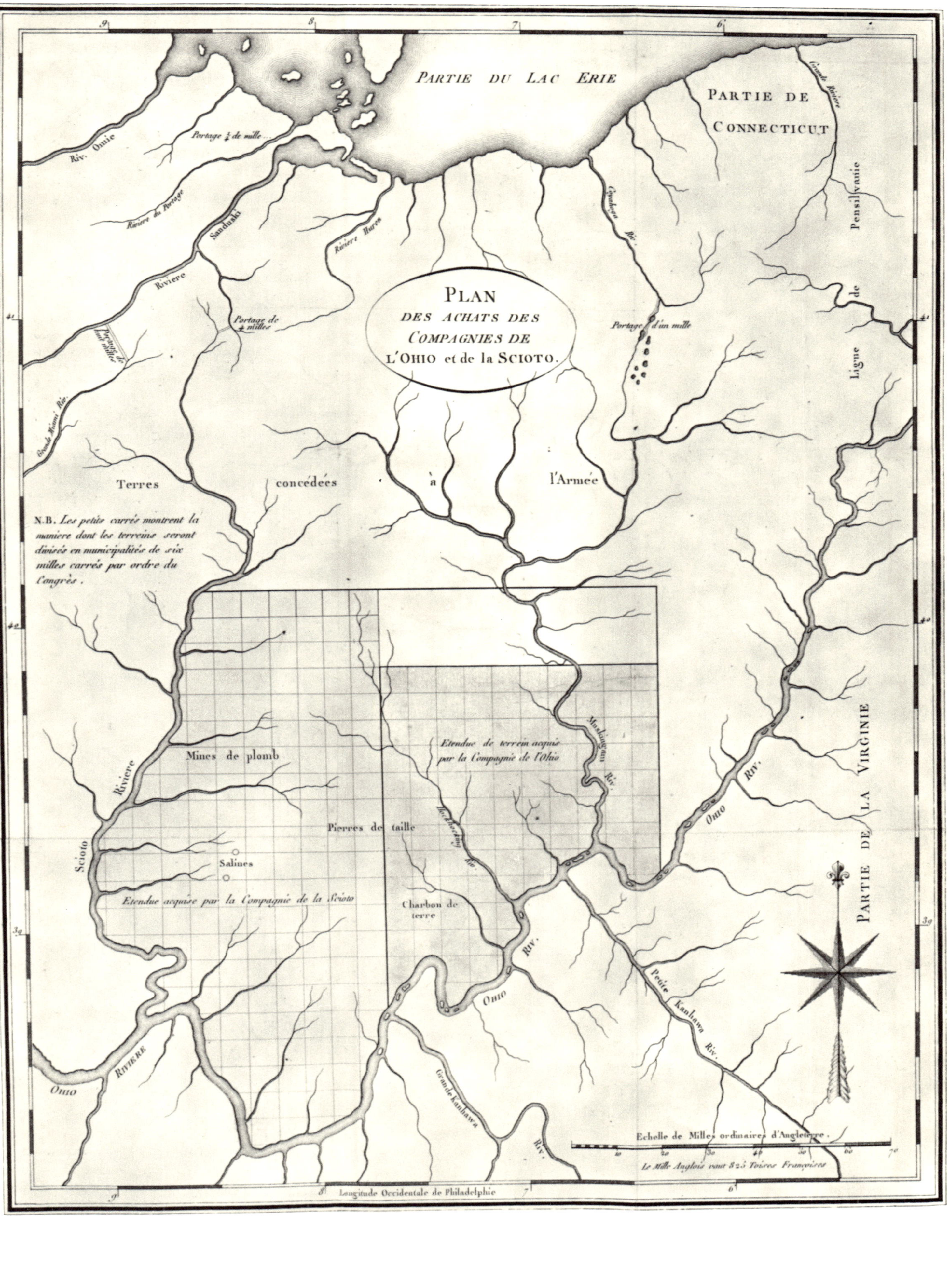
PLAN
DES ACHATS DES
COMPAGNIES DE
L'OHIO et de la SCIOTO.
PARTIE DU LAC ERIE
PARTIE DE
CONNECTICUT
Grande Riviere
Pensilvanie
de
Ligne
PARTIE DE LA VIRGINIE
Riv. Omie
Portage ½ de mille...
Riviere du Portage
Sandusky
Riviere Huron
Riviere
Portage de 4 milles
Portage d'un mille
Cayahoga Riv.
Grande Miami Riv.
Terres
concédées
à
l'Armée
N.B. Les petits carrés montrent la maniere dont les terreins seront divisés en municipalités de six milles carrés par ordre du Congrès.
Etendue de terrein acquis par la Compagnie de l'Ohio
Muskingum Riv.
Mines de plomb
Pierres de taille
Salines
Etendue acquise par la Compagnie de la Scioto
Charbon de terre
Hockhocking Riv.
Scioto
Riviere
OHIO RIV.
OHIO
RIV.
OHIO
RIVIERE
Grande Kanhawa
Riv.
Petite Kanhawa Riv.
Echelle de Milles ordinaires d'Angleterre.
10 20 30 40 50 60 70
Le Mille Anglois vaut 825 Toises Françoises
Longitude Occidentale de Philadelphie
9 8 7 6
41 40 39

laws of Congress are in force. The numbering of the ranges on this map conform to the laws prescribed by Congress.

The first town that is planned for this area will be located along the Ohio River, opposite the mouth of the Great Kanawha River and will be located in Section 3 of the eighteenth township if that is possible.

It is natural that the colonists would want some housing built for them before they arrive. To meet this situation, temporary housing has been built to house the colonists until they select for themselves the land they want to occupy and build their new homes.

The price of land will necessarily increase because of the improvements made and the agricultural development; consequently, the Scioto Company will not sell an acre of land in the future for the same price it receives today when the intrinsic value of an acre will increase. It will not, however, introduce any change in its price of land without first publishing a notice a month in advance. Once a colonist had selected his land, he will not be able to change to a more desireable tract. Changes in location of land in this tract will be made only through the Paris office; any other system would work against the interests and successful operations of the company.

Even though part of the land surveyed in the Seven Ranges was reserved for the purpose of retiring land bounties given as compensation to Revolutionary War veterans, it was soon apparent that those lands would not be sufficient to satisfy all the demands. After much consideration, the federal government, on June 1, 1796, set aside a 2,539,110-acre tract, known as the United States Military Tract, between the Ohio Company's land and that of the Western Reserve, and west of the Seven Ranges, to help fulfill its promise of compensation made to those who fought for the country's independence. The map "Plat of that Tract Country in the Territory Northwest of the Ohio appropriated for Military Services, and described in the Act of Congress, entitled 'An Act regulating the Grants of Land appropriated for Military Services, and for the Society of United Brethren for propagating the Gospel among the Heathen' surveyed under the direction of Rufus Putnam Surveyor General of the United States" (Plate IV) showed the location of this tract and the subdivisions made after

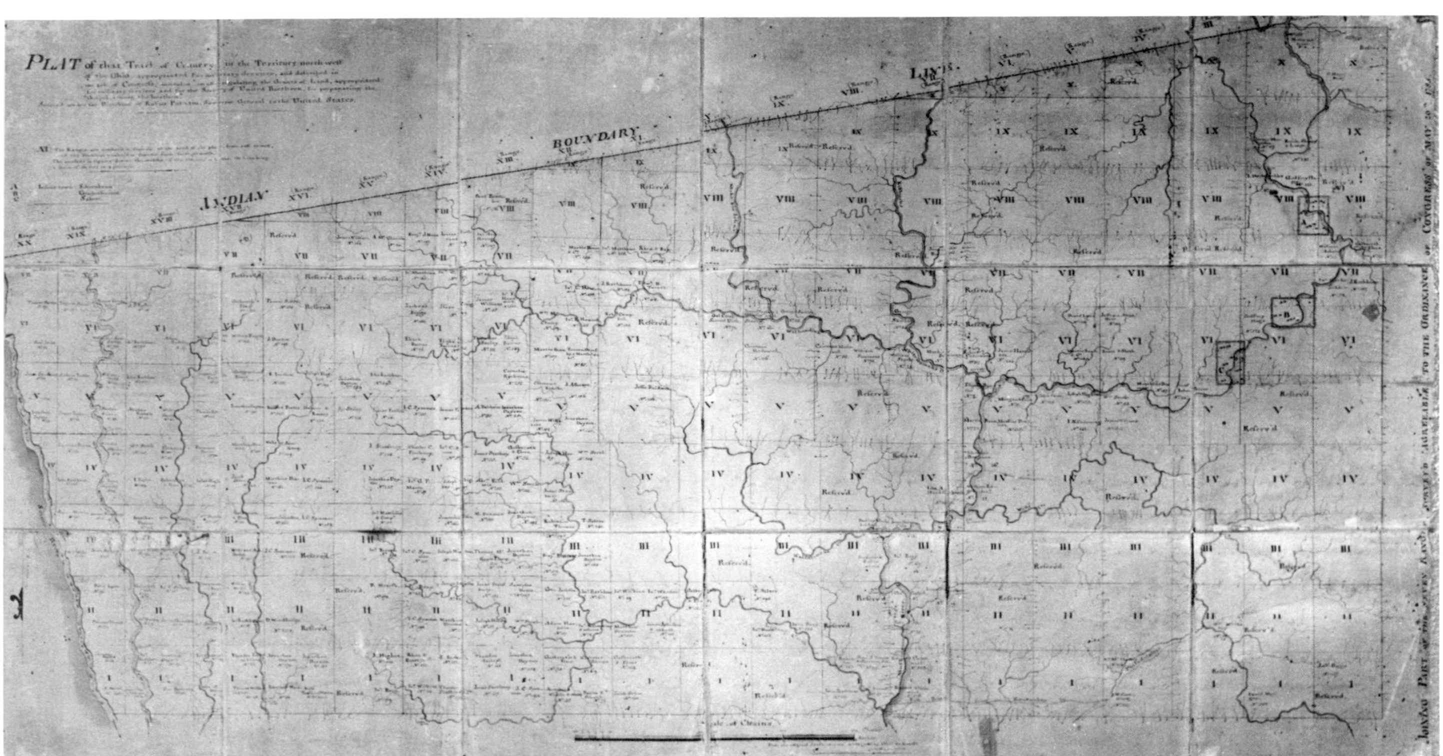

Plate IV: Putman, "Territory Northwest of the Ohio," (1800), **Dawes Memorial Library, Marietta College.** Established in 1796, the United States Military Tracts satisfied claims Revolutionary War veterans had against the United States.

the initial surveys. By the same act, Rufus Putnam was appointed Surveyor General of the United States, a position which replaced that of Geographer of the United States, held by Thomas Hutchins until his death in 1789. It was not until money was appropriated by Congress in 1797 and after its northern boundary, the famous Greenville Treaty Line, was established by Israel Ludlow in August that same year, that the actual surveying of the tract began.

The United States Military Tract was to be divided into townships 5 miles square rather than 6 miles square as prescribed by the Ordinance of 1785, and these townships were to be subdivided into quarter townships of 4,000 acres each for distribution. The 5-mile-square township was used in this tract because Congress had allotted land bounties to its soldiers as follows: a Major General received 1,100 acres, a Brigadier General 850, a Colonel 500, a Lieutenant Colonel 450 and so on, until a noncommissioned officer and an enlisted man received 100 acres. Since the land bounties were issued in multiples of 5, the 5-mile-square township permitted easier distribution. The smallest division was a quarter township or 4,000 acres. No one was given that large a bounty. Therefore, it was necessary for the separate bounties to be held by the government until their accumulative total equalled 4,000 acres before the land was granted to the bounty holders. The land within the quarter township was then divided privately between the various individuals. It was the duty of the Surveyor General of the United States to see to it that this land was divided properly. Once the necessary bounties were acquired, those who held such claims were anxious for their titles because in April 1806, Albert Gallatin, Secretary of the Treasury, ordered Jared Mansfield, the Surveyor General, to survey fifty townships within the Tract and cautioned that "the price of surveying must not exceed three dollars a mile, and you will be pleased to take measures for having the work speedily executed."[8] It is apparent from the map that some men, such as Jonathan Dayton, John Cleves Symmes and Matthias Denman had accumulated a number of land bounties because their names are conspicuous as large landholders. By 1803, the year Ohio entered the Union, the lands unclaimed by Revolutionary War veterans within this tract were placed for sale on the open market on the same terms as other public land, and the use of this tract to retire military bounties came to an end.

Besides reserving land for veterans, Congress also set aside for the Society of the United Brethren, or the Moravians, 3 tracts of land at Schoenbrunn, Gnadenhutten and Salem that totaled 12,000 acres. These mission villages had been established between 1772 and 1780 by the Moravians in order to provide Christian instruction for the Delaware Indians of the Tuscarawas Valley. The tragic history of these mission Indians is well known, but in recognition of their work with these primitive people on the western frontier, Congress provided that these tracts of land of 4,000 acres each were to be held in trust by the Moravians for the Indians. The 3 tracts shown on the Putnam map were held by the Society of the United Brethren until May 26, 1824, when the lands were returned to the government; and after pre-emptory claims were satisfied and the town of Gnadenhutten was laid out by the government, the remaining land was sold by Congress.

The map reproduced for this study is located in the Dawes Memorial Library at Marietta College, and was probably used by Rufus Putnam when he was Surveyor General of the United States in order to record those quarter townships exchanged for land bounties. The names that appear on the quarter townships are, of course, the names of the original landholders in the United States Military Tract; and the map illustrated the land ownership pattern much as it was on the eve of the passage of a land law on March 1, 1800, which liberalized land sales in the West and allowed the quarter townships yet unclaimed to be divided further into 100-acre lots. The small arrows on the map indicate the direction of the drainage flow into the streams that cut through the tract.

Although a fairly efficient system of land survey was prescribed by Congress for the division of the land in Ohio in its Ordinance of 1785, the Virginia Military District, which was bordered on the east by the scenic Scioto River and on the west by the Little Miami, was not divided according to the rectangular land survey. Because of its original land charter granted by James I in 1609, Virginia claimed all the lands "from sea to sea, West and Northwest," which included what is now Kentucky and Ohio and most of the Northwest Territory. In 1778 when it was proposed that the various states form a permanent political union, it was suggested by Maryland that all western lands claimed by the seaboard states should be ceded to the Congress formed by the country's first constitution, the Articles of Confederation. New York was the first to comply with Maryland's suggestion and was soon followed by Virginia, whose House of Burgesses ceded its western lands to Congress on January 2, 1781. Virginia, however, had to retain sufficient land to pay off the generous land bounties that the state legislature had promised to her own soldiers for their service during the American Revolution. The western regions of Virginia and Kentucky had been used for that purpose but Virginia feared that those areas had insufficient amounts of good land to meet her needs. Consequently, in her cession of lands to Congress, Virginia reserved about 4,000,204 acres in Ohio to be used in order to fulfill that state's obligation to its veterans.[9]

The method of land distribution used by Virginia was known as the indiscriminate survey system. Rather than having teams of surveyors systematically section off either 6- or 5-mile-square townships to be sold through a government land office, each individual holding a land bounty had the right to select a parcel of land for himself. The size of each individual tract depended on the soldier's rank; an enlisted man who served less than 3 years could receive 100 acres, while a Major General was entitled to 15,000 acres. At first, while large tracts of good bottom land were available, most of the land grants were rectangular; but as the better lands were taken, many peculiar shapes emerged as the latecomers wanted to obtain the best land remaining while eliminating the poor from their grants.

The survey books for the Virginia Military Tract in the Record Center of the Land Office in the State Auditor's Office in Columbus are filled with the plats of the original surveys that illustrate this point. For example, Survey No. 571 for 766⅔ acres found on page 65 of *Survey Book A* was a fairly early survey (Plate V). This particular tract was surveyed by Nathaniel Massie, who had been appointed deputy surveyor by Virginia to survey the land claims north of the Ohio River, and who had actually begun some survey work in the Virginia Military District in 1787, almost 3 years before Congress authorized Virginia to do so. The land bounty for this tract was owned by Albert Gallatin, who in 1792 was a member of the Pennsylvania General Assembly and who later served as Secretary of the Treasury between 1801 and 1814. Gallatin had moved to the western

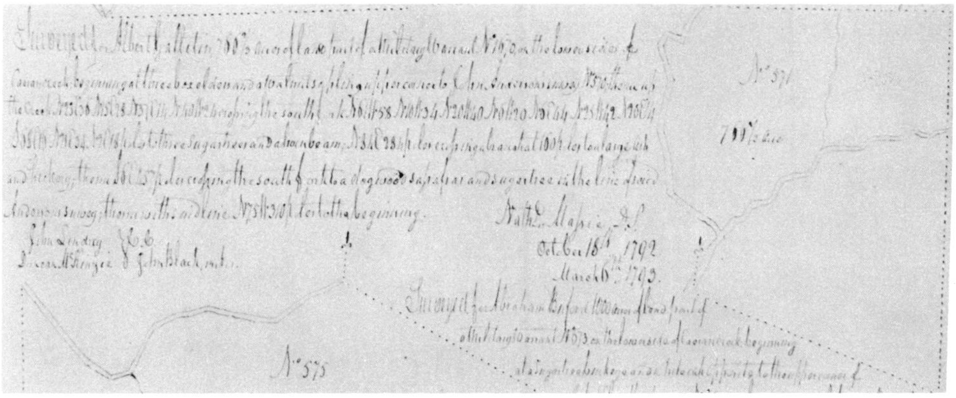

Plate V: "Plat of Survey No. 571," Virginia Military Tract, Land Records, State Auditor's Office, Columbus, Albert Gallatin, Secretary of the Treasury, owned this particular tract of land in the Virginia Military District.

portions of Pennsylvania in 1784 and had been successful as a land speculator. Apparently Gallatin, since he was not a member of the Virginia militia, had purchased this particular land bounty and had made his claim. The tract, located in Clinton County, was nearly rectangular, bordered by Caesar's Creek on the west and drained by Anderson's Fork. At the junction of these two streams, the village of New Burlington was built.

An example of the later surveys was the one made in 1822 by Deputy Surveyor Cadwallader Wallace for the representatives of Benjamin Harrison (Plate VI). This survey, Number 8867 and 10891, contained 1,040 acres near the North Fork of Paint Creek. The irregular shape of the tract was illustrative of the attempts to include only good land. The metes and bounds, or limits of the bounty, were described by the physical features and not the mathematical readings as used in the rectangular survey. This resulted in considerable confusion over the years as boundary lines were fixed. If, as this particular survey explained, a part of the plat line ran "thence S336 E 132 poles to two white oaks, thence S 40 W11 poles to two barren oaks," and years later the oaks were no longer standing, that particular part of the survey was a matter for dispute. Land continued to be surveyed in the Virginia Military Tract until February 18, 1871, when all claims on the land from Virginia were believed to be exhausted. In all, 76,745.44 acres remained unclaimed in this tract. In 1872, the Ohio General Assembly made it possible for all the unsold land from this district to be sold and the proceeds given to the Ohio Agricultural and Mechanical College, which later became the Ohio State University, for support of that school.

Although not a survey map, the historic "Heckewelder Map" of 1796 (Plate VII) is a very important one in the story of the early occupation of northeastern Ohio. John Heckewelder, a Moravian missionary and an assistant to Christian Frederick Post, first visited the Ohio country in 1764 when he and Post tried to section off a small parcel of land along the Tuscarawas River near the present site of Bolivar in order to establish a mission there. Hostilities on the part of some Indians forced these two missionaries to abandon their project at that time; but Heckewelder later returned, and in 1772 he, along with David Zeisberger, another Moravian missionary, established the Moravian villages of Schoenbrunn and Gnadenhutten on a tract of land along the Tuscarawas River. Because of the years spent in Ohio wandering over the frontier, particularly during the American Revolution when the British forced the Moravian Indians along with their white teachers, including Heckewelder, to abandon their homes in eastern Ohio and establish new ones in the western part of the state and around Detroit, probably no other white man knew this particu-

(Overleaf) Plate VI: "Plat of Survey No. 8867 & 10891," Virginia Military Tract, Land Records, State Auditor's Office, Columbus. Surveyed into irregular shapes, the land in the Virginia Military District illustrated the Southern pattern of land distribution.

David McArthur & Matthew Bonner M.k.r. — March 3rd 1822

Surveyed for the representatives of Benjamin Harrison 1040 acres of land on part of a military Warrant No 6014 on the waters of the north fork of paint Creek: beginning at a black Oak S65½ E80 poles from the most northerly corner of Robert Armstrongs representatives survey No 3964; thence with the several lines of said Harrisons Survey No 7806 of 600 acres S88 E110 poles to three bur oaks from one root; thence N70 E140 poles to a bur oak and hickory; thence N25 E44 poles to a large White oak; thence N24 W210 poles to a bur oak; thence N71 W90 poles to two bur oaks; thence S66 W60 poles to a hickory and bur oak; thence N56 W24 poles to a hickory and bur oak; thence N26 W92 poles to a stake in the line of Harrisons said Survey; thence East 63 poles to a black oak; thence S9½ E83 poles to three White oaks two from one root and a black oak, thence S60 E160 poles crossing a small branch to four bur oaks; thence N77 E80 poles to three bur oaks, two from one root; thence S31 E34 poles to three large white oaks corner to said Armstrongs survey No 4282; thence with his lines S75 E80 poles to two white oaks: thence S39 E72 poles to a black oak; thence S56 poles to three white oaks; thence S33 E 132 poles to two white oaks; thence S40 W100 poles to two barren oaks; thence S8 W78 poles to a large barren oak; thence S16 W200 to a stake northeast corner to said Harrisons Survey No 7805 of 100 acres, thence with his line S84 W167 poles to a stake corner to said Survey; thence with another line thereof S16 W115 poles to a black oak in the line of Survey No 3964; thence with the lines thereof N52½ W14 poles to three white oaks; thence S50 W26 poles to two bur oaks; thence with the line of entry No 9294 N28 W280 poles to a hickory corner to Survey No 3964 and corner to said Harrisons survey No 7805 of 150 acres; thence with said line East 144 poles to a stake corner to said Survey; thence with another of his lines North 100 poles to a stake; thence with another of his lines, West 220 poles to a White oak and three hickories corner to said survey and corner to said Armstrongs survey No 3964; thence N15 W80 poles to the beginning.

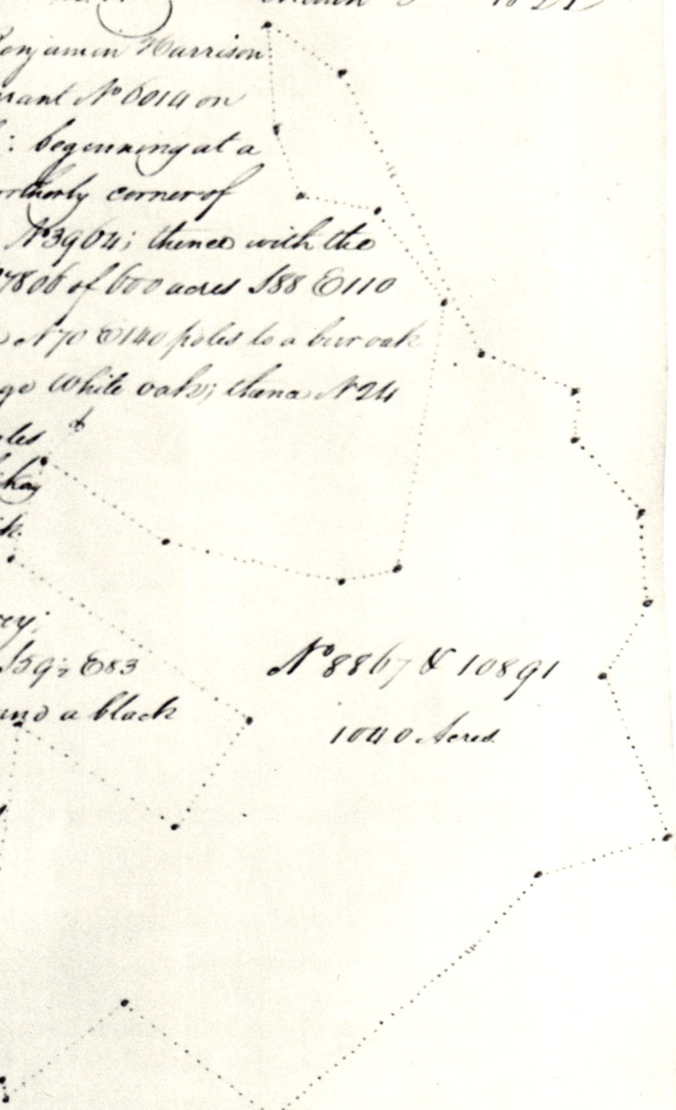

Cadwallader Wallace S.S.

Batteal Harrison
Robert Johnston } C.C.

January 18th 1822

A. C. Looker M.k.r. — March 5th 1822

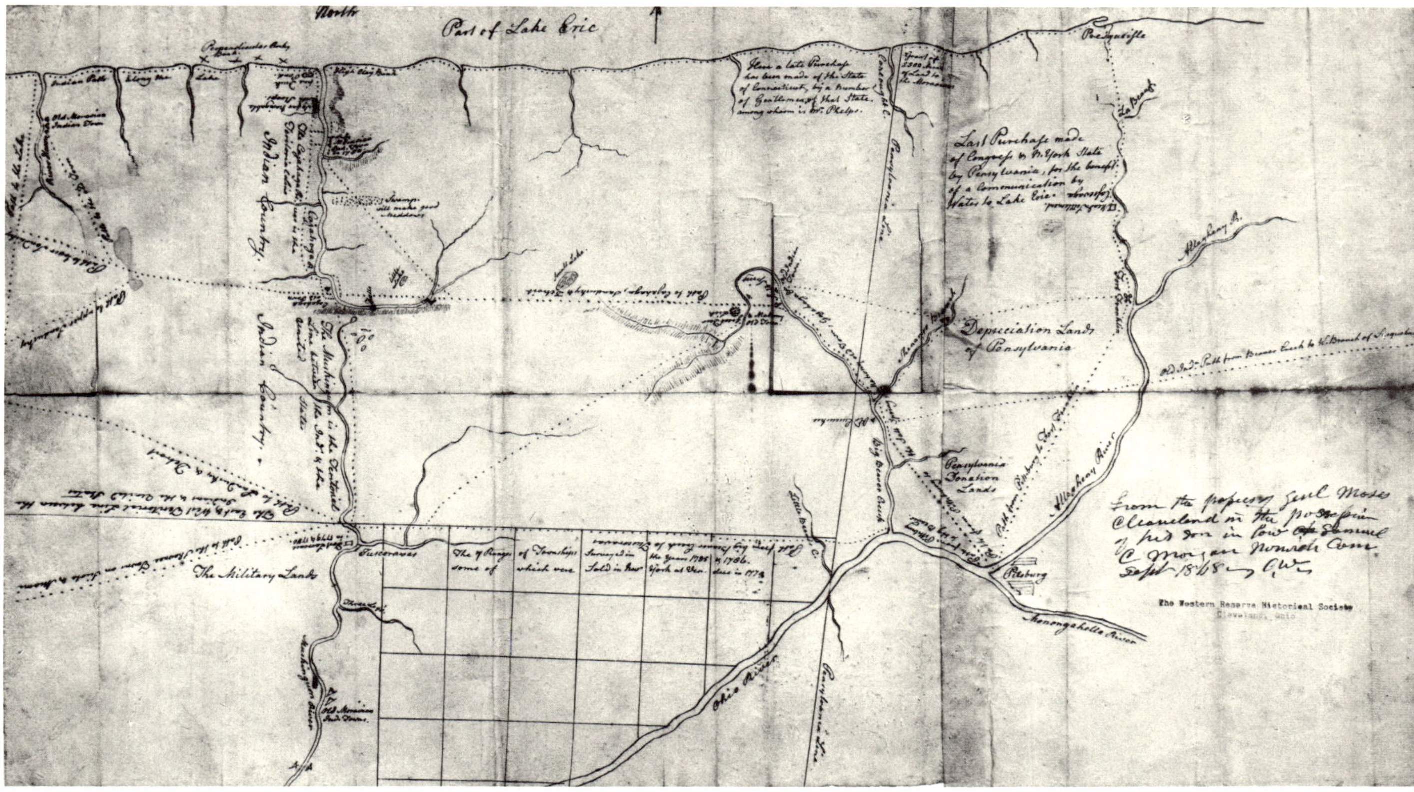

Plate VII: Heckewelder, "Heckewelder Map," (1796), Western Reserve Historical Society Library, Cleveland. Heckewelder recognized the potential of the land south of Lake Erie and his map was the first to detail information about the northeastern section of Ohio.

lar region better than Heckewelder. In a letter to John McNair, written on January 12, 1796, Heckewelder described the northeastern section of Ohio illustrated on his map:

> Altho the country in general containedth both arable Land & good Pasturage: yet there are particular Spots far preferable to others: not only on account of the Land being here superior in quality: but also on account of the many advantages presenting themselves.
>
> As the first place of utility between the Pennsylvania Line: (yea I may say between Presq' Isle) and Cujahaga; & in an East and West course as the dividing Ridge runs between the Rivers which empty into the Lake Erie; & those Rivers or Creeks which empty into the Ohio: (& which Ridge I suppose runs nearly Parallel with this Lake, & is nearly or about 50 miles distance from the same): Cujahaga certainly stands foremost; & that for the following reasons.
>
> 1) because it admits small Sloops into its mouth from the Lake, and affords them a good Harbour.
>
> 2) because it is Navigable at all times with Canoes to the Falls, a distance of upwards of 60 Miles by Water—and with Boats at some Seasons of the Year to that place—and may without any great Expense be made Navigable for Boats that distance at all times.
>
> 3) because there is the best prospect of Water communication from Lake Erie into the Ohio, by way of Cujahaga & Muskingum Rivers; The carrying place being the shortest of all carrying places, which interlock with each other, & at most not above 4 miles.
>
> 4) because of the Fishery which may be erected at its mouth, a place to which the White Fish of the Lake resort in the Spring, in order to Spawn.
>
> 5) because there is a great deal of Land of the first Quality on this River.
>
> 6) because not only the River itself, has a clear & lively current, but all Waters & Springs emptying in the same, prove by their clearness & current, that it must be a healthy Country in general.
>
> 7) because one principle Land Road, not only from the allegheny River & French Creek: but also from Pittsburg will pass thro that Country to Detroit, it being by far the most level Land path to that place.
>
> I will now endeavor to give an account of the Quality of the Soil of this Country: and will begin with the Land on the Cujahaga River itself.

Next to the Lake the Lands in general lay in this part of the Country, pretty high, (say from 30 to 60 feet high) except where there is an opening by a River or Stream. These banks are generally pretty level on the top, & continue so to a great distance into the Country. The Soil is good and the Land well Timbered either with Oaks & Hickory, or with lofty Chestnuts.

On the Cujahaga River are, I verily believe, as rich Bottoms, or intervals, as in any part of the Western Country. The timber in these are either Black Walnut, or White Thorn Trees, intermixed with various other Trees as Cherry, Mulberry, &c. The ground entirely covered with high Nettles.

In such Bottoms, somewhat inferior to the above, the Timber is principally lofty Oaks, Poplar, or Tulip tree, Elm, Hickory, Sugar Maple yet intermixed with Black Walnut, Cherry, Mulberry, Grape Vines, White Thorn, Haw-bush &c &c Ash &c Wild Hops of an excellent quality grow also plentifully on this River.

The richest Land on this River lieth from where the road crosseth at the old Town downwards. Within 8 or 10 miles of the Lake the Bottoms are but small, yet the Land rich, from here upwards they are larger & richer. At the Old Moravian Town as marked on my Map, they are exceedingly rich. Some low bottoms are covered with very lofty Sycamore Trees.

The Land adjoining those Bottoms within 10 or 15 Miles of the Lake, is generally ridgy, yet level & good on the top, excellently Timbered. Thro' these ridges run numbers of small Streams, & sometimes large Brooks; the water always clear and with a brisk current.

I have traced small Streams to their Sources, where I have found a variety of excellent Springs lying off in various directions. (see the run at the Moravian Town).

From these Lands upwards towards the old Town, & along the path towards the Salt Spring; the Country is in general pretty level; just so much broken as to give the Water liberty to pass gently off.

There is a remarkable fine Situation for a Town, at the old Cujahaga Town; & there can be no doubt of a large Trading Town being established here, as both a Road to Sandusky & Detroit crosses here: as also the carrying place between the two Rivers Cujahaga & Muskingum must be at this place.

Some miles above this Old Town there is a fall in the River. The Rock which runs across may be about between 20 & 30 feet high. No Fish can ascend higher up, or get over this Fall, tho there are Fish above it. Just under the Falls the Fish crowd together in vast numbers, & may be taken here the whole year round. At the more Easterly Crossing of this River as the Path runs; (the distance of which I do not exactly recollect, but think it between 15 & 20 miles) there is a most remarkable large Square Rock in the Middle of the Stream, which may at a future day, well answer the Pier of a Bridge. (see ^ this mark on the map) at this place there is a pretty large Plain on the Northwest Side of the River—and in several other places in this Country there are similar Plains or Flatts. On these the Land is rather thin in comparison to the other: yet not so that it would not bear good Grain.

There are also some Swamps in this Country, yet I have not seen one, which might not be cultivated, and make good Meadows.

Here and there I observed small groves of Pine, but never went to see of what kind they were. I supposed them only to border on some small Lake or Pond.

There are some beautiful small Lakes in this Country, with water as clear as Chrystall, & alive with Fish. In these Lakes as well as in Cujahaga River Water Fowl resort in abundance in Spring & Fall.

Between the head Waters of Beaver Creek & the head Waters of Cujahaga, the Country is rather more broken, yet not too much for tillage. The Land is good.

From the big Deer Lick on Beaver Creek to the Salt Springs (a distance of about 16 miles) the Country is rather of a colder Nature; but thinly Timbered, & much of a wet Clay ground. A Comp'y of gentlemen have obtained some Years ago a Title to this Tract of Country comprehending the Salt Spring.

I cannot leave Cujahaga without mentioning one Circumstance, viz. That when I left the Moravian Town on that River which was the Eighth day of October 1786, we had not then had one Frost yet, whereas all the Weeds & bushes had been killed by the Frost some Weeks before, on the dividing Ridge. Ind'n Corn, this year planted at the above mentioned place on the 20th day of June ripened before the Frost set in.

The Cujahaga Country abounds in Game, such as Elk, Deer, Turkey, Raccoons &c In the Year 1785, a Trader purchased 23 Horseload of Peltry, from the few Indians then Hunting on this River—

Of the Country to the Southward of Cujahaga & between the dividing Ridge & Tuscorawas, where the line strikes across, I cannot give a precise description, having only seen this Country in part, yet what I have seen has been pretty generally good, except it be some barren Plains, and Large Cranberry grounds. Otherwise off the River, and on the path from thence to Mahoning Old Towns, I saw vast bodies of very rich Upland, well Timbered, sometimes level Land, & then broken, especially the latter on the head Waters of the Beaver Creek towards Mahony.

From Tuscorawas Northerly for 12 or 15 Miles I thought the Land very good & observed extensive Meadows on the Banks of the Muskingum. But I think near the dividing Ridge the Country is rather Colder, The Country is in some places off the River interspersed with round Nobs or Hills, with short yet thick Trees upon them. The water of this Country is also clear & good.

I will insert the description the late Geographer to the United States gives to this part of the Country, copied from a Pamphlet he had printed in London in the Year 1778, which runs thus:

"The Muskingum is Navigable with large Batteaux or Barges to three Legs [Creek] and by small ones to a Lake at its head. From thence, (namely from three Legs) to Cujahaga, (the Creek that leads, to Lake Erie) the Muskingum is muddy, and not very swift, but no where obstructed with Falls or Rifts. Here are fine uplands, extensive Meadows, Oak and Mulberry Trees fit for Ship building, and Walnut, Chestnut, & Poplar Trees suitable for domestic service—Cujahaga furnishes the best portage between Ohio and Lake Erie: at its mouth it is wide enough to receive large Sloops from the Lake. It will hereafter be a place of great importance."[10]

Heckewelder's map, the first one to detail a large amount of information about the northeastern portion of Ohio from personal observations, was important to the increase in geographical information about the state. It located the several Moravian Indian villages and the numerous Indian trails and paths that passed through the area and identified the major streams at the time that the Connecticut Land Company had purchased this entire region from the state of Connecticut. There is little evidence to believe that Moses Cleaveland, a stockholder in the Connecticut Land Company and leader of the surveying party that divided the tract had access to the Heckewelder map at the time he began his surveys; he did not mention the map either during his preparation for the trip or after he arrived in the Reserve. The area indicated on the map included a large portion of western Pennsylvania to the exclusion of much of the Reserve's western territory. Also, on the lower portion of the map, Heckewelder noted that a purchase had been made "by a number of Gentlemen of that State [Connecticut], among whom is Mr. Phelps." Certainly this notation was not addressed to Cleaveland or Oliver Phelps, the founder of the Connecticut Land Company, since both already knew that. The map was probably sent by Heckewelder to John McNair along with the letter of January 12, 1796. McNair, a Philadelphian, along with others, was interested in forming a land company with the idea of speculating in regions in northwestern Pennsylvania and northeastern Ohio. Heckewelder probably made his map for them. For some reason, however, the McNair plans fell into Cleaveland's hands after he had begun his surveys.[11]

(Overleaf) Plate VIII: Pease, "Western Reserve," (1798), Western Reserve Historical Society Library, Cleveland. The famous Reserve was given to Connecticut to help satisfy its claim for western lands.

A Map of the
CONNECTICUT
Western Reserve
from actual Survey
By
Seth Pease
The Western Reserve Historical Society
Cleveland, Ohio
LAKE
Sandusky Bay
Huron River
Vermillion River
River
Rocky River
UNSURVEYED LAND
and
SUBJECT to INDIAN CLAIMS
FIRE LAND
FIRE LAND so called. It being a Grant of 500,000 Acres by the Legislature of the State of Connecticut to certain Sufferers in the late war with Britain; it is presumed that the Eastern Boundary of said Grant will fall near the mouth of Vermillion River, and is denoted by a broken line
N.B. The interior part of the Country west of the CUYAHOGA RIVER and subject to Indian Claims, has not been Surveyed.
The Townships are Numbered in the first Range at the Eastern boundary; the same Parild is the same Number in the different Ranges
A Scale of Miles
NORTHWEST TERRITORY
NORTH WEST TERRITORY

E R I E
Nº 13
Ashtabula Creek
Nº 12
Shaw
G R A N D R I V E R
Nº 11
Pierpont
Thomson
Trumbull
Nº 10
Leon
Nº 9
Euclid
Canton
Batavia
Windsor
Leffingwell
Nº 8
Williamsfield
Dwights field
Mesopotamia
Greenesborough
Gustavus
Nº 7
Kinsman
Tinkers Creek
Milan
Parkman
Nº 6
Aurora
Mennia
Nº 5
Streetsborough
Warren
Howland
Nº 4
Liberty
Nº 3
Salt Spring
Indian Path
Palmyra
Nº 2
Coitsville
Youngtown
Big Beaver River
Portage Path
10th Range
9th Range
8th Range
Randolph
7th Range
6th Range
5th Range
4th Range
Elsworth
3d Range
2d Range
Boardman
1st Range
Nº 1

The map, which included the western regions of Pennsylvania, showed the grant of 5,000 acres given to the Society of United Brethren in 1794 by the State of Pennsylvania "as an acknowledgment of their faithful labors among the Indians while these had resided in Pennsylvania"; the famous Erie Triangle, a tract of land just south of Lake Erie that New York had ceded to Congress and which was, in turn, sold to Pennsylvania in 1792 for $151,640, thus giving Pennsylvania access to the lake;[12] the Depreciation Lands, a tract of 720,000 which was sold after 1782 by the Quaker state for hard money and given to the veterans of the Pennsylvania line for their depreciated Continental Certificates;[13] and the Donation Lands which were set aside by the Pennsylvania General Assembly in 1783 and allotted to her veterans of the Revolutionary War.

By the time that Heckewelder drew his map, the Seven Ranges had been surveyed; the United States Military District had been set aside by Congress; the Connecticut Land Company had purchased the land east of the Cuyahoga River from Connecticut; and the Treaty of Greenville, which was signed by the Indians in August 1795 following the Battle of Fallen Timbers, had established the line which divided the lands that belonged to the Indians and those that were controlled by the United States. In 1796, Heckewelder and several others, including Rufus Putnam, journeyed to the Tuscarawas Valley to survey the Moravian Tract. Putnam, who was surveyor General of the United States, was charged with the responsibility of running the Greenville Treaty Line, which followed the Cuyahoga and Tuscarawas rivers southward to the crossing place above Fort Laurens and then turned westward. There, Heckewelder showed Putnam the exact location where the Indian trails from Pittsburgh to the Scioto and to Detroit reached the Tuscarawas at the famous crossing place. The Surveyor general marked a tree and placed "written directions, under a cover of bark, to this tree, for the surveyor who was running this line."[14]

As already mentioned, Heckewelder was at the Moravian lands along the Tuscarawas in 1796 to help supervise the surveying of the three 4,000-square-acre tracts which Congress had set aside in 1789 for the Moravians to hold in trust for the Christian Indians. It was here in 1772, Heckewelder himself explained, about 30 miles above the mouth of "One Legged creek," which he mistakenly marked as "Three Legs" on his map, that the site of an old Indian town, Gakalamukpeking, was selected for the founding of Schoenbrunn. The other two Moravian towns shown on his map were Petquotting on the Huron River and Pilgerruh, or Pilgrim's Rest, on the Cuyahoga River. Following the destruction of Gnadenhutten in March 1782, the Moravians lived for a time on the Sandusky River and later on the Huron River near Detroit. Determined to return to their beloved homes in the Tuscarawas Valley, they left Detroit in 1786 and, after a perilous journey, were forced to establish temporary quarters at an Ottawa Indian town on a high bank on the east side of the Cuyahoga River near the mouth of Tinker's Creek. They at once began to construct a village and to prepare the ground for cultivation. It was here, in June 1797, almost 10 years later, that Amzi Atwater, a member of Cleaveland's surveying party, was directed to build several houses in which the rest of the surveying team was to stay while working in that region of the Western Reserve. "Not far above the creek [Tinker's]," Atwater recorded, "we found the remnants of some old huts, partly overgrown with thorn and plum trees, one or more fragments of doors were fastened with nails, which to me was a curiosity to see in such a place."[15] From Pilgerruh, the

Moravians, along with Heckewelder, moved to Petquotting at the mouth of the Huron River in May 1787. Again they constructed shelters for themselves and called their village New Salem. Here they remained until 1790 when they again were forced to move, this time to the Thames River in Canada.

No doubt Moses Cleaveland and the other proprietors of the Connecticut Land Company studied the maps of North America made by Lewis Evans and Thomas Hutchins, along with Abraham Bradley's "Map of the Northern Parts of the United States of America," which appeared in Jedidiah Morse's *The American Universal Geography,* in order to determine the exact location and extent of their holdings in the Connecticut Western Reserve. The Connecticut Land Company had purchased 3,000,000 acres, or the larger part of the Connecticut Western Reserve, a tract of land that Connecticut kept when it surrendered its western lands to Congress in 1786, from the Connecticut General Assembly on September 2, 1795. Before this land could be sold it had to be subdivided, and between July 1796 and mid-September 1797, small parties of surveyors and chainmen tramped through this wilderness and laid out the land east of the Cuyahoga River into townships 5 miles square. As a result of their survey work, Seth Pease, an astronomer and surveyor who played a prominent role in the division of the Reserve, drew "A Map of the Connecticut Western Reserve" (Plate VI), which was engraved and printed by Amos Doolittle of New Haven, Connecticut, in 1798. Doolittle printed 500 copies of this map at a cost to Pease of $80.[16] This manuscript map is on display in the Western Reserve Historical Society's museum.

The Pease map of 1798, printed in two separate sections, illustrated the story of the political and economic problems faced by the Connecticut General Assembly, the diplomatic arrangements with the Indians in this region negotiated by the Connecticut Land Company, and the burdens and hardships suffered by the surveyors themselves as they subdivided this large tract of land. It indicated 3 separate sections of land within the larger Connecticut Western Reserve: the area between the Cuyahoga River and the western border of Pennsylvania, the land west of the Cuyahoga River to the eastern border of the Fire Lands, and the Fire Lands. Each has its own colorful history of land subdivision with its accompanying heroes, and the story of the surveying of the entire Western Reserve is part of the story of the mapping of Ohio.

The right half of the map represented that portion of the Western Reserve east of the Cuyahoga River for which Moses Cleaveland, who represented the Connecticut Land Company and was responsible for the division of its lands,[17] had first to negotiate with a delegation of Iroquois in Buffalo, New York, in order to satisfy their claims before moving his own surveying party westward. For that land, he signed a treaty on June 23, 1796, and gave to the Indians 500 pounds of money and goods, 2 beef cattle and 100 gallons of whiskey. This part was printed as a separate sheet and could be considered as an individual map since it dealt solely with the Company's first surveys even though the left half of the work was printed at the same time with the purpose that the two would make one map. It took Cleaveland's surveyors 2 years (excluding the winter of 1796-97, when all returned East except for a handful of hardy souls who braved the severe winters in that region) to survey the area into townships 5 miles square. The southern boundary or base line of the Reserve was to be the forty-first parallel extending westward from the Pennsylvania border 120 miles. On July 7 of the first year, 4 survey parties were organized to survey this east-west line and on July 23, Augustus Porter, the principal surveyor during the first year, located and marked the southwest corner with a chestnut post. Once this was completed, range lines that would be 5 miles apart were run north from this base line. The weather, mosquitoes, swamps and hunger delayed their work, and when the surveying stopped on October 17, 1797, only 4 range lines had been run.

After a brief but necessary stay in the East, Cleaveland and his party again returned to the Reserve to complete their surveys. Again, small survey parties were sent out, and their work was resumed on May 29, 1797, when Major Amos Spafford, one of the surveyors, began to survey "towns[ships] 10 and 11, in the 1st, 2d and 3d Ranges."[18] The hardships suffered by the surveyors were not lessened by the intervening winter. Tired and weary, the surveyors finished their work and left for home by November 1797.

The left or western half of the map represented both the area known as the Fire Lands, and those lands west of the Cuyahoga that belonged to the Connecticut Land Company but which were not surveyed during 1796–97 because of Cleaveland's arrangements with the Iroquois. This portion was printed at the same time as the other half but did not have the rule along its eastern edge. As Pease indicated on his map, this section of land was "Unsurveyed Land and Subject to Indian Claims." According to the Treaty of Greenville which was negotiated in 1795 following General Anthony Wayne's defeat of the Ohio Indians at the Battle of Fallen Timbers, the Indians had surrendered their claims to the land east of the Cuyahoga River, but they still retained control of the area west. Also, Cleaveland, in a conference with the representatives of the Six Nations, had promised not to venture across that river. He kept his word with the exception of the surveys and observations made by Augustus Porter who went as far as Sandusky Bay in order to survey the south shore of Lake Erie. It was not until the Treaty of Fort Industry, negotiated with the western Indians on July 4, 1805, that the lands between the United States Military Tract and the Western Reserve and that area between the Cuyahoga River westward to the meridian 120 miles west of the Pennsylvania line, or just about the western end of the Fire Lands, were ceded to the Government and opened for survey and settlement. The remainder of the Connecticut Reserve, which appeared only in outline form on the Pease map, was not surveyed until 1806, when Pease extended the southern boundary of the Reserve and Abraham Tappan ran the township lines for the Connecticut Land Company between the Cuyahoga River and the Fire Lands.

The area known as the Fire Lands, or the Sufferers' Lands, was the last to be subdivided in the Connecticut Western Reserve. It, too, as the western portion of the Connecticut Land Company's holdings west of the Cuyahoga River, appeared in outline form on Pease's map of 1798. This land, 500,000 acres in all, was set aside on May 10, 1792, following a protracted period of deliberations and considerations on the part of the Connecticut General Assembly to satisfy certain claims for damages first brought before that body on December 8, 1780. During the American Revolution, the British Navy selected certain Connecticut coastal towns such as Danbury, New Haven, Fairfield, Norwalk, and Groton for bombardment and destruction as punishment to their inhabitants for their participation in the fight for independence against England. The inhabitants of these coastal towns who did suffer heavily at the hands of the British petitioned the Connecticut legislative body for relief and assistance. Despite the fact that this area was set aside for their use in 1792, it was not until 1807, almost 30 years later, that the land in the western section of the Western Reserve was subdivided and made available to them.

In 1806, following the Treaty of Fort Industry, Maxfield Ludlow, an engineer for the United States government, began a survey of the Fire Lands in order to establish the exact location of its southwest corner so that a line could be run due north and thus determine the general outline of the reserve. Miscalculations were made, however, not by Ludlow, but by the surveyors working east of the Cuyahoga, and the results of his surveys were rejected. Undaunted, Connecticut hired Almon Ruggles, a surveyor who had a reputation of being skilled in woodcraft, in August 1807 to begin again the survey of the Fire Lands and directed him to divide it into townships 5 miles square and to subdivide each township into quarters. Again the wilderness took its toll on the surveyors, and Ruggles, while he was surveying in a large swamp near New Haven in Huron County, wrote in his field notebook, "We are in danger of our lives. 118th mile west. Sat a post in Hell. I've travelled the woods for seven years, but never saw so hideous a place as this." Once the division was made, however, those who claimed land there had by this time incorporated themselves into a company to best represent their interest. The task of the division of the land among the claimants began, and soon white settlements sprang up over the Fire Lands.

In 1824 Joseph Howard of Delaware County produced the first of at least four editions of his "Map of the Districts of Delaware in Ohio" (Plate IX). The District of Delaware, with its land office in Delaware, was created by Congress in order to provide a convenient location for land sales to those people who wished to settle there. Today this district includes the counties of Marion, Hardin, Crawford, Wyandot, Hancock, Seneca, Sandusky, Ottawa, Union, Wood, and parts of Lucas. This particular map, the "Fourth Edition" of Howard's work, is in the map collection at the Ohio Historical Society's library in Columbus. A copy of the "Third Edition" is at the Beegly Library at Ohio Wesleyan University in Delaware, and another is in the massive map collection at the Library of Congress. Howard's map illustrated the historic division and allocation of part of those lands in northwestern Ohio known as the Congressional Lands. All the territory north of the Greenville Treaty Line was acquired by the government through negotiations with the various Indian tribes living in that corner of the state. The numerous Indian reservations that were once in Ohio were located there and the many acres of land donated to Ohio for the construction of its canal system also lay within this section. Of significance is the fact that in this portion of the state the last of the experimental land survey systems was tried and accepted by the federal government. The pattern for land division used here was the one used throughout the remainder of the public domain.

The first parcels of land acquired by the United States in the Delaware District were through cessions made by the Indians in the Treaty of Greenville in 1795.[19] One such tract, the Twelve-Square-Mile Survey, located "at the British fort on the Miami of the lake at the foot of the rapids" was centered about Perrysburg on the Maumee River. In 1805 it was surveyed by Elias Glover[20] into four 6-square-mile townships and can be identified easily because it does not fit into the pattern of the surrounding surveys. It also extended beyond the ruled edge at the upper part of the map. By an act of Congress in 1816, Joseph Wampler, a surveyor,

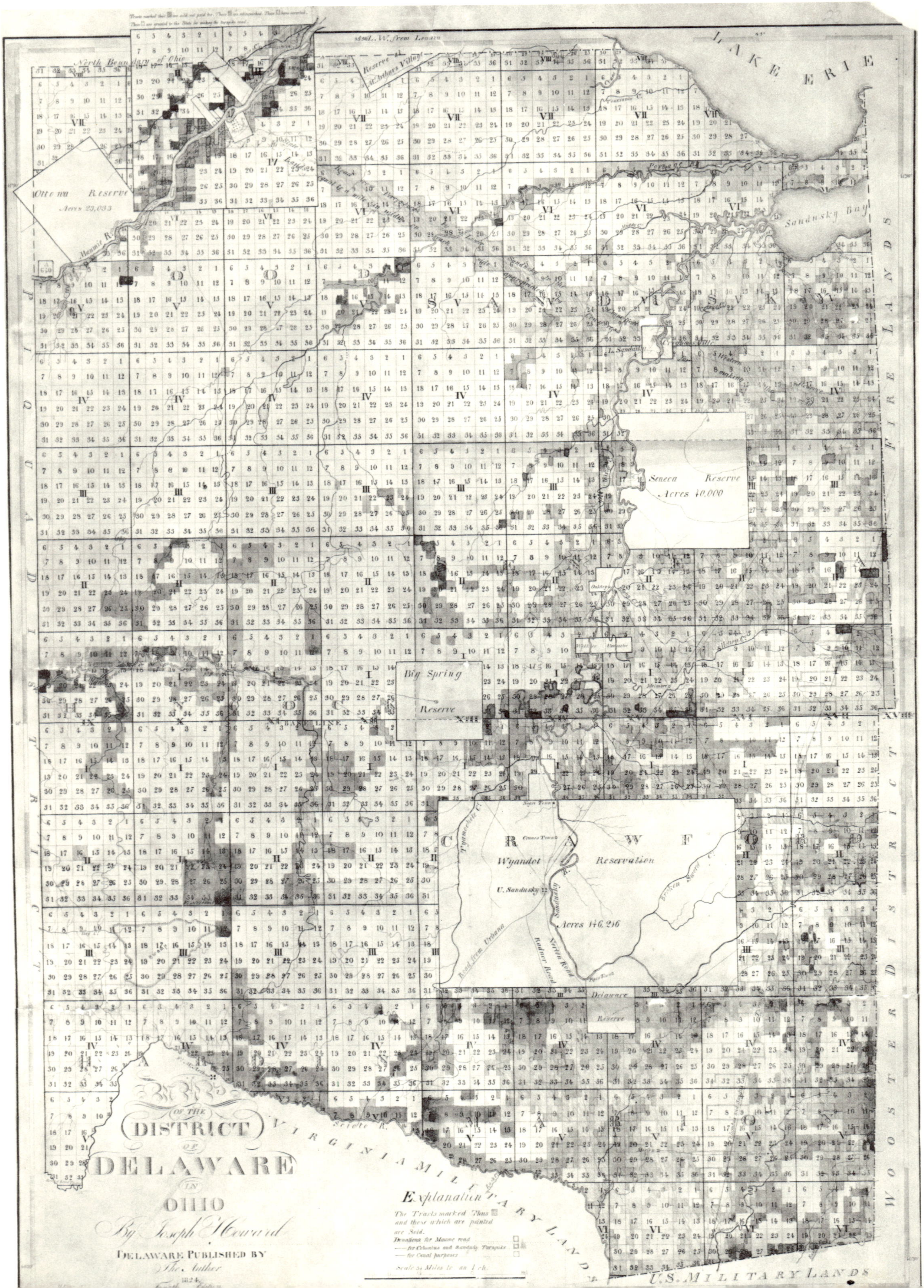
LAKE ERIE
Sandusky Bay
Seneca Reserve
Acres 40,000
Big Spring
Reserve
Wyandot Reservation
U. Sandusky
Acres 146,216
Delaware Reserve
FIRE LANDS
WOOSTER DISTRICT
VIRGINIA MILITARY LANDS
U.S. MILITARY LANDS
MAP OF THE DISTRICT OF DELAWARE IN OHIO
By Joseph Howard
DELAWARE PUBLISHED BY
The Author
1824
Explanation

divided the sections fronting the Maumee in this tract into ninety-three 160-acre plats which were known as the River Tracts. In April that same year, Congress directed that a town, not exceeding 2 sections, should be laid out along the Maumee, and thus Perrysburg was founded. Another tract within this district obtained in 1795 was the Two-Square-Mile Survey, located "at the lower rapids of Sandusky river." This tract centered around Fort Stephenson, which was built during the War of 1812 and heroically defended by the young Kentucky major, George Croghan. Here, also, in 1816 Congress directed that a town should be laid off, and soon the outlines of Croghanville appeared. This village, along with Perrysburg, Upper Sandusky and Gnadenhutten were the only communities in Ohio to have been laid out by the federal government. Later Croghanville was known as Lower Sandusky, and finally its name was changed to Fremont.

Another parcel of land in northwestern Ohio was ceded to the United States in the Treaty of Brownsville [Michigan] in November 1808 for the purpose of building a road through the Indian lands from the Twelve-Square-Mile Survey to the Two-Mile-Square Survey and then on to the western border of the Connecticut Western Reserve. This road terminated at the present site of Bellevue and was indicated by Howard on his map as "State Road from Perrysburg by Lower Sandusky to the Connecticut Western Reserve."[21] The mile wide strip on either side of the road was surveyed into long thin parcels of land running north and south the entire length of each section, but that division was not included in this map.

The land shown north of the Maumee River on this map was ceded to the United States in the Treaty of Detroit, which was signed with the Ottawas, Chippewas, Wyandots and Pottawatamies on November 7, 1807.[22] Within this cession the Indians reserved 12 tracts, 3 of which were in Ohio. The first was "six miles square, on the Miami of Lake Erie [Maumee River], above Roche de Boeuf, to include the village where Fondagame [or the Dog] now lives," and was shown along the upper left portion of Howard's map. The second was "three miles square on the said river [Maumee], (above the twelve-mile square ceded to the United States by the treaty of Greenville) including what is called Presqu' Isle." The third tract was at the mouth of the Maumee River which was not illustrated on this map.

In the famous "Treaty of the Foot of the Rapids of the Miami of Lake Erie," which was negotiated between the Wyandots, Senecas, Delawares, Shawanee, Pottawatamies, Ottawas and Chippewas and the United States in September 1817, all the Indian land that remained in Ohio was ceded to Congress with the exception of a tract of land between the Wabash and St. Mary's rivers and 9 tribal reservations and 14 individual grants.[23] Of these, all or parts of 4 reservations fell within the Delaware District, as did 8 of the individual grants. The largest of the tribal reservations was the 12-mile-square reserve kept by the Wyandots "at Upper Sandusky, the center of which shall be the place where Fort Ferree[24] stands." In a later supplementary treaty, the Treaty of St. Mary's, signed September 17, 1818,[25] this reservation was enlarged by two tracts of 55,680 acres. The larger was attached to the eastern line of the Wyandot Reserve and was shown as all one tract on Howard's map. The second was that square tract of land immediately north and adjacent to the northern boundary of the reservation. The Wyandots retained this land until 1842 when they relinquished most of their lands to the United States and moved westward. The second tribal reservation included on this map was the Seneca Reserve, which

(Opposite page) Plate IX: Howard, "District of Delaware," (1824), Ohio Historical Society, Columbus. This map illustrates the orderly sale of lands controlled by the federal government through local land offices, as well as land set aside for particular use.

contained 30,000 acres. This reserve also benefited from the supplementary treaty of St. Mary's in 1818, when an additional 10,000 acres was acquired.

A third tribal reservation was granted by the government "to Zee-shaw-au, or James Armstrong, and to San-non-do-you-ray-quaw, or Silas Armstrong, chiefs of the Delaware Indians," for the use of the members of their tribe, and was located adjacent to the southern boundary of the larger Wyandot reserve. The fourth and last reserve within the Delaware District was that portion of a 34-square-mile grant made to the Ottawas "on the south side of the Miami river of Lake Erie, and to include Tushquegan or McCarty's village." Part of this reservation was shown in the middle of the upper border of Howard's map, and he mistakenly labeled this tract the "Reserve at McArthur's Village."

The first individual grant of 1,280 acres, located on the west side of the Sandusky River below Croghanville, was given to Elizabeth Whitaker. She had been taken prisoner by the Wyandots and apparently lived most of her life with them. This grant was in what was then Seneca County and occupied most of sections 15, 22, 14, 23 of township V, Range XV, north and east of the first principal meridian. The second tract of 640 acres was granted to Robert Armstrong, who was also taken prisoner by the Wyandots and had married a Wyandot woman. The third grant of 640 acres was given to the children of William McCollock, who was killed at Maguaga, an Indian village south of Detroit, in August 1812. McCollock's children were one-quarter Wyandot. These last two tracts were adjacent but appeared as one on Howard's map and were located on the west bank of the Sandusky River just south of the large Seneca Reservation and opposite the site of Tiffin. Immediately north of the Whitaker tract was the Sarah Williams grant. This grant of 160 acres was given to the wife and children of Isaac Williams who was a half-blood Wyandot. It was located in the northern part of Sandusky County. To John Vanmeter, who was taken prisoner by the Wyandots and had married a Seneca woman, 1,000 acres were granted along Honey Creek, a tributary of the Sandusky. Adjacent to this tract, grants of 640 acres each were made to Catharine Walker and her son, John R. Walker, who was wounded in the battle of Maguaga in 1812. Both these grants are shown together and appropriately labeled on this map. Adjacent to the smaller Wyandot tract is the 640-acre grant given to Horonu, or the Cherokee Boy, a Wyandot chief. Within the smaller Ottawa Reservation, which resulted from the Treaty of Detroit in 1807, was a 640-acre tract granted to Peter Manor, the adopted son of the Ottawa chief Tontoq-sa-ny, or the Dog. Manor, known to the Indians as Saw-en-de-bans, or the Yellow Hair, was a French trader who lived among the Indians for years and was given this tract at the request of the Indians. He developed this well-timbered tract at the head of the rapids of the Maumee River and there laid off the town of Providence, which prospered as a mill town and enjoyed some growth during the canal period. Between 1850 and 1852, however, a cholera epidemic and a disastrous fire destroyed the town and it was never rebuilt.

As mentioned before, the Treaty of St. Mary's (September 17, 1818) was supplemental to the one at the foot of the rapids of the Maumee River. In addition to the acres added to the Wyandot Reservation at Upper Sandusky and the Seneca Reservation, 16,000 acres were set aside for the Wyandots "in a square form, on the head of Blanchard's fork, the center of which shall be at the Big Spring, on the trace leading from Upper Sandusky to Fort Findlay." Although the St. Mary's treaty negated many of the individual grants made in the treaty of 1817, many of them were subsequently granted. In a letter sent from Lewis Cass and Duncan McArthur, the two representatives of the United States who had negotiated the 1817 treaty, on September 30, 1817, to George Graham, Acting Secretary of War, both the commissioners argued that the individual tracts should be granted. "We have been compelled to admit the claims of a number of individuals, and to stipulate that patents shall be granted to them," they pleaded. "Almost all these persons are Indians by blood. Some have been taken prisoners in early life, have married Indian women, and have identified themselves in feeling, manners, and interests with the Indians. In the few instances only are they merely Indians by adoption. In every case, it was the urgent wish of the Indians that land should be granted to these persons. To have refused these requests would have embodied against us an interest, and created obstacles, which no efforts of ours would have defeated or surmounted."[26] Despite the negation of some of the individual grants, most were recognized as valid and appeared on Howard's map.

Most of the general surveying in this portion of the state was done between 1819 and 1821 under the direction of Edward Tiffin, Surveyor General of the United States. The border between Indiana and Ohio was surveyed in 1817 and that line became known as the First Principal Meridian for all the land surveyed in the rest of the United States. Between this line and the western boundary of the Fire Lands in the Connecticut Western Reserve, the land was divided into townships that were numbered from west to east. The forty-first parallel, which passed south of Carrothers in Seneca County and Ottawa in Putnam County and shown on Howard's map, was the base line from which tiers of townships were surveyed and numbered both north and south.

The survey procedures and patterns used here were based on an elaborate set of instructions set down for the deputy surveyors by Tiffin in 1815.[27] He cautioned the surveyors to check their chains constantly while in the field and explained that each surveyor's chain should be standardized with the one in the Office of the Surveyor General. He encouraged each surveyor to use a Rittenhouse compass, which had nonius divisions and movable sights. It had been developed by David Rittenhouse of Philadelphia, who had a reputation as an astronomer and mathematiciain during the latter part of the eighteenth century. Any violation or "departure from these instructions," Tiffin explained, would mean a "forfeiture of the conditions of contract," and that "loose inaccurate or precipitate work will not be admitted, either as it respects surveys in the field, or their return on paper." Tiffin's attention to detail and concern for accuracy paid dividends because the surveys in this area were very uniform and accurate. Charles E. Sherman explained that "the survey of this region was the last of the Government's experiments with systems of division in Ohio and the general plan of survey there adopted has since been used over the vast expanse of our public domain west to the Pacific Ocean."[25]

Thus, within Ohio are found examples of the various experimental survey systems employed by the United States to divide its public domain. The survey maps described in this chapter illustrate only part of these systems. The surveyors themselves played an important role in the division of land, and their selection of particular sites for settlement laid the foundations for the future growth of the state. Their maps form an important part of the historic material that illustrates the westward movement during the first half of the nineteenth century, and depicts Ohio's place in the settlement of the West.

1. For an excellent account of the surveying of the Seven Ranges, see William D. Pattison, *Beginning of the American Rectangular Land Survey System* (Columbus, Ohio: Ohio Historical Society 1970), pp. 105–68. For a description of this land, see Thomas Hutchins, "A brief account of the Soil and Timber in that part of the Western Territory through which and [*sic*] East and West Line has been Surveyed agreeable to an Ordinance of Congress of the 20th of May 1785." Papers of the Continental Congress, LX, 229–36.

2. For a history of Manasseh Cutler's map and pamphlet, see Philip Lee Phillips, *The First Map and Description of Ohio, 1787, by Manasseh Cutler: A Bibliographical Account* (Washington, D.C., 1918). This pamphlet is also reprinted in Appendix C. William Cutler and Julia Perkins Cutler, *Life, Journal and Correspondence of Rev. Manasseh Cutler, LL.D.*, II (Cincinnati, 1888), 393–406; also reprinted in Hélène Fouré-Selter, *Gallipolis Ohio, Histoire de l'éstablissement de cinq cents Francais dans la vallée de l' Ohio à la fin du XVIII*[e] *siècle* (Paris, 1939), p. 135.

3. Phillips, *The First Map,* p. 26.

4. *Ibid.,* pp. 27–28.

5. *Ibid.,* p. 36.

6. This pamphlet was printed in Paris in 1789. A copy of it is in the library of the Cincinnati Historical Society and a reprint appeared in Fouré-Selter, *Gallipolis Ohio,* pp. 151–79.

7. See Josephine E. Phillips, "The Naming of Marietta," "*The Ohio State Archaeological and Historical Quarterly,* LV (January-March 1946), 127, for a brief discussion of this map and its relationship to the naming of Marietta. Also see Henry J. Yeager, "Nouvelles du Scioto: The Story of a Fraud," *Ohio History,* LXXVIII (Autumn 1969), 261–72.

8. Letter from Albert Gallatin to Jared Mansfield, April 16, 1807, *Plats, U.S. Military Tract, 100 Acre Lots,* Vol. 10, Auditor of the State of Ohio, Columbus, Ohio.

9. For a brief discussion of this survey system, see Norman J. W. Thrower, *Original Survey and Land Subdivision* (Chicago: Rand McNally, 1966), pp. 16–18.

10. This letter was first published as John Heckewelder, "Map and Description of Northeastern Ohio," *Tract No. 64* of the Western Reserve Historical Society (November 1884); it also appeared as Charles C. Baldwin, "Map and Description of Northeastern Ohio," *Magazine of Western History,* I (December 1884), 109–14; and recently in Mildred Walmsley, *The Heckewelder Map: 1796* (Cleveland, 1968).

11. Mr. Meredith B. Colket, Jr., Director of The Western Reserve Historical Society, kindly supplied the author this information based upon his own research and that of Mary Lou Conlin. Letter, Meredith B. Colket, Jr., to author, November 21, 1970, possession of the author.

12. John E. Potter, "The Pennsylvania and Virginia Boundary Controversy," *Pennsylvania Magazine of History and Biography,* XXXVIII (1914), 419; Richard G. Wood, "Research Materials in the National Archives Pertaining to Pennsylvania," *Pennsylvania Magazine of History and Biography,* LXIX (1945), 98.

13. Robert H. Foster, "Depreciation," *Pennsylvania Archives,* 3rd Series III (1894), 759–71.

14. "Progress of the Society of United Brethren in Propagating the Gospel Among the Indians," No. 189, *American State Papers, Indian Affairs,* Vol. II, 1834, p. 389; John W. White, "Notes of Travel of William Henry, John Heckewelder, John Rothrock and Christian Clewell, to Gnadenhutten on the Muskingum, in the Summer of 1797," *The Pennsylvania Magazine of History and Biography,* X (1882), 125–57.

15. Charles Whittlesey, *Early History of Cleveland, Ohio* (Cleveland, 1867), p. 297.

16. For a detailed discussion of this map and its subsequent revision, see Russell H. Anderson, "The Pease Map of the Connecticut Western Reserve," *The Ohio States Archaeological and Historical Quarterly*, LXIII (July 1954), 270–78.

17. See Document XI, "The Mode of Partition of the Western Reserve, April, 1796," in *Tract No. 96* of the Western Reserve Historical Society (1916), 164–68. See also "Instructions of the Directors to their Agent," *Tract No. 96* of the Western Reserve Historical Society (1916), 186–87.

18. "Pease's Journal," quoted in Whittlesey, *Early History of Cleveland, Ohio,* p. 281. Pease's journal of the surveys also appeared in "Seth Pease's Journals," Part II, *Tract No. 94,* The Western Reserve Historical Society (1914), pp. 68–124.

19. *American State Papers, Indian Affairs,* Vol. I (Washington, D.C., 1832), 562–63.

20. *Commemorative Historical and Biographical Record of Wood County,* (Chicago: J. H. Beers & Co., 1897), p. 192.

21. *American State Papers, Indian Affairs,* I, 757.

22. *Ibid.,* I, 747.

23. *American State Papers, Indian Affairs,* II, 131–35.

24. Fort Ferree was built during the War of 1812 to help keep the line of communications open between northern and southern Ohio.

25. *American State Papers, Indian Affairs,* II, 166.

26. *Ibid.,* p. 139.

27. Charles E. Sherman, *Original Ohio Land Subdivision,* Vol. III of *Final Report, Ohio Cooperative Topographic Survey* (Columbus: Press of the Ohio State Reformatory, 1925), 197–201.

28. *Ibid.,* p. 131

CHAPTER FIVE

Mapping the State

Immediately following the close of the American Revolution, interest in the region north and west of the Ohio River heightened even before much was known about the geography of this vast territory. What knowledge there was had been obtained from traders and trappers who had passed through the region in pursuit of business with the Indians, and from a small handful of individuals who had entered the region on specific governmental assignments. The major maps of this region were constructed by John Mitchell, Lewis Evans, and Thomas Pownall, but all had gleaned their information from the reports of others, while Thomas Hutchins's map was based in great part on personal observations and visitation. It was probably one of these four maps that Congress used while discussing both the land ordinance of 1785 and the governmental ordinance of 1787. Even though Congress was concerned with the increasing number of squatters who had crossed the Ohio River and settled in the eastern parts of Ohio by 1783, that same year it passed an ordinance that required the Indians living in the region to retire westward behind a line drawn from the Maumee to the Miami, thus anticipating actual settlement in the Northwest. Certainly after 1785 and the beginning of the surveying of the Seven Ranges, maps of the Ohio country were necessary. As the demands from individuals and land speculators for lands increased, so did the need for accurate maps. Thus, in looking at the settlement patterns in Ohio and the political subdividing of the state, a study of its maps illustrates the growth of the seventeenth state during the nineteenth century.

At first, the quest for knowledge of the geography of the regions west of the mountains was closely connected with the development of American cartography, national literature and a growing national consciousness. One of the earliest maps of the region north of the Ohio River after the new nation was formed was Amos Doolittle's "A Map of the Northern and Middle States: Comprehending the Western Territory and the British Dominions in North America. Compiled from the best Authorities" (Plate I) which was "Delineated & Engraved" especially for the first edition of Jedidiah Morse's *The American Geography or, a View of the Present Situation of the United States of America.* This famous geography was published by Shepard Kollock in Elizabethtown, New Jersey in 1789 and clearly illustrated the author's pride in the new nation. With a sense of patriotism, Morse explained that his geography was to replace the one produced by William Guthrie, an Englishman who had produced a very popular textbook entitled *A New Geographical, Historical, and Commercial Grammar,*[1] and that it was not necessary for Americans to rely on a foreigner to inform them of their own country.

Doolittle's map, which included Ohio, was an interesting representation of the Northwest Territory at that time. It was probably made in 1788 and undoubtedly was the first map of the northern part of the country to show the several land subdivisions within the state. Part of the Seven

Plate I: Doolittle, "Northern and Middle States," (1788), in Morse, *The American Geography* (1789). Doolittle's map was probably the first to delineate land subdivisions in Ohio and showed Marietta as Adelphia.

Ranges had already been surveyed and the United States Military Tract had been set aside as well as the Virginia Military District. Land grants had been given by Congress to both the Ohio Company and to John Cleves Symmes. The only settlement represented in Ohio was Adelphia, the first community in the Ohio Company's lands. Its name was changed on August 14, 1788, to Marietta[2] and it is thus possible that Doolittle made his map from the plans of the Ohio Company before actual settlement. Morse, however, stated in his text that the settlement there consisted "of about 200 souls and is almost daily increasing."[3] He further explained that there was also a settlement in the Symmes lands, which, although not mentioned, was probably Columbia, and stated that "though [it was] very small at present, [it] is in prospect of a rapid enlargement." Much of Morse's material about the Northwest was taken from the reports produced by Thomas Hutchins.[4]

Morse's geography went through at least four editions, each one enlarged and improved from the last. In 1796, Abraham Bradley's "Map of the Northern Parts of the United States of America" (Plate II) replaced Doolittle's work and was published in volume two of the so-called third edition of Morse's greatly enlarged work, "*The American Universal Geography, or, A View of the Present State of All the Empires, Kingdoms, States, and Republics In the Known World and of the United States of America in Particular*," which was a continuation of the earlier geography and was published in Boston by Isaiah Thomas and Ebenezer T. Andrews. This geography was expanded to two volumes "at great pains and expense" to the author and included some 28 maps. Bradley's map also appeared in the later fourth edition of Morse's geography in 1802. Morse was certainly enthusiastic about the West and encouraged anyone in the East who was dissatisfied to journey west because there, he explained, the prospects for a bright future were great. The text concerning Ohio was a good capsule description of the state about the time of its birth. He estimated that there were approximately 65,000 Indians living in the entire Northwest Territory but did not

Plate II: Bradley, "Northern Parts of the United States," (1796), Morse, *The American Universal Geography* (1796). Accompanying a popular geographic text of the United States, Bradley's map was the first to illustrate county subdivisions.

specify how many were in Ohio; however, to show that the eastern regions, including Ohio, were not barbarous, Morse explained that "Chilicothe [*sic*] had been laid out like Philadelphia and that in a two-year period over 200 homes had been constructed." He assured his readers that a road had been cut through the wilderness from Wheeling to Limestone [Maysville] in Kentucky and that ferries had been established along its path at the Muskingum, the Hockhocking and the Scioto rivers. He also pointed out that the sale of western land would "greatly aid in extinguishing the national debt." A study of Morse's early geographies easily shows that in an effort to inform Americans about their own country, he summoned a sense of patriotism, added to the country's literary tradition through his descriptive narrative, and provided some of the earliest cartography of the Northwest Territory as well as Ohio.

Bradley's map of the Northwest Territory was important to the story of the mapping of Ohio because it was one of the earliest works that showed the political subdivisions for the region as specifically outlined in the Ordinance of 1787, and probably was the first to illustrate the county subdivisions within the proposed states. His map showed 6 counties within the territory, and only 3, Washington, Hamilton and Adams, in Ohio.

Very little is known about Bradley other than that he was born on February 22, 1767, in the Wyoming Valley in Pennsylvania, became a lawyer and judge, moved to Philadelphia, and later was appointed Assistant Postmaster General in 1800. He was removed from that position in 1829 for political reasons.

Without doubt, Bradley's best cartographic work was his "Map of the United States exhibiting the Post-Roads, the situations, connexions and distances of the Post-Offices, Stage Roads, Counties. Ports of Entry and Delivery for Foreign Vessels, and the Principal Rivers" (Plate III) which he drew in 1804. This map soon became the base or mother map of the United States, from which most others of the country were copied by many cartographers, and replaced in importance the ones made earlier by John Mitchell, Lewis Evans and Thomas Hutchins. The Ohio portion of his large map established some significant patterns that future map makers followed. Bradley showed the large land grants made by Congress along with some of the township surveys, and also included the names, but not the outlines, of the counties. The map was not accurate for the year it was drawn because he excluded Fairfield, Clermont, Franklin, and Scioto counties, which were formed between 1800 and 1803, while at the same time, he included Muskingum County, which was established in 1804. The location of the various roads within the state and the mileage between towns, of course, was important to immigrants wanting to go west. When compared with the works of contemporary cartographers, Bradley's map of Ohio was quite accurate, and his effort was a fine illustration of the state at the beginning of its statehood.

In 1805 there appeared *A New and Elegant General Atlas* by Aaron Arrowsmith and Samuel Lewis, published by the same Boston firm that had produced Morse's geography, and which was to accompany that famous work.[5] This particular atlas, however, had been first published the year before by J. Conrad and Company of Philadelphia as the companion volume to John Pinkerton's *Modern Geography.*[6] Lewis, whose map "Ohio" (Plate IV) appeared in both atlases, was a well-known map maker and engraver and had made the illustration for John Marshall's historic biography of George Washington. The Conrad atlas was without title and is listed by the Map Division of the Library of Congress as *Maps to Accompany*

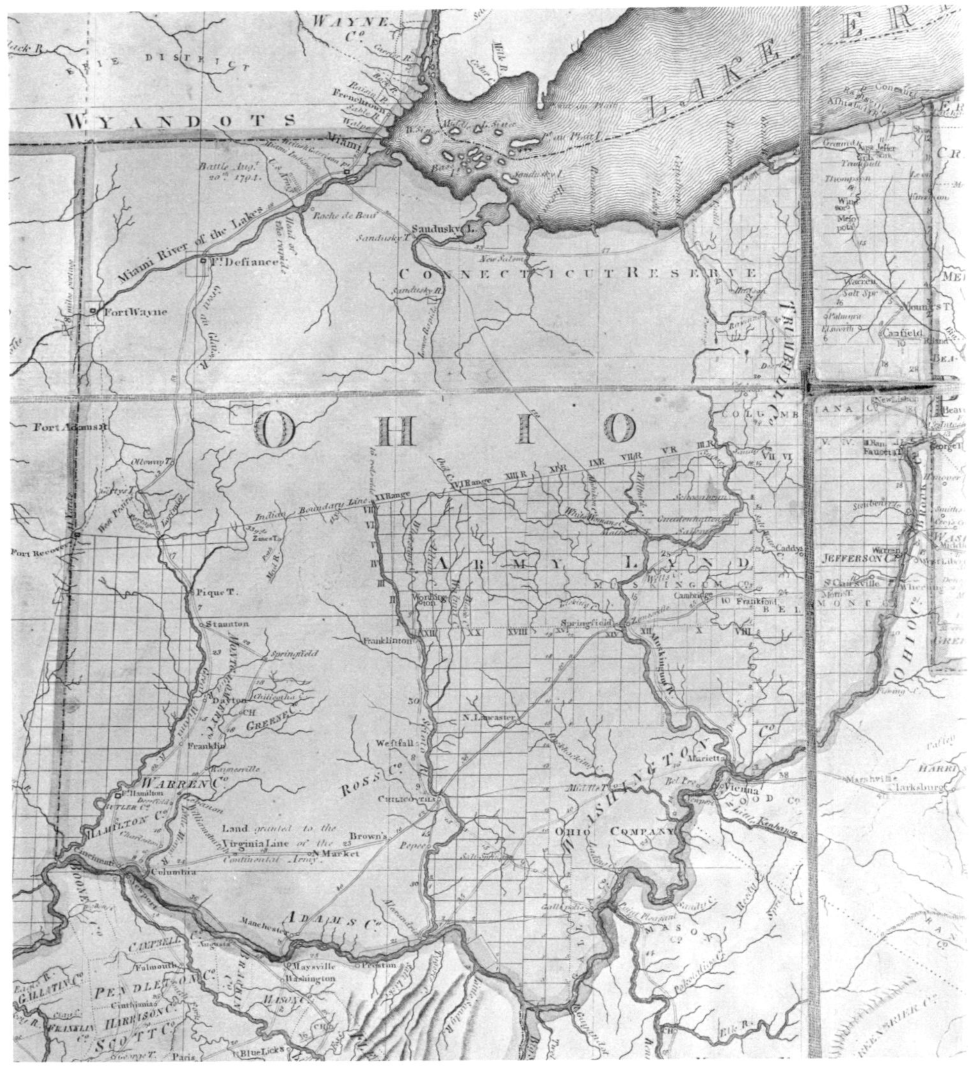

Plate III: Bradley, "Map of the United States," (1804), Cincinnati Historical Society, Cincinnati. While not entirely up to date about information concerning Ohio, Bradley's map became the base map of the United States for many years.

Pinkerton's Modern Geography (Philadelphia: J. Conrad and Company, 1804). The map itself does not show any county subdivisions but rather illustrated the larger land divisions made by Congress while the outline of lands held by the Ohio Company and the Connecticut Land Company were entirely omitted. No doubt Lewis was impressed with the military activity in the West because he located the important battlefields and posts in Ohio during the Indian Wars between 1790 and 1795. Even the site of the murder of Captain Alexander Trueman by the Indians in 1792 while he was on a peace mission was commemorated by the map maker. Also, Lewis included the historic Greenville Treaty Line. Because this particular map was published along with collections of other maps in atlases in both Philadelphia and Boston to accompany two separate popular geographies, while, at the same time, it also appeared in various editions of atlases prepared by Arrowsmith and Lewis between 1804 and 1819, many people during the early part of the nineteenth century received their first impressions of Ohio through this particular work.

Another map drawn by Lewis of interest in the story of the mapping of the state was "A Map of Part of the N: W: Territory of the United States compiled from Actual Surveys, and the best Information" (Plate V), which was made in 1796 but did not appear until 1802 when it was published by Mathew Carey in *Carey's American Atlas.* The map was basically an outline of the territory northwest of the Ohio River. It did contain the various land

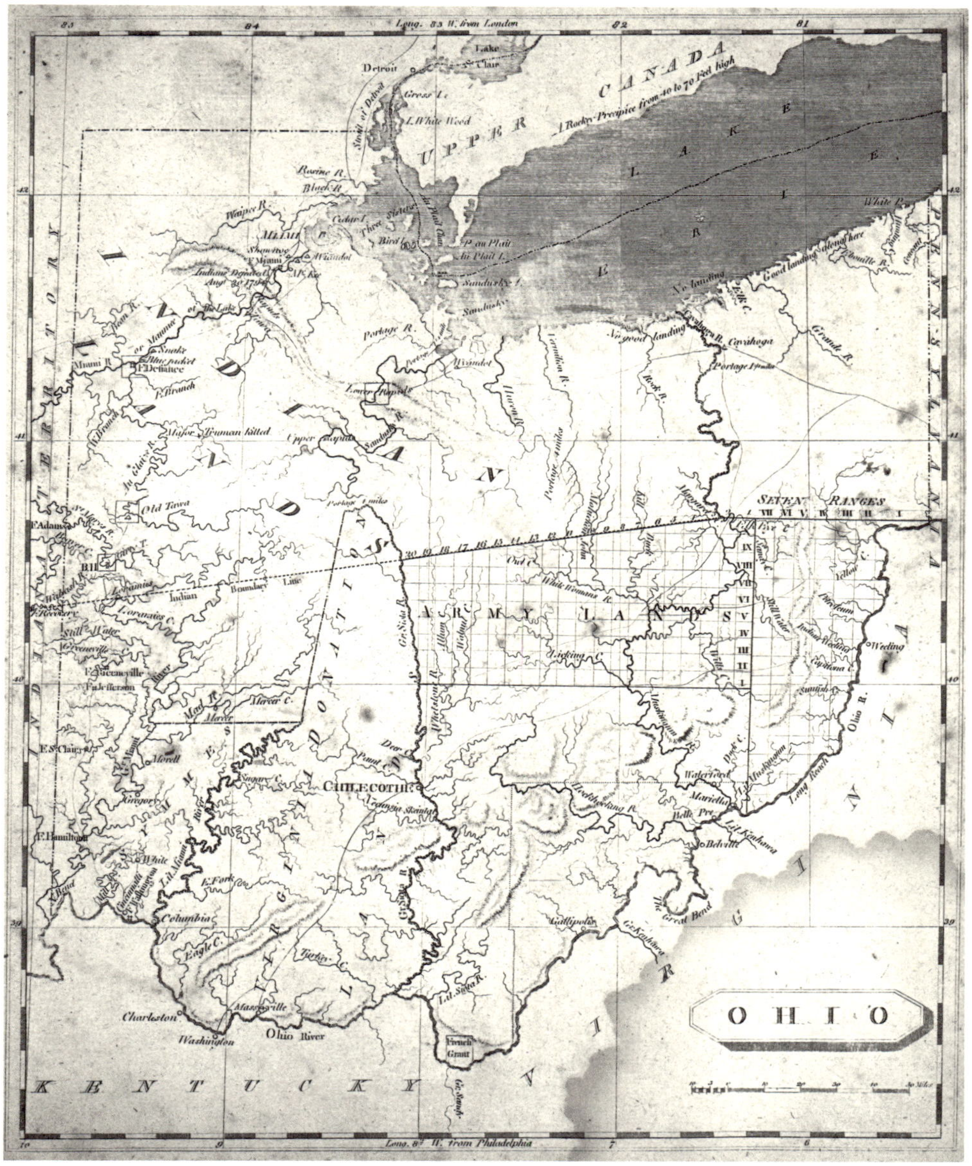

Plate IV: Lewis, "Ohio," (1805), Arrowsmith, *A New and Elegant General Atlas* (1805). Lewis's map of Ohio accompanied two very popular atlases published in the early nineteenth century.

subdivisions within Ohio. Of importance, the map showed the Greenville Treaty Line of 1795 and the numerous land reserves obtained from the Indians by the United States in that treaty. When compared with Bradley's map of the region, which was compiled about the same time, Lewis's work lacked both significant geographical information that apparently was common knowledge, and the cartographic artistry possessed by Bradley.

Mathew Carey, one of the most famous early book publishers in the United States, probably produced the first geographical atlas in this country at Philadelphia in 1795, to accompany *Guthrie's Geographical Grammar.* From about 1794 until the early 1830's, numerous atlases and individual maps came from Carey's press. One popular item was *Carey's American Pocket Atlas with a Concise Description of Each State,* which was first produced in 1796 for Carey by Land and Ustrick of Philadelphia. It was a pocket-sized, leather-bound volume that contained various maps of the states along with brief descriptions of each one. These small atlases served as travel guides to immigrants who sought new homes in the West and also were sources of information for easterners who wanted to learn more about the regions west of the mountains. Included in this atlas was a rather roughly sketched "Ohio and N.W. Territory," which was engraved by W. Barker and probably taken from Lewis's work. In the 1805 edition,

Plate V: Lewis, "A Map of Part of the N.W. Territory," (1796), Carey, *Carey's American Atlas* (1802). While lacking much specific information, Lewis's map was one of the earliest by the famous Mathew Carey.

Barker's Ohio map (Plate VI) was improved over the previous ones and it showed some of the land divisions and, more importantly, indicated the available roads within Ohio and the Indiana Territory over which immigrants could easily pass.

One of the most curious maps in the cartographic delineation of Ohio was Samuel Lewis's map "The State of Ohio with part of Upper Canada, &c." (Plate VII), which appeared in *Carey's General Atlas, Improved and Enlarged: Being a Collection of Maps of the World and Quarters, Their Principal Empires, Kingdoms &c.*, published in Philadelphia in 1814. The first of a series of Carey's atlases, under the general title *Carey's General Atlas,* was produced in 1804 to accompany a revised edition of Guthrie's geographical text, which was also printed by the Carey publishing house.[7] Lewis's map was similar to the one which appeared in the atlas that went with Morse's geography, but instead of showing the large land grants, the cartographer outlined 23 counties in Ohio. What made the map curious

Plate VI: Barker, "Ohio and N.W. Territory," (1805), Carey, *Carey's American Pocket Atlas* (1805). Probably copied from Lewis's map of Ohio, Barker's map depicted Ohio shortly after its entry into the Union.

was its inaccuracy. For instance, Gallia County was shown as Gallipolis and Columbiana as Columbia. Aside from these rather minor points, there appeared in the southwestern portion of the state, above Hamilton and Clermont Counties, Deerfield County, a political subdivision in the state that never existed. Lewis's Deerfield County occupied almost the same area as does the present Warren County, which was organized in May 1803. Perhaps Lewis knew of a county being created in that general location and assumed that it would be named Deerfield since Deerfield was one of the most important settlements on the Little Miami River north of Columbia, and Deerfield Township included most of the territory now contained in Warren County.[8] Lewis paid considerable attention to the exact location of the military activity that occurred in western Ohio during both the Indian Wars between 1790 and 1794 and the War of 1812. Apparently Carey had received some criticism that his maps had neglected to mention such a significant military area, and he doubtless instructed Lewis to detail the important points. Even with this caution, however, the map maker positioned Fort Adams and Fort Recovery too far west and outside the state, while he relied upon his ear rather than official reports

Plate VII: Lewis, "State of Ohio," (1814), Carey, *Carey's General Atlas* **(1814). While Lewis's map in this atlas** was inaccurate, it no doubt answered criticisms that had been received by Mathew Carey.

and spelled the site of Perry's victory in Lake Erie "Pudding Bay." Nevertheless, convinced that he had satisfied all demands and silenced his critics, Carey, in the introductory remarks to the volume, stated:

> One further remark will, I trust, be pardoned. A battle, a siege, a blockade, or some other remarkable event, gives a temporary celebrity to some insignificant place, never before known or thought of at the distance of ten or twenty miles from its local position. The guidances have recourse to their maps to discover its bearing and distance, and so forth. A *non est inventus* takes place. The enquirers are disappointed—they rail at the map, and abase the publisher.
>
> This case actually occurred with respect to Put-in-Bay, into which Commodore Perry retired after the battle of Lake Erie. It was not to be found in the maps of the Seat of War; which were of course condemned. Well may we exclaim with Sterne—"Excellent critics! Admirable connoisseurs!"

This same map, simply entitled "Ohio," however, with Gallia and Columbiana counties spelled correctly, but Deerfield County still shown, appeared in Fielding Lucas, Jr.'s *A New and Elegant General Atlas, Comprised of Maps of the United States,* which was published in Baltimore in 1816. By the 1817 edition of *Carey's General Atlas,* "The State of Ohio with part of

Upper Canada, &c.," was corrected and improved to show forty-six counties in the state, which reflected Ohio's growth during its first thirteen years of statehood and especially following the conclusion of the War of 1812.

The first map drawn of the state of Ohio from actual surveys was made in 1807 by John F. Mansfield and entitled "Map of the State of Ohio from the returns in the Office of the Surveyor General" (Plate VIII). Mansfield, who was described as a man with "distinguished scientific attainments and high promises," died quite young while commanding a company of light infantry volunteers during the War of 1812.[9] On his map he outlined 22 counties and indicated the 5 land sales districts as established by the Harrison Land Act of 1800, in addition to the Zanesville District, which was added in 1802. Mansfield's work was the first map of the state to show the actual township subdivisions; and in the Western Reserve, he not only showed the township subdivisions, but in Geauga and Trumbull counties, he tried to name the townships even though in some instances, the names are inaccurate. The copyright to this map was purchased probably in 1814 by B. Hough and Alexander Bourne of Chillicothe. These two surveyors revised, enlarged and reprinted the map in 1815. The famous Hough and Bourne map, based on Mansfield's map, was the first map of Ohio to show all the actual surveys within the inhabited part of the state. Because of its great size, no reproduction was attempted for this study.

During the three decades between 1810 and 1840, Ohio experienced rapid growth in population and agricultural wealth. Following the War of 1812 people began to settle in those regions in northwestern Ohio that had formerly been reserved for the Indians; and by 1820, the entire state was organized into some form of county subdivision. During this same period, canals were constructed and the wealth of the interior of the state was opened to the country's market places while, at the same time, products found their way to the growing communities in Ohio over the useful canal waterways. Because of this growth and promise, people in the East became more interested in Ohio, and during these three decades, some of the best popular descriptive accounts and the most accurate maps of the state were produced.

One of the most interesting descriptions of Ohio during the early part of the nineteenth century was written by John Melish, an Englishman of talent and ambition. After several attempts to establish a commerce house in the West Indies, Georgia and New York, Melish settled in Philadelphia in 1811. He was interested in geography and its relationship to the economic development of the United States. Before his first visit to this country in 1798, he read all the available travel journals in order to familiarize himself with the new land. Once he reached America, however, he found to his dismay that most of the accounts were filled with "effusions of ignorance and spleen," and was therefore determined to travel the entire country over to learn for himself of its promise. During his 2,400-mile tour of the United States, Melish kept extensive notes and published them in 1812 as *Travels in the United States of America, in the Years, 1806 & 1807, and 1809, 1810 & 1811,* which possibly could have been the first comprehensive descriptive geography of America. Melish explained that he had "spared no labour, nor expence, to have a good set of maps to illustrate this work. They have been drawn with great care from the best materials to which I could get access, aided by much local information; and

Map of the State of Ohio
Taken from the returns in the Office of the Surveyor General
By John F. Mansfield
TERRITORY
EXPLANATION
LAKE ERIE
UPPER
CONNECTICUT WESTERN RESERVATION
FIRE COMPANYS LANDS
CANTON DISTRICT
ZANESVILLE DISTRICT
STEUBENVILLE DISTRICT
MARIETTA DISTRICT
CHILLICOTHE DISTRICT
VIRGINIA MILITARY LANDS
PART OF CINCINNATI DISTRICT
INDIANA TERRITORY
PENNSYLVANIA
VIRGINIA
KENTUCKY
SANDUSKY BAY
Imaginary Northern Boundary
Division Line between the United States and Canada
Indian Boundary Line
Champaign
Miami County
Montgomery County
Butler County
Hamilton County
Warren County
Clermont County
Highland County
Adams County
Scioto County
Ross
Franklin County
Fairfield
Muskingum
Belmont County
Jefferson
Columbiana County
Trumbull County
Geauga County
Washington
Athens County
Gallia County
Greene County
OHIO RIVER

the engraving had been executed by the first artists of Philadelphia." His efforts produced a map entitled "Ohio" (Plate IX) which appeared in the first edition of his *Travels.* There is little question that Melish was influenced by Mansfield's earlier work of Ohio because he adopted the basic format used by the young map maker to illustrate his map, and also accepted Mansfield's interpretation of Lake Erie, which made the western end much too round. Nevertheless, the clear and simple features on this map of Ohio were representative of Melish's talents as a cartographer and observer. Melish's *Travels* glowed with enthusiasm about the United States. He reflected a confidence in the economic opportunities available in the new country, as well as what the author believed was the growth of democracy "unalloyed by the feudal system." Certain that the bond between the United States and England could never be dissolved even by war, he observed that the economic restrictions placed upon the American mercantile classes prior to the War of 1812 would make the United States an industrial nation and thus independent of Europe. While in Ohio, he recorded that the economic sanctions imposed by the national administration had had effects as far west as Marietta, the commercial center of Ohio which, he said, was "a dull place, though inhabited by a gay, lively people, most natives of Massachusetts" who were all "sober, industrious, intelligent, and discreet." The sparsely settled regions of the state up the Muskingum Valley attracted Melish. He was particularly impressed with the small village of Zanesville, which he visited for several days. "I found a large thriving town, with a great number of handsome brick houses," he reported. Because of its location on both the navigable Muskingum River

(Opposite page) Plate VIII: Mansfield, "Map of Ohio," (1807), Ohio Historical Society, Columbus (1807). This is the first map of Ohio drawn from actual surveys of the state.

Plate IX: Melish, "Ohio," in Melish, *Travels* (1812). His map was included with his descriptive travel narrative of North America which glowed with enthusiasm about the future of Ohio.

and the Zane's Trace, he said "this must certainly become a fine situation for manufacturers." The area surrounding Zanesville pleased Melish also, and he described it as "beautifully diversified," while the "whole is abundantly supplied with freestone, limestone, iron-ore, and inexhaustible beds of coal." He detailed the cost of labor, clothing, food and whiskey and suggested that a handsome profit awaited anyone willing to speculate in land in that area.

There were two other editions of Melish's *Travels*; the second published in 1815 and the third in 1818. The map of Ohio in the 1815 edition was the same as that of 1811 with the addition of information concerning the northwestern portions of the state. In the 1818 edition, his map of Ohio (Plate X) was vastly improved and considerable information added. For instance, even though Melish continued to show the various land sales districts, he also included 55 county subdivisions, as well as illustrating some townships. The author was interested in the land distribution system in America and exclaimed:

> Every industrious citizen of the United States has the power to become a freeholder, on paying the small sum of eighty dollars, being the first instalment on the purchase of a quarter of a section of land; and though he should not have a shilling in the world, he can easily clear as much from the land, as will pay the remaining instalments before they come due. . . . A farmer with an industrious family may become the proprietor of a whole section or more; and the land being *purely his own,* there is no setting limits to his prosperity. No proud tyrant can lord it over him.—He has no rent to pay—no game laws—nor timber laws—nor fishing laws to dread. He has no taxes to pay, except his *equal share* for the support of the civil government of the country, which is but a trifle. He can neither be gauged nor supervised.—He has no poor rates to pay; for he dwells in a country where government interposes not its greedy hand, to snatch the cup of industry from the lips of the

Plate X: Melish, "Ohio," in Melish, *Travels,* (1818). Melish's second effort to illustrate Ohio was more detailed than his first and showed county development shortly following the War of 1812.

> feeble.—He has no tithes to pay for there are no hireling priests in the community of which he is a member.
> Such are the blessings enjoyed by the American farmer. It might look invidious to compare it with the same class of society elsewhere.—It is sufficient to say, that here this class,—one of the most valuable in every community,—are prosperous and happy—and from a view of the whole subject, we are inclined to believe, that this is secured for ages to come. May the Almighty Father of the human race, pour down his choicest blessings upon the heads of those who planned, and carried into effect, such a benevolent system.[10]

To be sure, Melish's descriptive narrative, along with his excellent maps, reflected a spirit that encouraged those unsure of their own futures in the East to settle in the West where all had the opportunity to succeed.

This type of popular narrative geography was used to lure easterners, and Europeans newly arrived from the old world, to the frontiers of Ohio and to the Middle West. Aside from the more carefully prepared texts written by Melish and others, there were many shorter immigrant guides, hastily written to direct immigrants westward, either by land or by river. These works also were produced to secure quick fortunes for their authors. One such volume was Andrew Miller's *New States and Territories, or the Ohio, Indiana, Illinois, Michigan, North-Western, Missouri, Louisiana, Mississippi and Alabama in their Real Characters, in 1818* which was published in 1819 "for the benefit of emigrants, and others, intending to visit the western country." This particular guide described Ohio as being "220,000,000 miles square [*sic*], containing 25,500,000 acres, 400 thousand inhabitants, 3 thousand Indians, 48 counties, 230 villages, 33 printing-offices, 27 banks, 6 furnaces, 8 forges, 5 glass works, 8 woollen mills, 12 cotton mills, 8 paper mills, and three nail factories, 3 colleges, styled universities, 9 academies, 8 public libraries, an almost infinite number of stores, grist, merchants, and sawmills, with schools in all parts, and churches for all denominations."[11] The narrative continued to discuss the topography of the land, the conditions of the soils, the types of timber, the climate, and the various diseases encountered, including "fevers and agues" which prevailed near swamps and marshes. After a brief discussion of several counties, Miller stated that Cincinnati had streets almost 4 rods wide and contained "about 1300 buildings, mostly of wood and brick and elegantly painted." Addressing himself to the social conditions of the state, he explained that "in the country of Miami there are no slaves, and very few free blacks.—Slavery in every shape is prohibited throughout the state; and our laws interdict the residence of free blacks, unless under very special circumstances. There is not any prospect that the constitution of this state will be altered so as to allow slavery."[12] The guide included a foldout map of Ohio (Plate XI) with explicit instructions for the use of the information it contained. This, no doubt, is one of the most unusual maps of the state ever printed.

There was no question that people were interested in Ohio during the decades immediately following the close of the War of 1812. Ohio's population increased from 230,760 in 1810 to 581,295 in 1820, to rank it fifth in the nation in population, and enough people moved into the state during the next decade to raise it to fourth. Although the immigrant guides were useful, they were too general. People wanted more specific information about Ohio, and, as the state's population and wealth increased and its rough frontier society changed to one of respected permanency, a need was created for a specific fact book. In May 1816, therefore, John Kilbourn

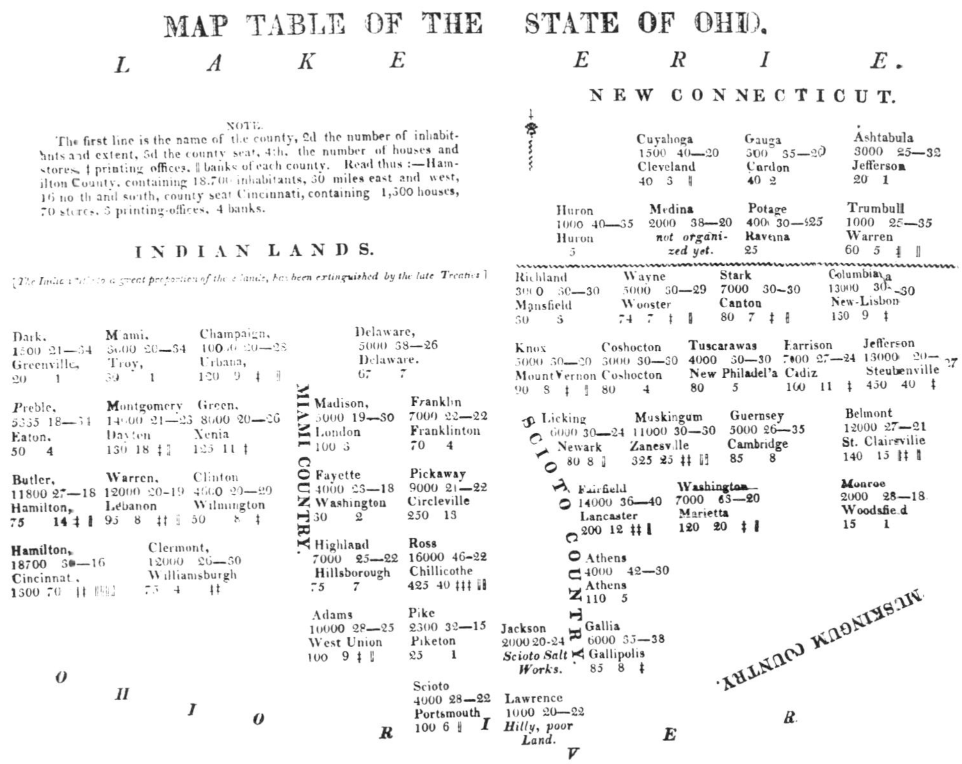

Plate XI: Miller, "State of Ohio," (1818), in Miller, *New States and Territories* (1818). This map accompanied an emigrant guide and in a unique fashion illustrated Ohio's development after fifteen years of statehood.

of Columbus published the first edition of *The Ohio Gazetteer or Topographical Dictionary* to meet this demand. Because of "a rapid and extensive sale of the first edition . . . with in a few months," Kilbourn produced a second edition of his *Gazetteer* the same year. He explained that such a publication was necessary in order to satisfy the "eyes of the Eastern members of our Columbian Republic" and, at the same time, to inform the citizens of the state about their own environment. No other publication of this kind about Ohio had been published before. It was gratefully received by the people, its popularity reflecting a widespread curiosity about the state as well as indicating the development of a sense of pride in it.

Kilbourn was born in Connecticut in 1787, graduated from Vermont University and came to Ohio to join his uncle, Colonel James Kilbourn, in Worthington, a town the Colonel had laid out in 1804. In Worthington, John Kilbourn became the principal of the Worthington Academy. Apparently he taught geography along with other subjects because in 1813 he wrote *Compendious System of Universal Geography Designed for Schools and Youth in General,* which was printed by Putnam & Israel of Zanesville. Kilbourn explained that his geography was the result of 6 years of labor, but it had been published several years before in pamphlet form for the use of his young students. It was compiled, he stated, to correct certain errors he had found in Morse's popular geography. For example, the 1810 edition of Morse stated that Sandusky was the principal settlement in the northwestern section of Ohio, and the venerable geographer had placed Chillicothe on the Great Miami River. Two years later, in 1815, Kilbourn wrote *Columbia Geography, or A Description of the United States of America To Which is Prefixed A Concise Introduction to the Science of Geography in General,* which was published in Chillicothe. This work was descriptive, as most of the geographies were at that time, and apparently was quite popular. Included was a chapter on Ohio, which presented a good description of the state at that time. With an interest in education and 2 fairly good geography textbooks already on the market, Kilbourn left his teaching post in Worthington and opened a bookstore on High Street in Columbus, just "a few doors south from the state house." It was here in 1816 he produced the first edition of his *Ohio Gazetteer.*[13]

The *Ohio Gazetteers* were filled with facts concerning the state's antiquities, canals, topography, colleges, government, literature, public lands, roads, newspapers, banks and many other topics. Today they are valuable sources of information about the state's past between 1816, when the first edition appeared, and 1841, when the last edition was published. The information was gleaned from many sources, including questionnaires sent to officials of each county seat, about which Kilbourn complained that he had never received complete cooperation; the author's personal observations; Daniel Drake's *Picture of Cincinnati*, which was produced in 1815; Return Jonathan Meigs, who was Postmaster General between 1814 and 1823; state legislators such as Alfred Kelley of Cuyahoga County and Robert Lucas from Pike and Scioto counties; and interested individuals who volunteered information about their own locales. The first several editions of the *Gazetteer* were rough volumes indeed, but, as its popularity increased, so did the quantity of information and the quality of the printing.

The first map to appear in an *Ohio Gazetteer* was the "Plat of the town of Columbus" (See Chapter VI, Plate IV) which was published in the 1817 edition. The first map of Ohio to appear in Kilbourn's work was included in the 1818 edition and it was drawn by the author himself (Plate XII). The foldout map was a rough outline of the state that showed 54 counties and the major rivers. After this edition, each *Gazetteer* contained a map of Ohio that increased in quality every year. The 1821 edition of the map, "A Map of Ohio" (Plate XIII), showed the many internal improvements in the state, the growth in county organization, especially in northwestern Ohio, and the improvement in Kilbourn's cartography. His work indicated some discipline in drawing and attention to detail. The map illustrated 71 counties, the main roads and the distances between the county seats. The last edition of the *Ohio Gazetteer* to appear under Kilbourn's editorship before his death was in 1831. That year he dedicated the edition to General Duncan McArthur and explained in the preface that in its first 9 printings, the *Gazetteer* had sold approximately 14,000 copies.

After Kilbourn's death, the *Gazetteer* was published by C. Scott and I. Whiting of Columbus. Even though it carried Kilbourn's name on the title page in the 1833 edition, it was explained that it was "a continuation of the work originally compiled by the late John Kilbourn" but that it had been "enlarged by a citizen of Columbus." The next editions, until the final one in 1841, were compiled by Warren Jenkins, who printed *The Ohio Register and Anti-Masonic Review* between 1830 and 1833, and published by Scott and Whiting. The maps that appeared with these volumes were made by William Woodruff of Cincinnati and entitled "The Travellers Guide Through Ohio with its Canals, Roads & Distances" (Plate XIV) and were superior in craftsmanship to those made by Kilbourn. This particular map accompanied the 1836 edition of the *Ohio Gazetteer.* It indicated an increase in population in the eastern part of the state, especially in the Western Reserve, with the addition of counties in that region. Woodruff included the population figures in the upper part of each county. Jenkins changed the name of the Kilbourn publication to *The Ohio Gazetteer and Traveler's Guide*; and, although it retained much of the data found in the earlier editions, its additional travel information made it serve almost as a road atlas for the state. Thus, between 1816 and 1841, the *Ohio Gazetter* chronicled the growth and development of the state during the first half of the

Plate XII: Kilbourn, "Ohio," (1818), in Kilbourn, *Ohio Gazetteer* (1818). Crude but technically well-done, Kilbourne's wood engraving of Ohio used metal type-set to illustrate the counties.

Plate XIII: Kilbourn, "Ohio," (1821), Kilbourn, *Ohio Gazetteer,* (1821). Kilbourne's map in this edition showed better cartographic skill as well as more information about the state.

Plate XIV: Woodruff, "Travellers Guide Through Ohio," (1836), in Jenkins, *Ohio Gazetteer* (1836). Maps found in the *Ohio Gazetteer* served as road maps to thousands of migrants who came to Ohio between 1814 and 1841.

nineteenth century; and the maps contained in these volumes made by two Ohio cartographers accurately illustrated that growth.

The maps of the state that appeared in the earlier *Gazetteers,* however, were not Kilbourn's only cartographic effort. In 1820 he and Alexander Bourne combined their talents and produced a map entitled "Map of the State of Ohio Drawn by A. Bourne Including the Indian Reservations Purchased and laid out into Counties and Townships in 1820 Drawn by J. Kilbourne" (Plate XV) which appeared as a foldout map in Caleb Atwater's *Description of the Antiquities Discovered in the State of Ohio and other Western States,* in Volume I of the American Antiquarian Society's *Archaeologia Americana. Transactions and Collections* in 1820. Alexander Bourne, who was already well known for his cartographic work, which he and B. Hough produced in 1815, was a civil engineer with some reputation in the state. He was born in Wareham, Massachusetts, in 1786, and in 1810 floated down the Ohio River and probably landed in Marietta. He settled first in Zanesville and later moved to Chillicothe, and still later, to Columbus. In 1842 he moved back to his place of birth and died there six years later. Aside from his map work, he was appointed by Duncan McArthur as a member of the Ohio Canal Commission and helped to survey the canals within the state. On April 27, 1816, Congress directed that a town should be established at the site of the fort on the Maumee River and that same year, Bourne helped to survey Perrysburg.[14] He also published several articles in Silli-

Plate XV: Bourne and Kilbourn, "Map of the State of Ohio," (1820), in Atwater, *Description of the Antiquities* (1820). The effort of these two cartographers appeared as a fold out map in Atwater's book of the prehistoric sites in Ohio published by the American Antiquarian Society.

man's *American Journal of Science* and in 1834 wrote *The Surveyor's Pocket-book,* which was published in Chillicothe. The map drawn by Kilbourn and Bourne showed seventy counties, the major roads and the most important settlements in each county. Of importance were the clear topographical features, including the Black Swamp in northwestern Ohio and the "Unbroken Sandstone Region" in the southeast, illustrated by Bourne. This particular map is also important in the story of the mapping of Ohio because it appeared in Atwater's classic description of the western antiquities, which was the first systematic study of the prehistoric Indians in North America, and because it was the result of the combined efforts of two of Ohio's leading early nineteenth-century map makers.

Of course, not all the maps made of Ohio were included in geographies, gazetteers or travel guides. Some remained in unpublished manuscript form and were used for purposes not entirely apparent. Several interesting manuscript maps of the state were drawn by William Creese Pelham of Zanesville, Ohio. The Pelham family moved to Ohio from Williamsburg, Virginia, and first settled in the Virginia Military District before going on to Zanesville. The record is not clear about this family, but both the father, William Pelham, and his son, William Creese Pelham, had very active public and private careers, and both were concerned about the elevation and perfectability of society. William Pelham became the editor of the *Ohio Republican,* a Whig newspaper in Zanesville, and on July 1, 1818, was appointed postmaster of that Ohio community. He also operated a publishing firm which printed at least 75 separate titles, and helped to found a private library, the Zanesville Atheneum. Pelham was also an enthusiastic contributor to Kilbourn's *Ohio Gazetteer*. In 1824 he left Zanesville for Robert Dale Owen's experimental community of New Harmony, in Indiana, where he became an accountant and editor of the *New Harmony Gazette*. William Creese Pelham, who had attended Joseph Neef's experimental school in Philadelphia, succeeded his father as postmaster in Zanesville and as editor of the *Ohio Republican*. In 1827, however, he was persuaded by his father to join the New Harmony colony. He died in 1847.

The two maps drawn by William Creese Pelham were accurate, considering that they were done by an apparently amateur cartographer. On the Pelham map "Ohio" of October 1817[15] (Plate XVI), the state was shown

Plate XVI: Pelham, "Ohio," (1817), William L. Clements Library, Ann Arbor, Michigan. Pelham was a newspaperman in Zanesville and a contributor to Kilbourn's *Gazetteer*.

with 48 counties, while the territory northwest of the Greenville Treaty line was subdivided and temporarily attached to already organized counties. This was the method by which new counties were created and governed by the parent county until sufficient population was available to warrant a new county government; the Pelham maps are the only ones to illustrate this county satellite system. During the latter part of 1817 and the first month in 1818, Ohio's General Assembly created 5 new counties. These new counties, Perry, Clark, Brown, Logan and Hocking, were to assume legal status on March 1, 1818. Pelham's map, therefore, was quite accurate insofar as it indicated the general geographic location for these new political subdivisions with upsidedown lettering 5 months before their actual organization. He was not flawless, however, and showed Hocking County as Hawkin, and, apparently not knowing that Clark was the name given to the new county created from Champaign, Madison and Greene, he simply marked it New county. Pelham's map of June 26, 1818[16] (Plate XVII), showed the state with the 5 new counties and included the major transportation routes between principal communities and, in some instances, indicated the distances. There was a third Pelham map, dated November 1817,[17] which was quite similar to that of October 1817. The purpose for which Pelham made his maps is not known; his work, however, was accurate and clear, and the maps are valuable documents for studying early nineteenth-century Ohio.

In 1828 Robert Thomas Anderson of Cadiz made an interesting map of the state that was printed by James Turnbull of St. Clairsville, Ohio. This map, "Ohio"[18] (Plate XVIII), showed the various federal land sales districts through which the public lands in Ohio were sold and the major land tracts set aside by Congress. Anderson outlined 72 counties and included at the top of each its population according to the Census of 1820, while at the bottom, he listed the number of white males over 21 years of age. The cartouche was carefully executed and showed that Anderson was an artist as well as a fairly good cartographer.

It is not certain why this particular map was printed. Single sheet maps were printed by several enterprising publishers to be sold to the public. Perhaps in this instance, the publisher, James Turnbull, had wanted to include it in Humphrey H. Leavitt's *The Ohio Officers' Guide and Clerks' Companion,* which he published in 1832 and again in 1833, or in *The Ohio Officers' and Justices' Guide,* by the same author, which Turnbull printed in 1835 and in 1843. The map, however, was not included in any of these volumes.

Turnbull went to Steubenville from Philadelphia in 1816 and established himself as a bookbinder and publisher and soon became a prominent citizen in his new home. Since Philadelphia was the center of the publishing industry in the United States, it can be assumed that Turnbull learned his trade there. Aside from publishing Leavitt's guides, he also published *The History of Joseph and his Brethren* in 1829. Several other maps of Ohio were also published by Turnbull. One, "A topographical and statistical map of the state of Ohio, drawn and engraved by H. Anderson, Cadiz, 1832," was a large and good representation of the state. Its cartographer, Hugh Anderson, who had engraved the plates for the American edition of the *Edinburgh Encyclopedia,* also produced other state and Ohio county maps.[19] In 1835 Turnbull published a very fine color cartographic work, "Map of the State of Ohio, Indiana, Illinois with the Principal parts of Missouri and Michigan by the latest authority," which was made by Anderson. This map went through several editions, and by 1847 it was

Plate XVII: Pelham, "Ohio," (1818), Cincinnati Historical Society, Cincinnati. Pelham's map clearly illustrates the growth and development of counties in Ohio more so than any other piece of cartography.

reproduced as a Samuel August Mitchell map sold by Thomas and Cowperthwait of Philadelphia.

The same year that Turnbull printed his cartographic work in St. Clairsville, a very important map in the history of the mapping of the state of Ohio in the nineteenth century was produced. This piece, "Topographical Map of the State of Ohio" (Plate XIX), which was engraved by William Woodruff of Cincinnati and printed by Horton Howard of Columbus, became the base map for the state. The original was compiled by Alfred Kelley, legislator, land speculator and one of the central figures on the Ohio Canal Commission. In a letter to Benjamin Tappan, a lawyer

Plate XVIII: Anderson, "Ohio," (1828), Ohio Historical Society, Columbus. Detailed and finely drawn, Anderson's map is an example of early Ohio cartography that was produced at the local level in the state.

from Steubenville and a federal judge who was also a member of the Commission, and to Micajah T. Williams, of Cincinnati and chairman of the famous Canal Commission, Kelley explained that he had made the map from the results of the surveys of the state authorized by the commission between 1822 and 1824 when that body was trying to determine the most suitable canal routes through Ohio.[20] Kelley wanted his work to be accepted by the General Assembly as the state's official map, but no action was taken. Woodruff, who had engraved Kelley's map, used it for his own map of Ohio that was included with the *Ohio Gazetteer* after 1831. The engraver also ran a copper plate printing business and his shop was located on Third Street, between Main and Walnut streets, in the Queen City. He was a very careful craftsman and produced some fine cartography of the state during the 1830s, including plans of Port Clinton, Cincinnati and Portsmouth. Woodruff was known for his excellent portrait engraving and he produced the likenesses of John Adams, Benjamin Franklin, Marquis de La Fayette, George Washington and others during the first quarter of the nineteenth century.[21] By 1837, Woodruff's publishing firm offered for sale some of the best examples of midwestern cartography, including Morrison's "Map of Indiana," Munsell's "Map of Kentucky" and Peck's "Map of Illinois."[22] Horton Howard, the printer, published the *National Enquirer* during the late 1820s. He later was a member of the board of directors for common schools in Columbus in 1831, but died during the cholera epidemic in 1833.

Revealing 73 counties, towns and major roads in the state, the Woodruff map's most interesting features are the comments made by either the surveyors or by Kelley concerning certain topographical characteristics across Ohio. The outline of Lake Erie more accurately resembled its actual configuration and the earlier Mansfield interpretation of the lake was now discarded by most cartographers. The cartouche used for this map was most fitting. It depicted an aqueduct over one of Ohio's rivers along a canal route, with a mule pulling a sluggish canal boat through its waters. This map was the most accurate one drawn of the state to that date and remained one of the most popular maps of Ohio until Charles E. Sherman produced his "Map of Ohio showing Original Land Subdivisions" for the Ohio Cooperative Topographical Survey in 1922.

The period between 1820 and 1850 has been termed the golden age of American cartography and it was during this period that some of the most beautiful maps and elaborate atlases were produced by some of the country's finest map makers.[23] Using Woodruff's map of Ohio as the base map, the famous cartographers of this period, including Anthony Finley, David Burr, Henry Schenck Tanner and Samuel August Mitchell, produced maps of the state almost every year. During this same period, Ohio experienced a rapid population growth, as the number of people in the state increased from 581,295 in 1820 to 1,980,329 in 1850, ranking it third in the nation. With this growth during the 3 decades, members of the Ohio General Assembly created 29 new counties, or one-third the state's total, in order to provide local government for the inhabitants. In 1851, with the establishment of Noble County, the 88-county political subdividing of the state stopped. Also, by the half-century mark, Ohio's agrarian-dominated economy began to give way to a growing industrialization and urbanization; the development of the canals, roads and railroads within the state encouraged this change. The maps produced by these cartographers during these 30 years, therefore, recorded graphically the state's growth and development.

Probably the rarest, as well as the most accurate, map made of Ohio during the nineteenth century was "Ohio," compiled by Charles Whittlesey and engraved by Curtis M. Doolittle and Samuel B. Munson of Cincinnati (Plate XX). This map, printed in 1838, was the result of the partially completed first geological survey made of the state,[24] conducted during the summer of 1837. The act that provided for this survey instructed the head geologist to make mineral surveys, measure the height of mountains and collect examples of rocks, fossils, ores and soils, and at the same time, to "construct and prepare for engraving a complete geological

(Overleaf) Plate XIX: Kelley, "Topographical Map of the State of Ohio," Ohio Historical Society, Columbus. This map, made by surveys conducted by Alfred Kelley between 1822 and 1824, became the base map for the State of Ohio.

State Line as claimed by Ohio
WILLIAMS
HENRY
WOOD
PAULDING
PUTNAM
HANCOCK
VANWERT
CRAWFORD
HARDIN
MERCER
ALLEN
SHELBY
LOGAN
UNION
DELAWARE
DARKE
MIAMI
CHAMPAIGN
CLARK
MADISON
FRANKLIN
PREBLE
MONTGOMERY
GREENE
FAYETTE
PICKAWAY
BUTLER
WARREN
CLINTON
ROSS
HAMILTON
CLERMONT
HIGHLAND
BROWN
ADAMS
PIKE
SCIOTO
LAWRENCE
KENTUCKY
Fort Wayne
Columbus
Dayton
Urbana
Perrysburg
Chillicothe
Portsmouth
Greenville
Winchester
Hillsborough
Xenia
Sandusky Bay
Scale 10 Miles to one Inch

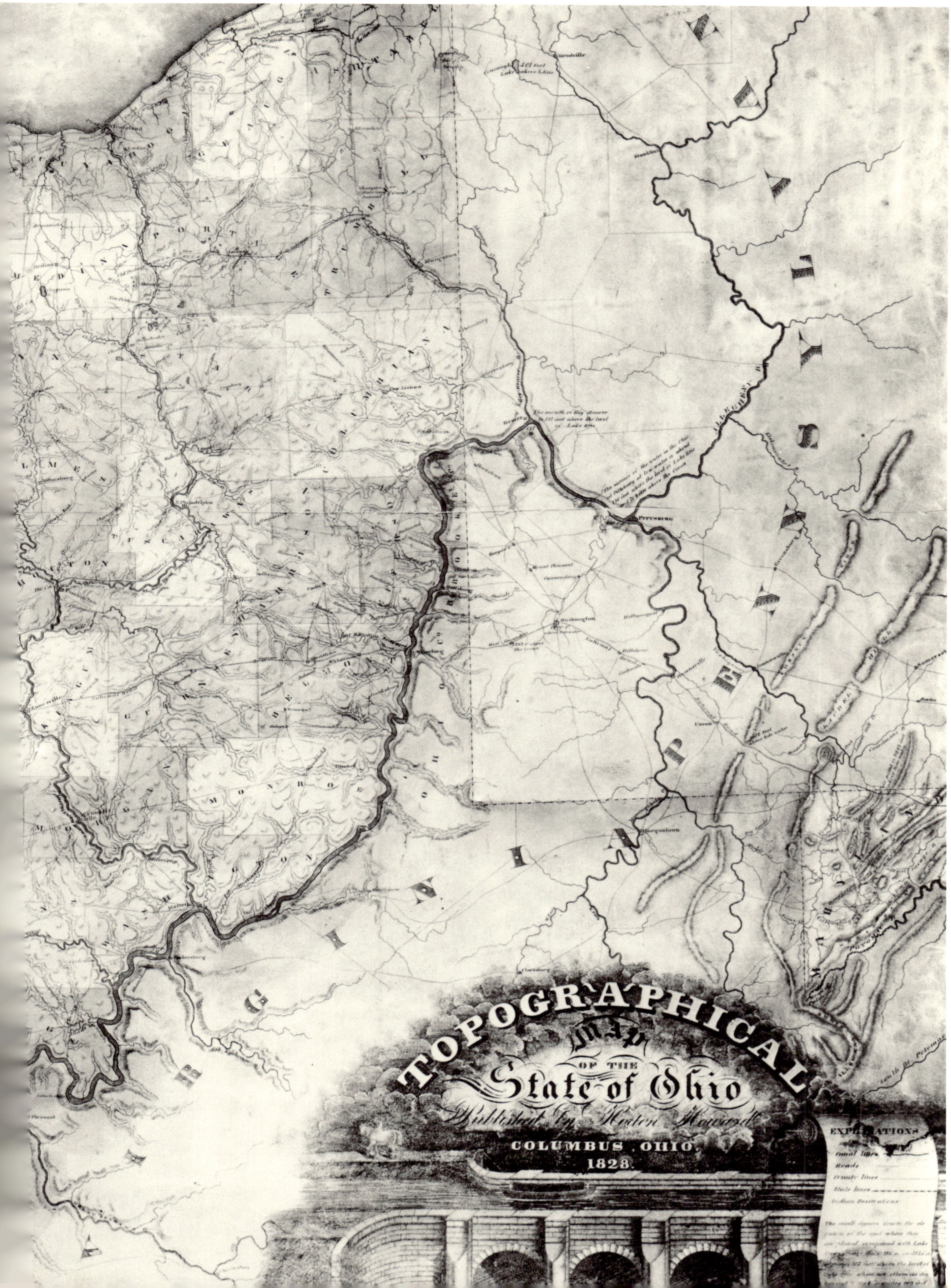
TOPOGRAPHICAL
Map
OF THE
State of Ohio
Published by Horton Howard
COLUMBUS, OHIO.
1828.
PITTSBURG

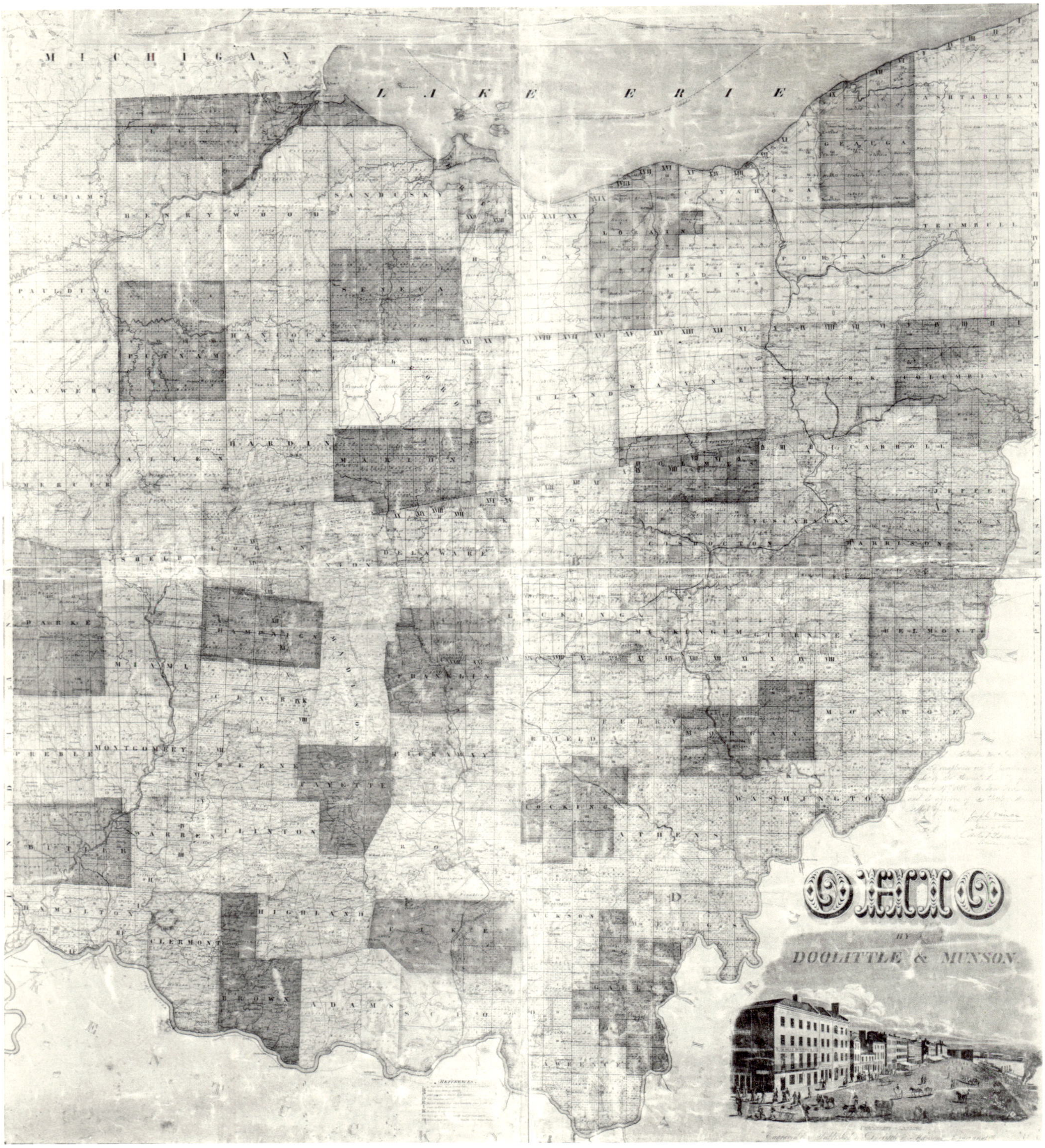

Plate XX: Doolittle and Munson, "Ohio," (1838). Ohio Historical Society, Columbus. Probably the most accurate map of Ohio drawn during the nineteenth century, this cartographic piece was the result of the first geological survey made in 1837.

map of this State." The legislation was the result of several years of activity on the part of many who were interested in learning the full extent of the physical and natural resources of the state,[25] including Governor Robert Lucas, who urged such a survey in his annual message in December 1835. Plans were laid for the actual survey during 1836 and the first part of 1837. Twelve thousand dollars were earmarked for the survey team, which was eventually headed by W. W. Mather, a geologist from New York. Because of the depression of 1837, the legislature retrenched its finances and withheld support from the project before it was completed.[26]

Principal Geologist Mather had 5 assistants, including Samuel P. Hildreth of Marietta, John Locke, who surveyed prehistoric earthworks in western Ohio, and Charles Whittlesey of Cleveland. Whittlesey was in charge of the Topographical Department of the survey, and aside from keeping the records for the project, he was to "collect all accessible information from the Land office and public records, in relation to the boundaries of counties, towns, sources of streams, &c., preparatory to the construction of a topographical map of the State."[27] Mather explained in his report that the map was to contain "the whole contour of the ground, and the physical character of the country, as if it were actually spread before the eye of the observer, showing, at a glance, its plains, and mountains, hills, valleys, streams, bottom lands, swamps, marshes, woods and cultivated fields." He argued that the available maps of Ohio were "erroneous" in many instances and urged that the legislature should provide for a trigonometrical survey of the state at an estimated cost of $100,000.[28]

Whittlesey worked diligently and tirelessly to collect all available data. He provided skeleton maps of each county and township to the other geological assistants who, in turn, were to fill in the important geological information and return it along with the maps to his office. Also, he collected data from the various land offices across the state and from many other useful sources. In his report to Mather, he suggested that the state should take advantage of the material he had amassed and produce maps of each county within the state. He commented that there was a "great deficiency of county maps," and estimated that such an enterprise would not cost the state more than $150 per county. Cost estimates had been obtained from the engraving and printing firm of Doolittle and Munson in Cincinnati. The naked sheets would cost "nine and seven-tenths cents" each, while colored sheets would be "thirteen and seven-tenths cents." A full set of county maps would cost approximately $12.75, and Whittlesey was sure that the counties would willingly pay the state for their own bound copy.[29]

Because of the economic setback during 1837, none of the maps proposed by either Mather or Whittlesey materialized. On January 27, 1838, however, the Ohio legislature provided for the purchase of "one hundred and fifty copies of Doolittle and Munson's Map of the State of Ohio, at the rate of not more than six dollars per copy."[30] The Secretary of State was in full charge of these superb maps. He was to distribute them in a "convenient and safe mode," and to ensure that "one copy [was sent] to the Auditor of each county in this State, for the use of such county; one copy to the Executive of each of the States and Territories in this Union, for the use of such States or territories; one copy to each of the public officers of the State, at Columbus, for the use of such office; and one copy to each of the colleges and universities within this State, for the use of such institutions; and retain the balance, if any, for the future disposition of the General Assembly."

The map itself was engraved by Doolittle and Munson, a firm that specialized in copper-plate printing, whose shop was located in 1836 on the northeast corner of Main and Fifth streets in Cincinnati.[31] Little is known about Curtis M. Doolittle other than that he was an engraver, a partner of Munson and was no longer in Cincinnati by 1850. He apparently was not related to Amos Doolittle, the famous engraver and map maker from New Haven, Connecticut. Samuel B. Munson lived in New Haven, Connecticut, in 1835 and probably moved to Cincinnati the following year.[32] He was a portrait engraver of some note and continued business on his own into the second half of the nineteenth century.

The map failed to illustrate the topographical detail anticipated by Mather. Basically it served to update the 1815 Hough and Bourne map of the state. County subdivisions were shown over the entire state, including those in the northwestern portion, which, of course, the map of 1815 did not show. Townships were detailed except in the Western Reserve and, in most cases, they were identified by name. The original large tracts and land sales in the state were also shown. Roads were illustrated, but the distance between major points was not noted as on other maps of the same period. The state's canal system was outlined and a profile of the Ohio-Erie Canal appeared across the top of the map. An interesting scene of "Cincinnati Landing" decorated the lower right hand corner. As prescribed by the Ohio legislature, the map was endorsed by Governor Joseph Vance and Secretary of State Carter B. Harlan. Over the years copies of this map have been discarded, and those that are identified are located at the Ohio Historical Society Library, the Ohio State Library and the Library of Congress.

Included in the story of the mapping of the state during this period should be the episode of the Michigan Border War, which brought Ohio and the Michigan Territory close to an actual confrontation in 1835 over the boundary line between the two states from the western end of Lake Erie to Indiana. Because of faulty geographical information and political misunderstanding and jealousies, Ohio and Michigan were involved in an almost three-year dispute over their common border. According to the provisions of the Ordinance of 1787 for the division of the Northwest Territory, which were based on the earlier maps of Lewis Evans and John Mitchell, a line dividing Ohio and Michigan to a point north of Maumee Bay on Lake Erie. Lake Michigan, however, was further south than at first believed. The survey line reached Lake Erie south of Maumee Bay, thereby placing the mouth of the Maumee River within Michigan. This fact was clearly understood by the Ohio Constitutional Convention in 1802 because that body included in the state's basic law, which was ultimately accepted by Congress, a resolution that stated that when the actual survey was run, the northern boundary of Ohio should be located at the northern cape of Maumee Bay. In order to carry out this provision of the state's constitution, William Harris surveyed that particular line in 1817 for the state of Ohio. The government of the Territory of Michigan, however, refused to recognize the line, conducted its own survey, which placed the boundary south of Maumee Bay, and exerted political control over the disputed territory. The issue climaxed in 1835 when Michigan applied for statehood and Ohio refused its permission for admission until the boundary question was settled to its own satisfaction. To be sure, Ohio was granted the territory in question, but not until a comic episode nearly resulting in open warfare was enacted throughout the Maumee River region.

The story of this dispute is easily related through cartography. Most of the maps of Ohio prior to 1832 and 1833 showed that the state claimed political control to the northernmost survey line claimed by Ohio, but indicated political organization only to the southern survey line as provided by the Ordinance of 1787. Of particular interest to this story was the "Map of Ohio and the settled parts of Michigan, 1832" (Plate XXI) which appeared in the 1833 edition of Anthony Finley's *A New General Atlas, Comprising a Complete Set of Maps Representing the Grand Divisions of the Globe, Together with the Several Empires, Kingdoms, and States in the World* published in Philadelphia. The map, which showed Ohio with 73 counties, illustrated

Michigan's political control over the disputed area through its county subdivisions, thereby placing Port Lawrence, later Toledo, and the Maumee Bay within Michigan. There is little doubt that Finley used the Woodruff map of Ohio as a guide, although his own work was of exceptionally fine detail. Finley began to produce atlases as early as 1824, and in July of that year, the *North American Review*[33] publicized his first atlas as a work of some competency. He also made the map for the *New Jersey Gazetteer* of 1834.

Plate XXI: Finley, "Map of Ohio," (1832), in Finley, *A New General Atlas* (1833). Michigan's control of Maumee Bay and the Toledo area before the end of the Michigan Border War is illustrated by Finley's map.

David H. Burr, a surveyor and cartographer, also was involved in the Ohio-Michigan controversy. Burr, who was born in 1803, the year that Ohio entered the Union, was geographer to the national House of Representatives and apparently charged with providing Congress with correct geographic information concerning the disputed border. His map of the area between Lake Michigan and the western portions of Lake Erie was printed in the House Executive Documents of the 23rd Congress, 2nd Session on June 14, 1834. It is a greatly reduced and simplified version of his original survey map entitled "Map Exhibiting the Country between

Lake Michigan and Erie and the Contested Boundary Lines," which is in the National Archives in Washington, D.C.[34]

Aside from his work for Congress, Burr also produced several atlases and individual maps. Several editions of his *A New Universal Atlas Comprising Separate Maps of all the principal Empires, Kingdoms and States Throughout the World and Forming a Distinct Atlas of the United States* were published by D. S. Stone of New York. The maps of Ohio contained in these atlases were not as expertly done as work by others and probably did not reflect Burr's true ability as a cartographer. His best representation of the state was produced by J. H. Colton and Company of New York in 1836 (Plate XXII). There were several other editions of this map printed. Burr's "Ohio" showed the state with 75 counties, including Lucas, which was the county through which Ohio extended its political control to the northernmost boundary line. He did not extend the county line to the Harris survey. His map of Ohio detailed the townships by name in the eastern half of the state, and a comparison of his map with any earlier one readily indicates the extension of roads, canals, proposed canals and the newly constructed railroads in the state. The decorative border and inset of a plan of Cincin-

Plate XXII: Burr, "Ohio," (1836), *Colton's Atlas* (1836). Burr, who surveyed the Ohio-Michigan border in 1834 for the national House of Representatives, indicated on this map Ohio's control over the disputed territory with Michigan.

nati was part of the cartographic style used during the second quarter of the nineteenth century. The clarity of detail in this work establishes Burr as a cartographer of some ability. It should be noted, however, that in 1831, Burr produced a small foldout traveller's guide to Ohio that was similar to Woodruff's map.

Burr drew a series of maps of the states showing post roads and post offices, which he sold to the United States Post Office. In 1835, he complained that Congress had not appropriated enough money with which to purchase his maps and requested that those interested in such matters should write to that legislative body and encourage it to act accordingly. Apparently his method of lobbying was successful because on March 27, 1837, the Postmaster General purchased 1,810 maps of the United States and the various states from Burr for $6.00 each or $78.00 per set of 13 maps, a total of $10,800. These maps were distributed to the post offices throughout the country. In 1841, Burr petitioned Congress to relieve him from paying the tariff duty on a set of maps that he had had printed in England by J. Arrowsmith, to whom he owed $9,000. Congress refused Burr's petition; the Postmaster General explained to Congress that he had enough of the cartographer's maps and, in fact, the Post Office was willing to sell them to the public at cost. Burr also did some illustrating for books, but his cartographic work made him one of the leading map makers of the first half of the nineteenth century.

Without doubt, the foremost cartographer and most prolific map publisher of the first half of the nineteenth century was Henry Schenck Tanner of Philadelphia, who produced in 5 installments between 1818 and 1823 the magnificent work, *A New American Atlas: Containing Maps of the Several States of the American Union, Projected and Drawn on a Uniform Scale from Documents Found in the Public Offices of the United States Government and Other Original and Authentic Information.* In this atlas, which contained individual maps of all states and territories, Tanner included Ohio and Indiana on the same sheet, entitled "Ohio and Indiana" (Plate XXIII). Only the Ohio portion is reproduced here. In the introductory "Geographical Memoir" to his atlas, Tanner explained that his sources for the Ohio map were the Alexander Bourne and B. Hough map of Ohio of 1815 and a manuscript map of the state compiled by Bourne alone that "was projected and drawn expressly for my [Tanner's] use" and contained information that had not been published before. Bourne also wrote a short geographical description of the state for the famous cartographer. Tanner used a manuscript map, "Map of the Country about the West end of Lake Erie," prepared for him by Samuel Williams of Chillicothe, and another manuscript map entitled "Map of the North-West Part of Ohio," which illustrated the most recent surveys in that portion of the state.[35] Aside from these sources, Tanner also relied on the works of Daniel Drake, John Kilbourn and Zadock Cramer, the author of many works concerning navigation on the Ohio River, to piece together his map of Ohio. Aside from the fact that no history of the mapping of Ohio would be complete without the work produced in this landmark atlas, Tanner's reproduction of the northwestern portions of Ohio is important to the delineation of the state. Here the various forts and blockhouses constructed during the War of 1812 are pinpointed, along with the network of roads that kept open the lines of communications between them.

In 1829 Tanner produced a rather large and detailed map of the United States, and to accompany this piece he published his *Memoir on the*

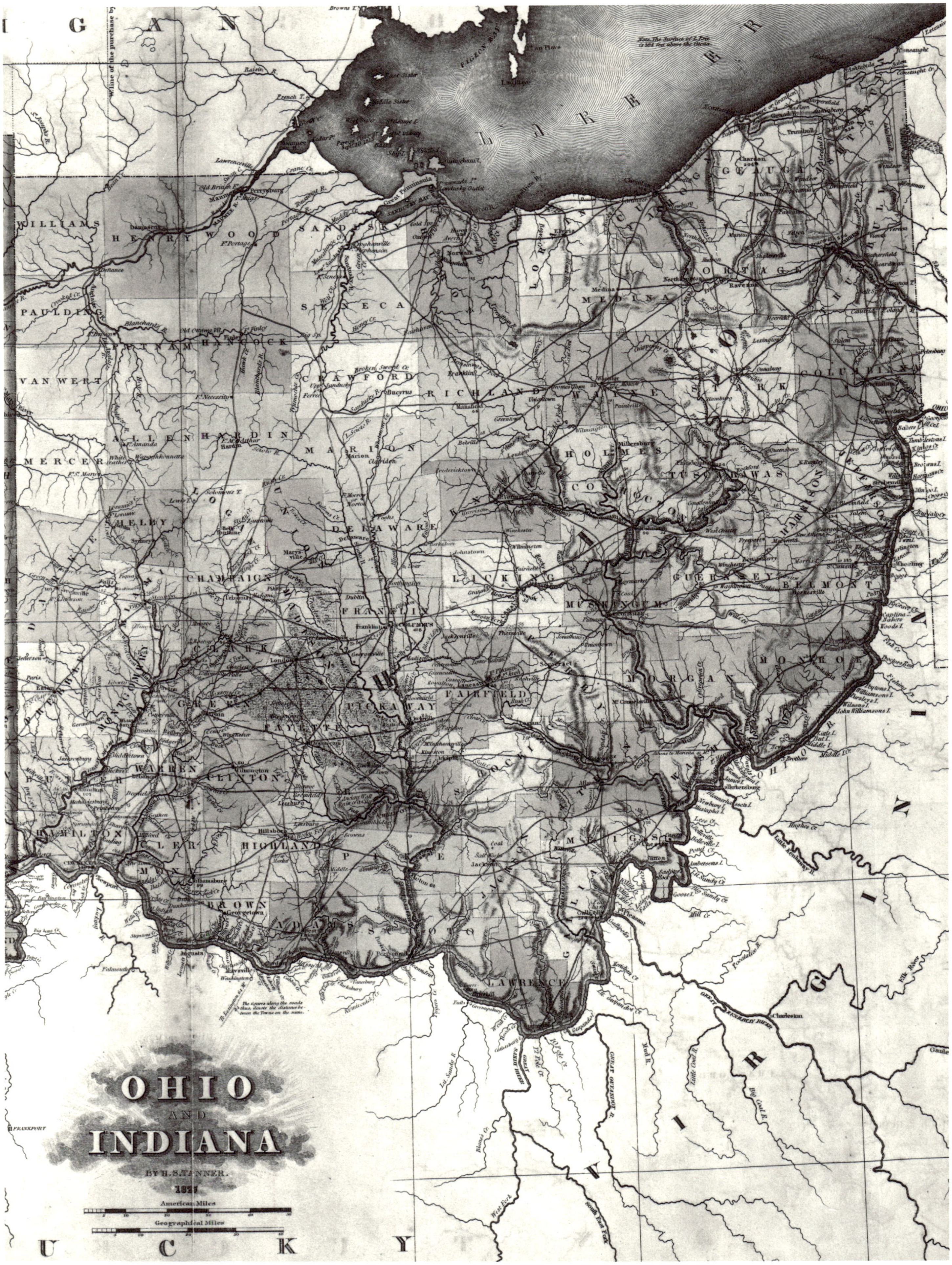
OHIO
AND
INDIANA
BY H. S. TANNER.
American Miles
Geographical Miles
The figures along the roads thus, denote the distance between the Towns on the same.

Recent Surveys, Observations and Internal Improvements in the United States. In order to gather information for this narrative, he sent letters of inquiry to knowledgeable individuals in each state, and from their replies he gleaned details for his map and information for his *Memoir.* Alexander Bourne of Chillicothe, the author of several maps of Ohio, supplied Tanner with information about his adopted state. In his summary of the state, Tanner remarked that "whilst we find its [Ohio's] physical geography there permanent, no other state in the union, nor perhaps any district of equal extent in the known world, affords an instance of such rapid increase in population and extension of settlements as this interesting state presents. As a necessary consequence of this increase of population," he continued, "towns, innumerable have sprung into existence as if by magic."[36]

Tanner's publishing house in Philadelphia printed a variety of items from prints of landscapes, boats and flowers to maps pasted on cardboard ready for use in the classroom. In fact, in 1836, he offered 104 separate items for sale. Travelers' guides were still popular among those seeking new homes and opportunities in the West, and that year he produced *The American Traveller: or Guide Through the United States* and offered many of the maps contained in this volume for sale as separate sheets. The one for Ohio cost $1.00.

In 1830, Tanner published a small foldout map contained in a leather-bound cover entitled, "The Traveller's Pocket Map of Ohio with its Canals, Roads and Distances." This same map was frequently used by Tanner in his atlases or sold as an individual sheet. It later appeared as "A New Map of Ohio with its Canals, Roads & Distances" (Plate XXIV) in the 1845 edition of his *A New Universal Atlas, Containing Maps, of the Various Empires, Kingdoms, States & Republics of the World.* This map was updated from the earlier ones to show 79 counties in Ohio. It also reflected a change in the style of map publishing and gave a rather cluttered appearance. Aside from the map itself, it had a very elaborate border, an inset of Cincinnati, a small framed cartouche, a profile of the Ohio Canal and a chart for distances from ports along the Ohio River to important centers of commerce on the interior waterways of the United States. This map, however, appeared to be very similar to one that William Woodruff produced in Cincinnati for the *Ohio Gazetteer.*

Another famous nineteenth-century map maker was Samuel Augustus Mitchell, an artist, geographer and cartographer, who had been a close associate of Tanner. By the mid-nineteenth century, he had taken over the operations of Tanner's Philadelphia publishing firm and continued to produce maps and atlases almost until the end of the century. Mitchell, like John Kilbourn, had been a geography teacher who became dissatisfied with the material available for teaching the subject, and had begun to write and publish geography textbooks. He made a substantial reputation for himself and expanded his publishing interests to include maps and atlases. In 1834 he began to publish a series of tourist pocket maps to meet the demands for more geographical information of a mobile population that was constantly pushing westward. The next year he produced "The Tourist's Pocket Map of the State of Ohio Exhibiting the Internal Improvements, Road Distances, &c." (Plate XXV), which was engraved by James H. Young.[37] Young was an engraver active in Philadelphia between 1817 and 1866. This map illustrated 73 counties and depicted the Ohio-Michigan border contest in the northwestern corner of the state. The engraver, Young no doubt, had been influenced by Woodruff's earlier work of Ohio.

(Opposite page) Plate XXIII: Tanner, "Ohio and Indiana," in Tanner, *A New American Atlas* (1818-1823). Based upon the work of other cartographers, this magnificent map of Ohio was produced by Henry Schenck Tanner, one of the foremost cartographers of the nineteenth century.

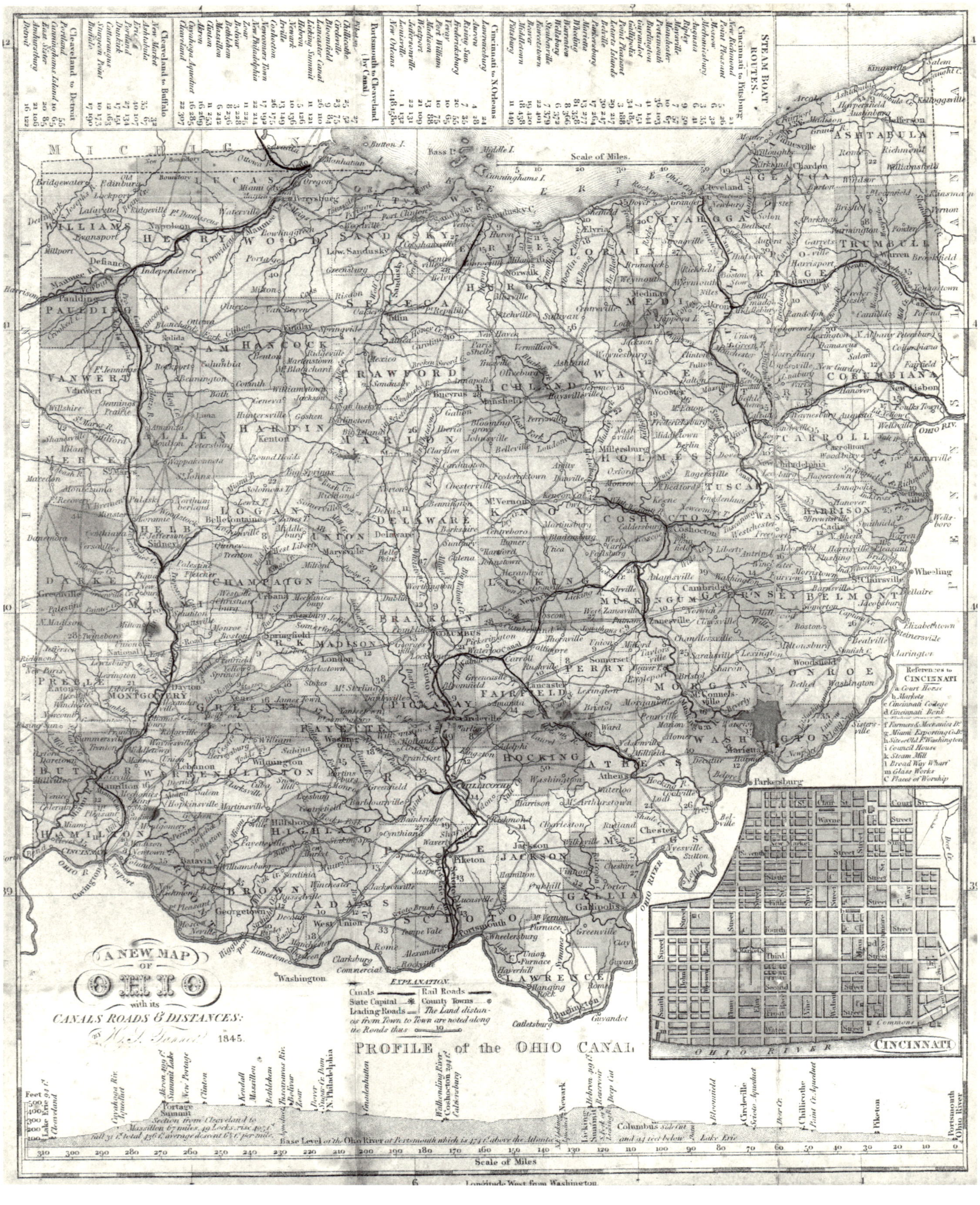

Plate XXIV: Tanner, "A New Map of Ohio," (1845), in Tanner, *A New Universal Atlas* (1845). First used as a fold-out map to a travelers' guide, this map appeared in many atlases produced by Tanner.

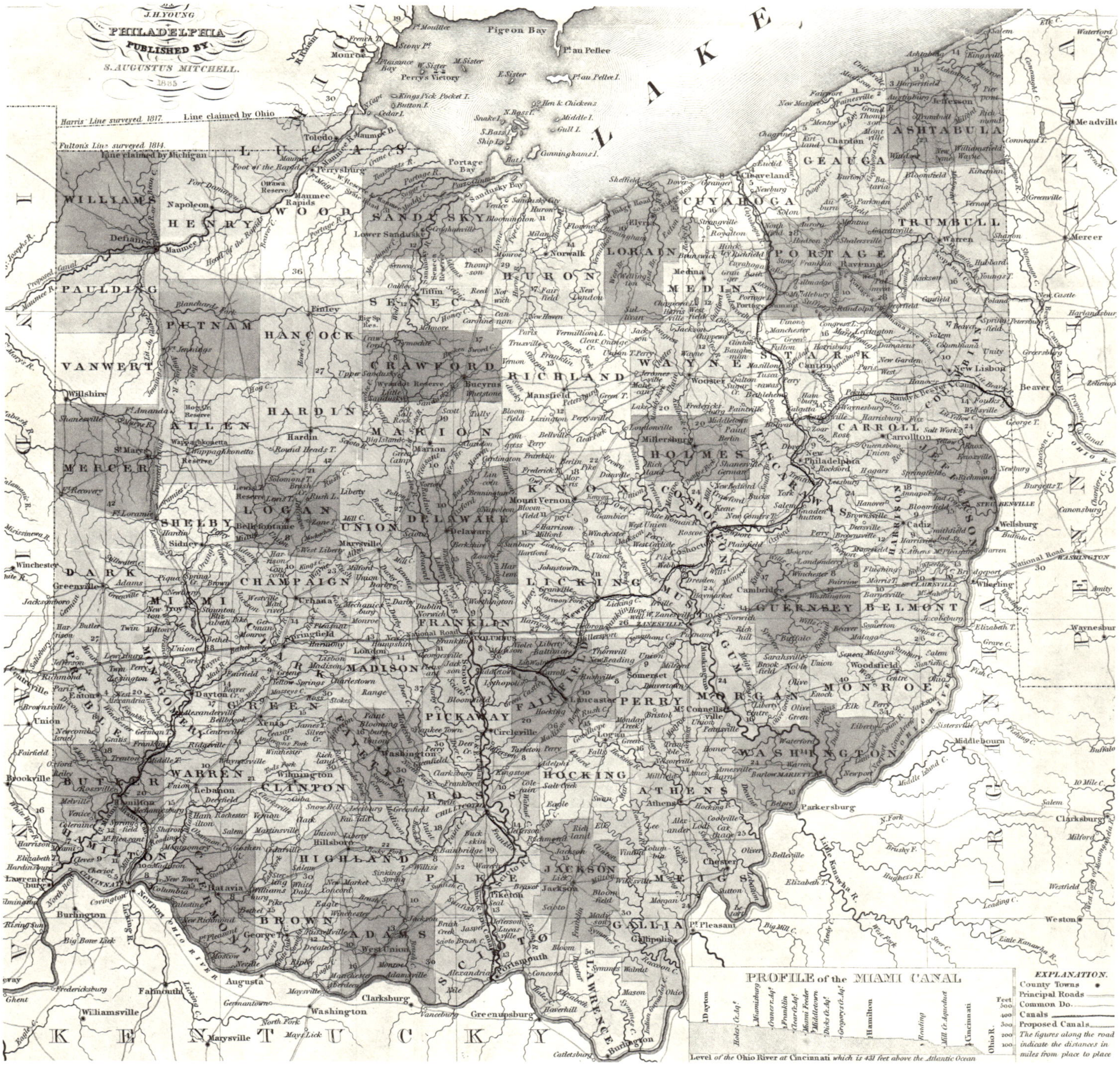

In 1847, Mitchell published *A New Universal Atlas Containing Maps of the Various Empires, Kingdoms, States & Republics of the World.* It became quite popular and was known simply as *Mitchell's Atlas.* It earned for him a greater reputation as a map publisher than Tanner's. "A New Map of Ohio with its Canals, Roads & Distances," which appeared in that atlas, closely resembled the one made by Woodruff for the *Ohio Gazetteer* as well as the one used by Tanner. In fact, the profile of the Ohio Canal that appeared at the bottom of the map could have been copied directly from Woodruff. This atlas appeared in various editions between 1847 and 1858. In 1851, however, and until 1858, although still published as *Mitchell's Atlas,* various editions were printed by Philadelphia publishing firms such as Thomas, Cowperthwait and Company, Cowperthwait, Desilver and Butler; and Charles Desilver. In the 1851 edition, which was printed by Thomas, Cowperthwait and Company, appeared "A New Map of the State of Ohio"

Plate XXV: Mitchell, "Tourist's Pocket Map of the State of Ohio," (1835), Ohio Historical Society, Columbus. The production of this and similar fold-out maps illustrated the demand for information about Ohio from migrants moving West.

Plate XXVI: Thomas, Cowperthwait, "Map of the State of Ohio," (1851), in Mitchell, *Mitchell's New Atlas* (1851). This map was used by Mitchell for many years and frequently updated. It shows Ohio with all eighty-eight counties.

(Plate XXVI). That year, J. F. Desilver, a publisher and bookseller in Cincinnati, acted as the western agent for the publishing house of Thomas, Cowperthwait and Company and distributed the famous *Mitchell's Atlas.*[38] This map was the same one used by Mitchell in 1835 and entitled "The Tourist's Pocket Map of the State of Ohio Exhibiting the Internal Improvements, Road Distances, &c." "A New Map of the State of Ohio;" it was, however, updated and included all 88 counties. For the first time in a Mitchell atlas, it was shown that Ohio controlled all the territory to the Harris survey line. It also illustrated the expanded growth of roads and railroads throughout the state and each of the railroads was clearly identified.

During the second half of the nineteenth century, many national and county atlases were produced by publishers across the country. Such important publishing houses as C. W. and C. B. Colton, A. J. Johnson, Henry Francis Walling, George F. Cram and George W. Gray, to name only a few, produced atlases that contained maps of Ohio almost every year from the 1860s until the end of the century. County atlases, comprised of township and city maps, illustrations of houses and businesses, election and population statistics and short histories of the county and state were produced during the last third of the century by private publishing firms that sought to make profits by marketing a product that appealed to provincial interests. During this same period, too, cadastral maps, or land ownership maps, were produced by enterprising men who mapped each county in the state literally inch by inch. Since a systematic survey of the United States was never conducted by the federal government, the mapping of the country fell to those who could put together atlases that appealed to local interests and that could be sold for profit.[39] Many maps, geographies and traveler's guides produced during the early part of the nineteenth century served as guides to eastern and European immigrants alike, who journeyed west to seek a new life. These same works, since they described locales for the immigrant and encouraged migration to a particular spot, reflected the ambitions of the national drive westward as well as the population and territorial growth of the country. It should not be overlooked that the descriptive geographies and guides, along with the accompanying maps that were produced, are part of America's literary tradition. These works represent what was happening in America during its westward movement, and their appeal for national as well as local pride and patriotism was extremely clear. The cartography also reflected part of America's cultural development. While it pictured America's business during the century, it also was American art, pure and simple. It should be recognized that while the geographers and cartographers performed a definite service to the state and to the nation, they were artists, and their works are a major contribution to the cultural heritage not only of Ohio, but of the United States as well.

1. Guthrie's geography was first published in 1770 and went through many revisions until 1827.

2. Archer Butler Hulbert, ed., *The Records of the Original Proceedings of the Ohio Company,* Vol. I, *Ohio Company Series* (Marietta, Ohio, 1917), 50, fn. 59.

3. Jedidiah Morse, *The American Geography or a View of the Present Situation of the United States of America* (Elizabethtown, New Jersey: Shepard Kollock, 1789), p. 458.

4. *Ibid.,* p. vi.

5. Aaron Arrowsmith, *A New and Elegant General Atlas, Comprising All the New Discoveries, to the Present Time; Containing Sixty-Three Maps, Drawn by Arrowsmith and Lewis. Intended to Accompany the New Improved Edition of Morse's Geography, but Equally Well Calculated to be Used with any Former Edition, with His Gazetteer, or Any Other Geographical Work* (Boston: Thomas and Andrews, 1805).

6. John Pinkerton, *Modern Geography. A Description of the Empires, Kingdoms, States, and Colonies: with the Oceans, Seas, and Isles; in all Parts of the World; Including the Most Recent Discoveries, and Political Alterations* (Philadelphia: J. Conrad & Co., 1804).

7. William Guthrie, *A New System of Modern Geography; or, A Geographical, Historical, and Commercial Grammar; and Present State of the Several Kingdoms of the World* (Philadelphia: M. Carey, 1794–1795).

8. W. H. Beers & Co., *The History of Warren County, Ohio* (Chicago, 1882), p. 244.

9. Edward D. Mansfield, *Memoirs of the Life and Services of Daniel Drake, M. D.* (Cincinnati: Applegate & Co., 1860), pp. 72, 98–100. John F. Mansfield lived in Cincinnati with Jared F. Mansfield, who was Surveyor General of the United States between 1803 and 1812 and who through astronomical computation established an accurate east-west base line which provided for a more accurate subdivision for the Public Domain. Without doubt, Jared Mansfield's influence was felt during the production of this map.

10. John Melish, *Travels in the United States of America in the Years 1806 & 1807, and 1809, 1810 & 1811,* II (Philadelphia, 1818), 57–59.

11. Andrew Miller, *New States and Territories, or the Ohio, Indiana, Illinois, Michigan, North-Western, Missouri, Louisiana, Mississippi and Alabama in their Real Characters, in 1818;. . .* (Printed for the Benefit of Emigrants, and others, Intending to Visit the Western Country, 1819), pp. 8–9.

12. Miller, *New States and Territories,* p. 25.

13. Kilbourn had acquired a printing press from James E. Gardner in Franklinton, who had published *The Freeman's Chronicle* in 1812. Kilbourn published only 2 issues of his own newspaper, the *Columbus Gazette.*

14. Alexander Bourne, "The Siege of Fort Meigs Year 1813, An Eye-Witness Account," *Northwestern Ohio Quarterly,* XVII (October 1945), 139–54; XVIII (January 1946), 39–48.

15. This map is in the Map Division of the William Clements Library, Ann Arbor, Michigan.

16. This map is in the map collection at the Cincinnati Historical Library, Cincinnati, Ohio.

17. This November 1817 edition of the Pelham map is at the Cincinnati Historical Society Library.

18. Thomas Anderson's map is at the Ohio Historical Society Library, Columbus, Ohio.

19. John Kent Folmer, ed., "Pre-Civil War Sentiment from Belmont County: Correspondence of Hugh Anderson," *Ohio History,* LXXVIII (Summer 1969), 202. For a partial list of Hugh Anderson's engravings, see David McNeely Stauffer, *American Engravers upon Copper and Steel,* II (New York, 1907), 15.

20. Letter from Alfred Kelley to Benjamin Tappan and Micajah T. Williams, October 2, 1826, attached to map, Map Room, Ohio Historical Library, Ohio Historical Society.

21. Stauffer, *American Engravers upon Copper and Steel,* pp. 560–62.

22. From advertising in J. W. Woodruff's *The Cincinnati Directory Advertiser for the Year, 1836–37* (Cincinnati, 1836).

23. Erwin Raisz, "Outline of the History of American Cartography," *Isis,* XXVI (March 1937), 373–90; see also Raisz, Chapter 4, "American Cartography," *General Cartography* (New York: McGraw-Hill Book Co., 1948), pp. 43–53.

24. "An Act Providing for a Geological Survey of the State of Ohio, and other purposes," March 27, 1837, *Acts of a General Nature Passed at the First Session of the Thirty-Fifth General Assembly of the State of Ohio* (Columbus, 1837), p. 85. This map is located at the Ohio Historical Society Library, Columbus, Ohio.

25. See letter from G. W. Featherstonaugh to Lewis Cass, January 12, 1833, printed in "Report from the Secretary of War, January 18, 1833," *Senate Documents,* 22nd Congress, 2nd Session, (1833), Document No. 35, Serial No. 230. Featherstonaugh was editor and publisher of *Geological Journal,* which was printed in Philadelphia. It was Featherstonaugh's idea to encourage each state to conduct an extensive geological survey to determine its mineral wealth. One result of such a project, he pointed out, would be a complete set of topographical maps of every state in the Union. Ohio's effort to conduct such a survey several years later was due partly to Featherstonaugh's encouragement.

26. For a very brief description of this first geological survey, see Edward Orton, *First Annual Report of the Geological Survey of Ohio* (Third Organization), (Columbus, 1890), pp. 1–2.

27. W. W. Mather, *First Annual Report of the Geological Survey of the State of Ohio* (Columbus, 1838), p. 21.

28. *Ibid.,* pp. 22–23.

29. W. W. Mather, *Second Annual Report of the Geological Survey of the State of Ohio* (Columbus, 1838), pp. 43–44.

30. "An Act to provide for the purchase and distribution of one hundred and fifty copies of Doolittle and Munson's Map of Ohio," *Acts of a General Nature Passed by the First Session of the Thirty-Sixth General Assembly of the State of Ohio* (Columbus, 1838), pp. 15–16.

31. *The Cincinnati Directory Advertiser for the Year 1836–37* (Cincinnati, 1836).

32. Stauffer, *American Engravers upon Copper and Steel,* I, 67.

33. *North American Review,* XIX (1824), 261–62.

34. Record Group No.77, 137–1, Map Division, National Archives, Washington, D.C.

35. Henry Schenck Tanner, *A New American Atlas* (Philadelphia: H. S. Tanner, 1823), pp. 13–14.

36. Henry S. Tanner, *Memoir on the Recent Surveys, Observations and Internal Improvements in the United States* (Philadelphia, 1829), pp. 83–84.

37. Stauffer, *American Engravers Upon Steel and Copper*, I, 299; II, 296–97. This map is located at the Ohio Historical Society Library, Columbus, Ohio.

38. Charles Gist, *Sketches and Statistics of Cincinnati in 1851* (Cincinnati, 1851), p. 233.

39. See Appendix A for a list of Ohio county atlases.

CHAPTER SIX

Mapping the City

Surveyors not only measured and mapped the township and section lines, but they laid out as well the towns and villages that became the commercial, industrial and cultural centers of Ohio. In many instances they were gamblers, men who could stand on a piece of ground at the confluence of several streams or at the place where a major Indian trail crossed a deep, clear river, surrounded by the dense forests that covered Ohio, and determine the location of a city that would one day grow and prosper. They had help, to be sure. The Indians before them had selected key geographical sites for their own homes. Many cities in Ohio today have their foundations on the ruins of some prehistoric culture or the remains of a village site that was occupied by Indians when the trappers and traders first set foot on the western slopes of the Allegheny Mountains. The record of urban Ohio has been well kept through its cartography. Maps of settlements were used to encourage immigrants from the East to move to sites in the West that offered new opportunities and a bright future. Urban cartographers illustrated the development and expansion of Ohio cities which reflected national population and economic growth throughout the nineteenth century. Both the work of the imaginative surveyor and the art of the map maker have left a brilliant and transparent portrait of the development of the urban industrial and cultural centers of the state.

Marietta, the first permanent settlement in Ohio, was also the first planned community in the Old Northwest Territory. The founding fathers of this historic settlement knew the value of planning for prosperous economic development and social improvement. The "Plan of Marietta agreeable to the Resolves of the Ohio Company" (Plate I) showed the careful and intelligent considerations that the founders of the famous Ohio Company gave to their first settlement along the east bank of the Muskingum River in 1788.

Without doubt, much deliberation went into the planning of the city of Marietta. The plat of the town was first accepted during the third meeting of the Ohio Company in Mr. Brackett's Tavern in Boston on August 30, 1787, before the original group of settlers began their journey

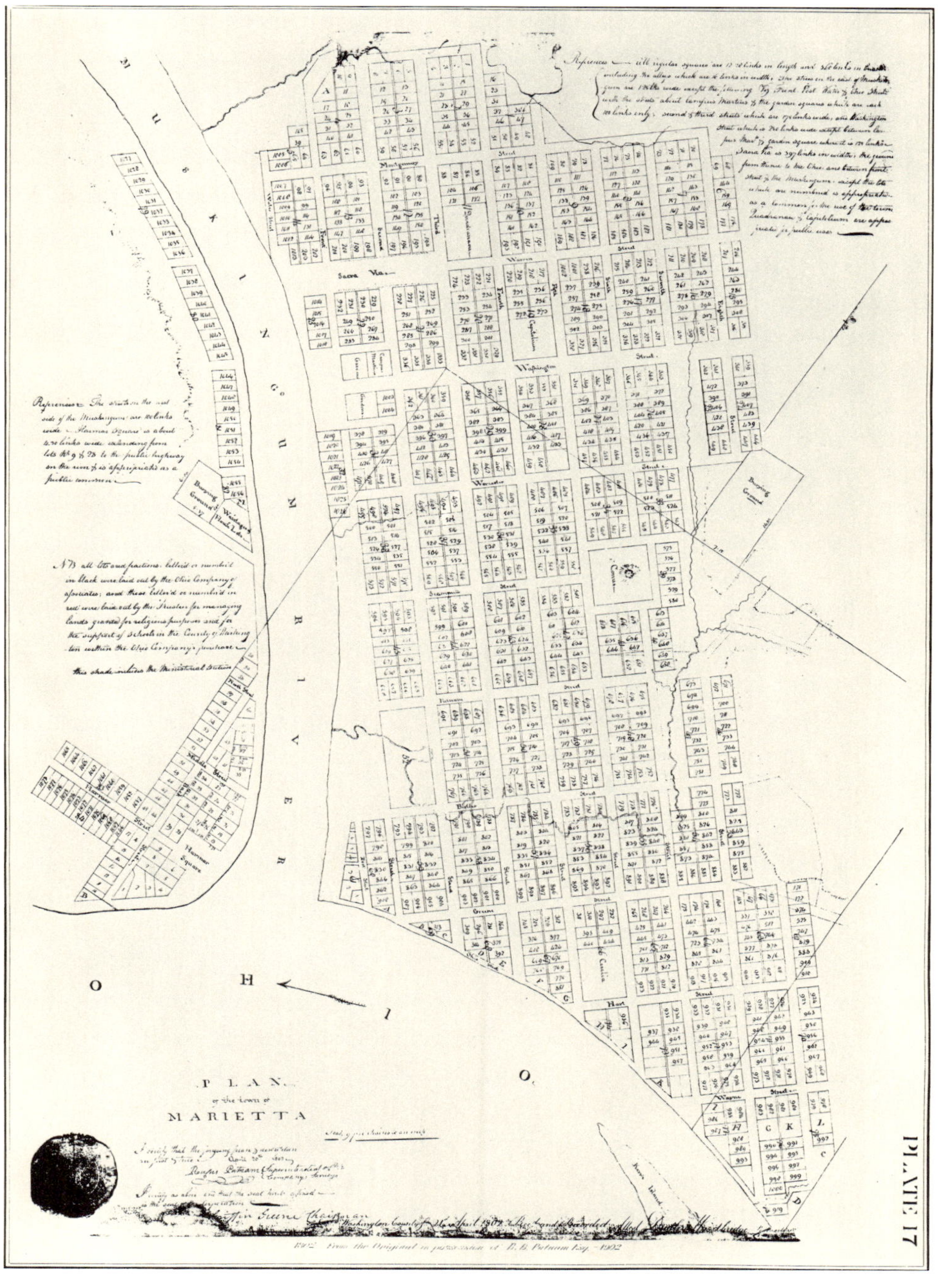

Plate I: Putnam, "Plan of Marietta," (1802), in Roe. *Atlas of the City of Marietta* (1902). The original map is in possession of the Putnam family. It illustrates Ohio's first permanent settlement fourteen years after its founding.

west. At first, the Company resolved that a tract of 5,760 acres "near the confluence of the Muskingum and Ohio Rivers, be reserved for a City & Commons," and prescribed "that within the said Tract and in the most eligible situation there be appropriated for a City, sixty squares of three hundred & sixty feet by three hundred & sixty feet each, in an oblong form, of ten squares in front and six in depth, with streets of one hundred feet in width through each range."[1] Four squares were also reserved at that time for public use.

Nearly 3 months later, on November 21, the members of the Ohio Company, meeting this time in Cromwell's Head Tavern in Boston, revised the initial plan for their city. The town tract was reduced in size from 5,760 to 4,000 acres, with an additional 8,000 acres adjacent to the town to be divided into 1,000 lots of 8 acres each. The town lots were increased from "sixty feet front and one hundred feet in depth" to "ninety feet in front and one hundred and eighty feet in depth." The 4 squares were still retained for the public, but "eight house lots in the city at the Mouth of Muskingum, [were] reserved for publick use."[2]

On July 3, 1788, the members of the company met for the first time in their new home along the Muskingum. The community's name was changed from Castrapolis to Adelphia; finally, as a friendly gesture to France in recognition of that country's timely assistance to the United States during the American Revolution, the proprietors of the Ohio Company named their settlement Marietta, in honor of Queen Marie Antoinette. The city itself sprawled out over the remains of a great Hopewell earthwork. In order to preserve that heritage and the memory of their own beginnings, the Marietta fathers reserved for public use the land around their stockade called *Campus Martius*; the right half of square 11, which contained a large truncated mound called *Quadranaou;* the right half of square 19, which included an elevated square named *Capitolium*;[3] the entire of square number 66 called *Cecelia*; and *Conus,* a square that was to be presented to the French Queen, was called Marie Antoinette Square. The great earthen way leading from *Quadranaou* to the Muskingum River was also set aside and named *Sacra Via.*[4] Educated in the classical tradition, and truly men of science and progress, the proprietors of this first settlement in Ohio tried to create an environment that was not only conducive to pleasant living and productivity but culturally and historically rewarding as well.

On December 3, 1788, however, not long after Marietta was first laid out, the Committee on the Lease of Public Squares was established to lease the public squares to individuals until such time as the squares could be used for the purpose for which they were reserved. All due care was taken that the mounds be preserved in their original state, and the Committee was particular "to point out the Mode of improvement for Ornament and in what manner the Ancient works shall be preserved," while the land was in private hands.[5] Eventually, *Conus, Capitolium, Quadranaou*[6] and *Cecelia*[7] were leased to private individuals. For instance, on July 29, 1790, a little over 2 years after their arrival, the Ohio Company leased to Captain Jonathan Heart for 20 years the square called *Cecelia* with the understanding that "he shall make and keep open sufficient Dreins to take off the waters on said Squares and the street running between squares 61 and 62, which Dreins are to Communicate with a Drein to be opened from the River Muskingum, to the said streets running between S^{d} squares 61 and 62."[8] On March 7, 1791, the committee leased *Conus* to Rufus Putnam for

12 years. In order to fulfill his lease, he was to surround the entire square with mulberry trees, at each corner of the square plant elms, set in more mulberries and honey locust, weeping willows and evergreens at particular points over the lot and, finally, to ensure that the entire square was in grass.[9] Similar landscape provisions were included by the committee when it leased *Capitolium* to Dudley Woodbridge[10] and *Quadranaou* to Benjamin Tupper.[11] It is certain that the proprietors of the Ohio Company were practical men who were willing to make use of land when it was necessary, but at the same time they realized that certain lands had intrinsic value. In their own way, they sought to conserve these prehistoric monuments from the past ages as cultural features which, when added to their own heritage, would create attractions that were unique and would reach beyond those possessed by their eastern cousins.

The Marietta settlement prospered under the keen direction of the leadership of the Ohio Company. In November 1789, a year and a half after the first party of settlers reached the banks of the Muskingum, General Josiah Harmar wrote to Thomas Mifflin of Philadelphia:

> It would afford me great happiness if you could steal three or four months from the Atlantic, and spend them with me. I am now on the wing, expecting to move down the Ohio in a few days, and to fix headquarters opposite the mouth of the Licking river, [Cincinnati] about three hundred miles below this garrison [Fort Harmar], where I should be proud of being honored with your company. Venison, two or three inches deep cut of fat, turkeys at one pence per pound, buffalo in abundance, and catfish of one hundred pounds weight, are stories that are by no means exaggerated. I am going to a country where there is a much greater plenty of game than there is here at present. Our New England brethren are a very industrious set of people. With the protection afforded them by the troops (which they acknowledge with great gratitude), they have converted the wilderness into a fine state of cultivation. Their settlement bids fair to be very flourishing. Cornfields, gardens, &c., now appear in places which were lately the habitation of wild beasts. Such are the glorious effects of industry.[12]

At about the same time that the Ohio Land Company began its initial settlement in the Ohio Country, other land schemes were nurtured and soon numerous settlements blossomed along the Ohio. One such scheme was John Cleves Symmes's Miami Purchase, noted on cartographic works as Between the Miamis. The most important settlement in this tract was the one established on the north shore of the great river on 800 acres opposite the mouth of the Licking River in Kentucky. Here, in December 1788, Colonel Robert Patterson, Israel Ludlow and Matthias Denman, along with approximately 23 other men, landed after floating down the river from Limestone, Kentucky, and began erecting crude wooden shelters, built from the wood of their flatboats. Ludlow, who was familiar with the Miami Country from a previous expedition, was chief surveyor for the Miami Purchase and laid out the town of Losantiville in 1789. The following year its name was changed to Cincinnati.

The map "Plan of Cincinnati including all the late Additions and Subdivisions" (Plate II) was drawn by Thomas Danby for Daniel Drake's *Natural and Statistical View, or Picture of Cincinnati and the Miami Country* which was published by the Looker and Wallace publishing firm in Cincinnati in 1815. Drake, a physician by profession, was representative of those men whose native intelligence, education and natural curiosity propelled them into numerous fields of interests where, in a new country, their success was limited only by the amount of their own energy and imagination. During his 67 years, Drake, among other things, received a medical

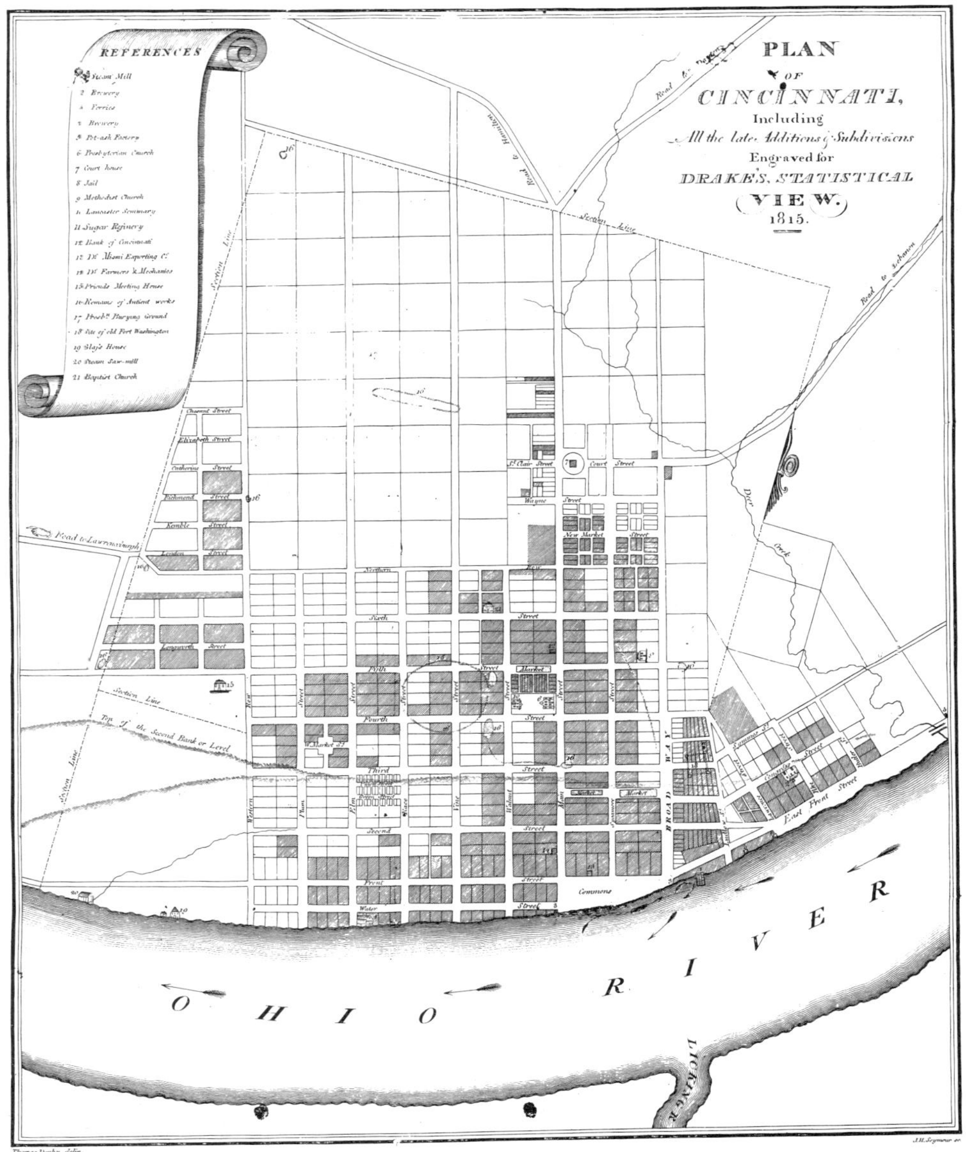

Plate II: Danby, "Plan of Cincinnati," (1815), in Drake, *Natural and Statistical View* (1815). Danby's map was used to induce immigrants to settle in Cincinnati and the Miami Valley.

certificate from the University of Pennsylvania in 1806 at the age of nineteen, practiced medicine, taught at Transylvania College, helped to found an experimental school, helped to organize a museum, established medical colleges, directed a temperance movement and wrote some 3,000 published articles. His *Natural and Statistical View*[13] followed his famous descriptive book, *Notices Concerning Cincinnati,*[14] which, published in 1810, had become popular among those who desired to settle in the Miami Valley. Many, however, complained that his work was too technical to be of much use. To satisfy these critics, Drake made his *Natural and Statistical View* "more extended, and less professional," and explained that he wanted to inform those who "desire to know what there is in a new country, that can recommend an emigration thither: [and] in a city, we seek for that which is worthy of imitation or adoption."

The map of Cincinnati included in this volume was drawn by Thomas Danby "from materials obtained chiefly at the office where the surveys of the town plat are recorded; and has therefore all the accuracy which can be conveniently given to such a work."[15] Situated on a site previously occupied by prehistoric Indians who had constructed their ceremonial earthworks on the second bank overlooking the Ohio, Cincinnati, because of its location on the river and at the head of the Miami Valley, soon became the Queen City in the West. In 1815, there were nearly 1,500 structures in the city, including 20 that were made of stone, 250 of brick and 800 of wood. Of these, 660 were used as dwellings, while the remaining were either shops, manufactories or offices. The houses were usually unpainted, 2-story buildings that were simple in design and decorated with Corinthian or Tuscan cornices. "Several [homes]," Drake noted, "have lately been erected with an additional [third] story, and exhibit, for a new town, some magnificence. A handsome frontispiece or balustrade occasionally affords an evidence of opening taste; but the higher architectural ornaments—elegant summer houses, porticos and colonnades, are entirely wanting."[16] Danby's "Reference" scroll in the upper right corner of his map provided the immigrant with an interesting little directory to the city. No doubt it can be assumed that Drake supervised the drawing of this map and ensured that particular information was prominently illustrated. His work was aimed at an eastern audience eager to learn of the virtues and promises of the West and to those who were willing to leave the East for a new home along the Ohio. Also, the map and Drake's descriptive material provided a faithful portrait of the Queen City 25 years after its founding.

Barely visible on Danby's map were the outlines of the prehistoric earthworks that were once located where Cincinnati now stands. The works and mounds were identified by number 16. The large earth ellipse situated between Race and Walnut, and Fourth and Fifth streets, enclosed approximately 11 acres. Its wall was nearly 2 feet high and nearly 30 feet wide at its base. The wall that ran from near the corner of Fourth and Walnut to the northeast corner of Third and Main streets was 1 foot high, 9 feet thick and 120 feet long. Oblong mounds 8 feet high and 60 feet wide were located at both sides of the eastern opening of the ellipse and at the end of the long wall. From in front of the Methodist Church to the corner of Broadway and Third Street was what appeared to be a partially completed semicircular wall. To the east of this wall was a small circle that enclosed nearly 2,826 square feet. North of the main part of the city was situated an elongated enclosure, the parallel walls of which were 45 feet apart and 760 feet long. Several small mounds were also found on this site. One, at the western end of Fifth Street, was a large mound nearly 35 feet high. When General Anthony Wayne and his army occupied Fort Washington in 1793, he had the summit of this mound removed and a sentry box constructed in its place. An excellent plan entitled, "Ancient Works, Cincinnati, Ohio, Now Obliterated," prepared by Daniel Drake in 1815, was published in Charles Whittlesey's article, "Descriptions of Ancient Works in Ohio," which appeared in Volume III of the *Smithsonian Contributions to Knowledge* in 1850. This article was the result of the labors of Whittlesey and others on the first Geological Survey of Ohio during 1837 and 1838 and was considered to be a supplement to the famous Squier and Davis survey of Ohio's prehistoric earthworks that appeared in Volume I of the same series.

The stream mill shown along the Ohio between Broadway and Ludlow Street was constructed between 1812 and 1814 by William Green, a local stone mason, for George Evans. This structure, built on the river's edge, was the tallest edifice in the city and rose some 110 feet on 9 stories above the water. Certainly at that time it was an engineering marvel. Drake noted that the mill had 24 doors and 90 windows and was made from 90,000 bricks, 14,800 bushels of lime and 81,200 cubic feet of timber and weighed an estimated 15,655 tons. Part of the building was used for the milling of flour, while another part housed machinery for the processing of wool, cotton, and flax seeds, a fulling mill and other equipment. "It is equally creditable to the prudence of the superintendent and the temperance of the laborers," Drake boasted, "that during the erection of this house, not one serious accident occurred."[17]

That religious facilities were available was important to Drake, and he believed that it was necessary to point out to prospective inhabitants of the city that in Cincinnati there were places of worship if the soul so desired. Consequently, he indicated on the map the locations of the Presbyterian Church and cemetery, the Methodist Church, the Friends Meeting House and the Baptist Church. The first Presbyterian Society in Cincinnati was founded in 1791 by the Reverend David Rice. In 1815 its congregation, the largest one in the Miami Valley, numbered 160 souls, and boasted a nearly completed church on Main Street. Drake was enthusiastic about this new brick building and the small sketch on the map surprisingly resembled its actual style. The church was 68 feet by 85 feet and its cupola rose 80 feet into the air from the pavement. Inside the church, Drake explained, were 120 pews and 5 aisles. The large stone Methodist Church at Fifth Street was built in 1805 and 1806. Its congregation grew from 10 members in 1804 to 400 in 1815 and was the largest in Cincinnati.[18] On Sixth Street was a "handsome and commodious brick" First Baptist Church. This church was founded during the latter part of 1813 by 11 Baptists; by the summer of the following year, John Stites Gano, a prominent local businessman, had donated the land for the church and construction had begun. Evidently this church had adequate financial support because Drake said it was "well furnished with doors and windows, ornamented with a balustrade, and finished inside with taste."[19] The Society of Friends, the smallest church in Cincinnati, was founded in 1813 by a small group of Quakers from the East. At that time it was part of the Ohio Yearly Meeting. Its wooden meeting house had previously been a modest private dwelling and its membership numbered about 32 families.[20]

The Cincinnati Lancaster-Seminary on Fourth Street was founded in 1814 upon the educational principles advocated by Joseph Lancaster, a Quaker from England. Like many schools in early Ohio, it was financed through private subscription, each share being worth $25. The Presbyterian Church granted a tract of land with a 99-year lease to the school without cost, asking only that the school permit 28 poor children to attend tuition free. The bankers of the town, concerned for quality education within their city, saw to it that funds at exceptionally easy terms were available for the construction of the building. Tuition costs were low, only $8 a year, and, by April 1815, 420 students were enrolled in the school. That same year, the Board of Directors of the school decided to establish a separate school for the Negro children of Cincinnati.[21]

In 1815 when Drake's *Natural and Statistical View* went to print, the Cincinnati Lancaster-Seminary was nearly completed. Drake, who was one of its founders, was very pleased with the school and included in his book the following description of the new building:

> The Cincinnati Lancaster-Seminary, on Fourth street, in the rear of the Presbyterian church, is an extensive two story brick edifice, built, with some alterations, on a plan furnished by Isaac Stagg. It consists of two oblong wings, extending from Fourth-street, 88 feet deep. Near the front, they are connected by an apartment, for stair cases, 18 by 30 feet, out of which arises a dome capped peristyle, designed for an observatory. The front of this intermediate apartment is to be decorated with a colonnade, forming a handsome portico, 12 feet deep and 30 feet long. The front and each side are ornamented with a pediment and Corinthian cornice. The aspect of the

building is light, airy, and might be considered elegant, were the doors wider, the pediments longer, and divested of the chimnies [*sic*], which at present disfigure them. One wing of this edifice is designed for male, and the other for female children; and between them there is no connecting passage, except through the portico. The lower stories are finished entire, and calculated for the reception of 900 children. Each upper story is to be divided into three apartments, two in the ends 30 feet square, and one in the centre of 25, with a sky light, and the appurtenances of a philosophical hall. When completed, the whole building can receive about 1100 scholars.[22]

Since Drake was interested in promoting Cincinnati and the Miami County to prospective businessmen, laborers and farmers alike, he included in his book descriptions of the local banks, manufacturers and markets. For instance, the oldest banking house in the Miami Valley and in Ohio, the Miami Exporting Company located on Front Street opposite the commons, was founded in 1803 as a local effort to promote the exportation of agricultural products from the Miami Valley. Given the right to issue specie by the first session of the state legislature, the company suspended commercial activity in 1807 and began full-time operation as a bank. In 1815 its stock, worth $150,000, was owned by 190 people who received between 10 and 15 percent interest annually on their investment.[23] The second banking institution listed on the map by Drake was the Farmers' and Mechanics' Bank. Founded in 1812, the bank had a capitalization of $200,000, a little more than its competitor, and paid annual dividends of 8 and 14 percent. Drake pointed out that of the 13 directors on the managing board of the Farmers' Bank, "one third . . . must be practical farmers, and the same proportion practical mechanics."[24]

So that the farmers of the Miami Valley would have a place to dispose of their produce and livestock, two open-air markets were provided for them within the Queen City. The first was located between Main and Sycamore streets, while the second, adjacent to the first but not yet opened in 1815, was between Sycamore Street and Broadway. There the townspeople could find sides of beef and pork and quantities of veal and mutton for sale. Fish, too, were available and everything from perch to eel was displayed. A great abundance of produce, both native and cultivated, was there also. From the woods and fields surrounding Cincinnati, wild blackberries, crabapples, pawpaws, wild cherries, cranberries, mulberries, and fox or wild grapes were gathered while farmers brought peaches, apples, strawberries, melons, culinary roots and herbs, and pears from their own fields and orchards. The market was open to all every day except the Sabbath, and the city fathers tried to regulate its activities, although it was admitted that "violations are constantly suffered to pass unnoticed." Drake confessed that the quality of the beef as well as the cheese and butter offered for sale was not equal to that found in the maritime states,and pointed out that the better breeds of beef cattle and dairy cows had not yet been introduced. He was not pessimistic in his report, but rather suggested that the future for anyone interested in farming in the Miami Valley was bright.[25]

At the extreme right edge of the map was located the Cincinnati Manufacturing Company, an expanding firm which Drake said would produce between 6 and 7 tons per week of both white and red lead. This company was the third such industry established between the Mississippi and the mountains.[26] Along the river, Cincinnati boasted of 2 breweries, one of which was built in 1811, that consumed 30,000 bushels of barley annually in their manufacture of beer, porter, and ale, the quality of which was "equal to that of the Atlantic states."[27] One used water from the Ohio River in its process, while the other exploited well and cistern water. The sugar refinery shown was only under construction at the time the map was prepared, but Drake believed that it would commence sugar production soon.[28] It should be noted that to those who sought a new life in the West, Drake pointed out that Cincinnati "at no time had surplus of laboring population or of capital." Certainly, through his map and book, he offered the hope of prosperity and good fortune to those who were willing to build a new life for themselves in a fertile and growing country.

Not all settlement activity west of the mountains was confined to the new land adjacent to the great Ohio River. By the spring of 1796, a survey party of 37 men under the supervision of Moses Cleaveland, one of the principal stockholders in the Connecticut Land Company and agent for that Company, began to survey in the Connecticut Western Reserve, the land west of the Pennsylvania border and between Lake Erie and the forty-first parallel. Cleaveland worked the entire summer and explored most of the company's tract himself. Apparently he had some difficulty in selecting a site for what was to be considered the capital of the Company's holdings; on August 5, 1796, he wrote to Oliver Phelps, director of the Board of Directors of the Connecticut Land Company, "It is impossible at the present to determine on the place for the capital. . . . I believe it will be on the Cuyahoga [.] it must command the greatest communication, either by land or water of any other place on the purchase, or on any ceded lands west of the head of the Mohawk."[29] Despite his apprehension, however, by the middle of September a square-mile tract of land was set aside along the south shore of Lake Erie at the mouth of the Cuyahoga River for a city, and surveys began on September 16 of that year by Seth Pease and Amos Spafford, both surveyors, under the direction of Augustus Porter, the principal surveyor for the entire Reserve. By October 1, the original surveys were completed and the city of Cleveland was founded.

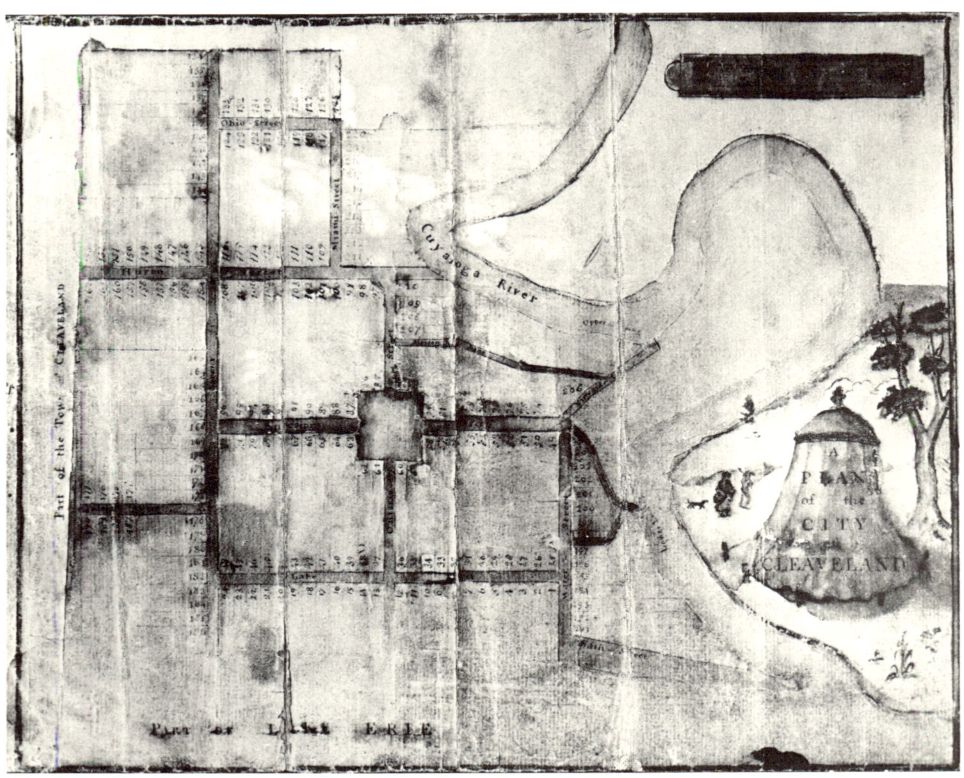

Plate III: Pease, "City of Cleaveland," (1796), Western Reserve Historical Society Library, Cleveland. The site for settlement at the mouth of the Cuyahoga River had been suggested by Benjamin Franklin and John Heckewelder long before the formation of the Connecticut Land Company.

The first map of the new city was made by Spafford in 1796;[30] though the so-called first official map, which accompanied the report to the proprietors of the Connecticut Land Company, was "A Plan of the City of Cleaveland" (Plate III) drawn by Pease himself from the results of the original survey. Several copies of the Pease map were made in 1796 and 1797, but it was never engraved for wide popular distribution. In 1816, the map and Pease's field notes were brought to Cleveland by James Root, brother of Ephraim Root, who was the secretary of the Connecticut Land Company.[31] This map is displayed at the Western Reserve Historical Society in Cleveland. Cleaveland's idea for the selection of this site at the mouth of the Cuyahoga River for a city was not wholly original. Even before Cleaveland set foot within the Reserve, Benjamin Franklin predicted that this was a suitable site for a thriving metropolis, and John Heckewelder, the Moravian missionary, too, pointed out that this location was suitable for a major settlement.

The Pease map showed 220 two-acre lots that were originally laid out in 1796.[32] The next year, 1797, 10-acre lots were surveyed east of the original site but included as part of the town. Starting with lot Number 1, every fourth lot was offered for sale to the general public at $50 a lot, a price that was considered high.[33] Each purchaser had to settle on his land almost immediately. Lots 58 through 63 and 81 to 87, those bordering the large square, were set aside for public use. Fourteen streets were laid out. The largest, Superior, which originally was named Broad, was 40 chains or 2,640 feet long and 200 links or 132 feet wide. The other streets were 99 feet wide. Because the site was surveyed before the forest was cleared, the lots and streets were laid out through the woods. The northwest corner at the west end of Superior was marked by a white oak on which was carved the letter "D" and the eastern end was specified by a letter "F" marked on the bark of a similar tree. Originally, Ontario was named Court, and Miami, Deer Street. Federal Street was later renamed St. Clair, in honor of Arthur St. Clair, the governor of the Northwest Territory, and no doubt reflected the favorable consideration given the governor from that sector. Pease's map is actually upside down, as Lake Erie is at the bottom or southern edge.

At the time Cleveland was laid out several structures were already located on the original town site. During the summer of 1796, Amzi Atwater, an employee of the Connecticut Land Company, built a cabin on lot 53. It was occupied by the Job Phelps Stiles family. The Connecticut Land Company promised to Talitha Stiles, the first woman settler in Cleveland, a 2-acre lot, a 10-acre lot and a 100-acre lot. This promise was never fulfilled.[34] On Lot 203, between Union Lane and the Cuyahoga River, Pease constructed a shelter to house the members of the survey team while they subdivided the Western Reserve. It was known as Pease Hotel. Also, on the same lot, and Lot 202, the one immediately north, were built sheds in which were housed the tools and instruments needed by the surveyors. Ezekiel Hawley, one of the original settlers of Cleveland, purchased town lots 49, 50 and 51. He built a cabin on Lot 49 that was later used by Elisha Norton as a trading post and residence. In 1806 it became the first post office in Cleveland.[35] On June 4, 1797, Pease selected the northern portion of lots 97 and 98 for the settlement burying grounds. That day, David Eldride, one of the company's employees who had drowned in the Grand River several days before, was buried "with as much decency and solemnity as could possibly be expected."[36] Because the land along the shore was not the most desirable for farming in the Connecticut

Reserve, the early settlers preferred to build their farms in the interior. Consequently, the city of Cleveland had to wait several decades until the economic conditions were right within the state before it became the prosperous city envisioned by those who had the imagination and faith to locate a city in a forest at the mouth of the Cuyahoga River.

Almost midway between Lake Erie and the Ohio River, along the east bank of the Scioto River, some 90 miles from its mouth, the city of Columbus was located in 1812. Two years earlier, in 1810, the Ohio General Assembly had appointed a committee to select a site for a state capital and stipulated that it should be located approximately in the center of the state. This committee met in Franklinton, a small village situated on the west bank of the Scioto in the northern portions of the Virginia Military Tract which had been laid out in 1797 by Lucas Sullivant. On September 12, 1810, the committee recommended that the state capital be situated where the present village of Dublin is located. The General Assembly was not enthusiastic about this recommendation and no immediate action was taken. The issue rested until February 1812 when an imaginative group of men from Franklinton proposed an ingenious scheme to the state legislature. This plan, which was accepted, not only proved the inventiveness of the land speculator in search of profit, but provided the state of Ohio with a capital in a manner that doubtless was unique in American history.

Land speculators at Franklinton formed a land company and offered its holdings in Township 5, Range 22, on the eastern side of the Scioto, directly opposite their city, to the state legislature for its capital. The company proposed that if the Ohio General Assembly met on their site by December 1817 and continued to hold its sessions there until at least 1840, the company would, in turn, lay out a town before July 1, 1812, deed to the state a 10-acre square for its capitol building and a similar tract of land for its penitentiary, and construct for the state an office building and a penitentiary. If the total value of the buildings, the proposal continued, did not amount to over $50,000, then the company would pay to the state the difference between the appraised value and the fixed sum. If, however, the total value of the buildings exceeded $50,000, then the state had to compensate the company in a manner agreeable to both parties. The state legislature accepted this rather unusual proposal on February 14, 1812. Thus Columbus, as the capital of Ohio, was founded.

The "Plat of the town of Columbus" (Plate IV) was the first map to be published in John Kilbourn's *Ohio Gazetteer* and appeared in the third printing of the 1817 edition. Although this rather crude map lacked detail and clarity, the explanation that accompanied it more than made up for the obscurity of the cartography. Kilbourn tried to present a picture of Ohio's capital city to those who were interested in the latest developments in the West and to those who might have been interested in the city as a place to settle. The Ohio geographer and cartographer explained his map as follows:

> Explanation
> Of the plate, on the preceeding page.
>
> The plate represents only the in lots; beyond or east of which is Fourth street; after which are reservations by the original proprietors, extending east about 2000 feet; then comes Seventh street parallel with High & Third streets. Immediately beyond which are the out-lots, 130 in number, (a few of which however lie to the north and west) and all containing from 2¼ to 3 acres each.

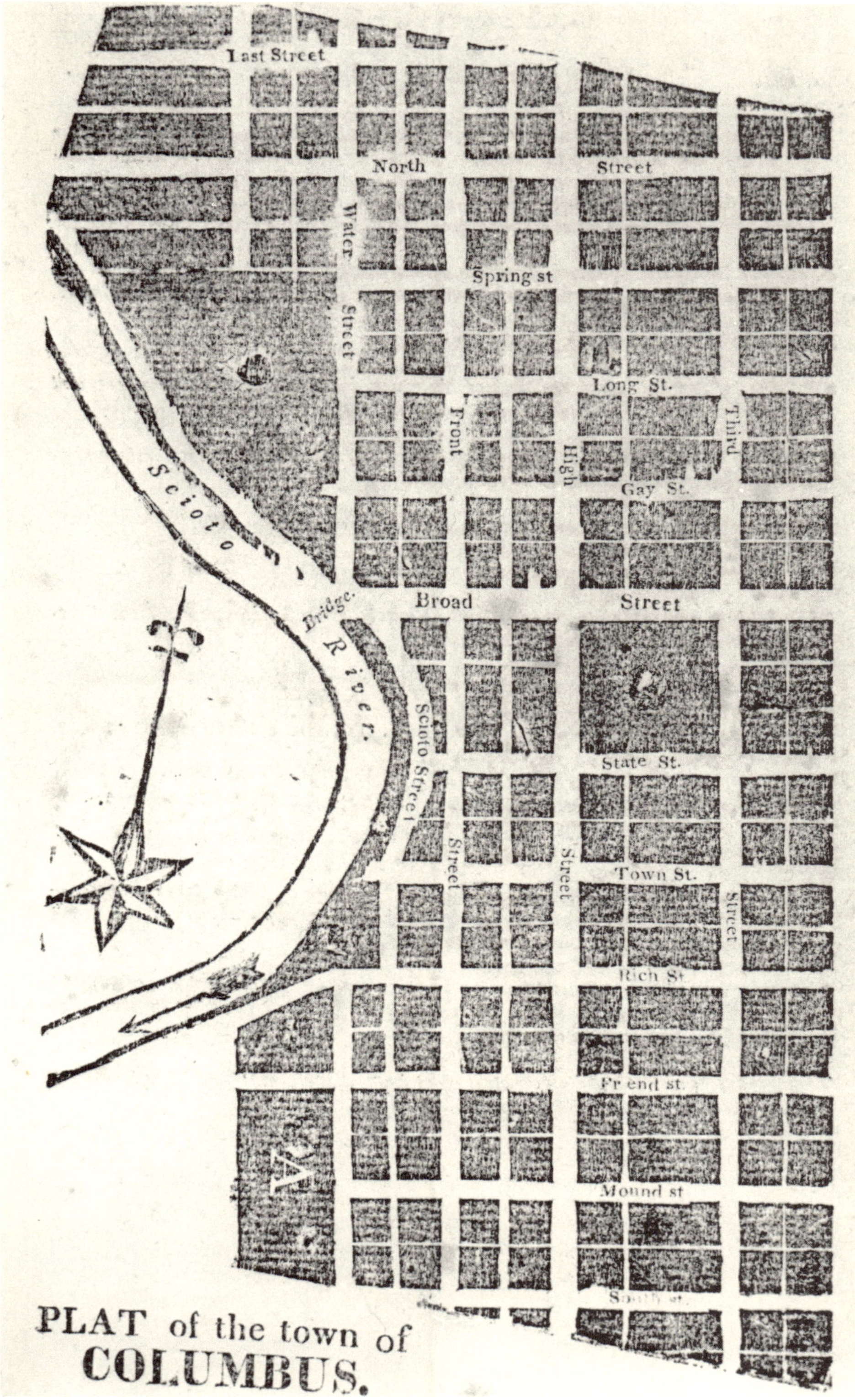

Plate IV: Kilbourn, "Columbus," (1817), in Kilbourn, *Ohio Gazetteer* (1817). Kilbourn's wood engraved map is a curious piece of cartography, but depicted Columbus shortly after its founding.

The ground included between Broad & State, High & Third streets is a Public Square, of 10 acres; on the southwestern corner of which stands the state house.

A—Is also a lot of 10 acres, for the penitentiary.

The slip of ground along the bank of the river, between it and Scioto street, and Broad and Town streets, is the property of the corporation of Columbus.

The street running parallel with the Scioto river from Broad street northwardly, is the present road leading to Worthington.

The road to Chillicothe is a continuation of High street southwardly.

The turnpike road to Newark leads out from the east end of Broad Street.

The bridge across Scioto river is built obliquely, with Broad street, in the direction of the word bridge on the plate.

High and all streets parallel therewith run north 12 degrees west.

Broad street is 120 feet wide, High street 100, and all the rest 83½. The alleys are each 38 feet broad.

The scale on which the plate is laid down is that of 800 feet to an inch.

The number of in-lots, including fractions, is 660.

All the mercantile business is now [1817], done on High street, between Broad and Rich streets and on Broad street between about opposite the middle of the Public square and the Scioto river.

On the Scioto river, about 12 or 15 rods below the intersection, of North street therewith, is a sawmill from which large quantities of lumber are obtained for building in the town.

The market house stands on State street immediately west of High street, and extending westwardly toward the river.

The second map of the state capital, "Plat of Columbus, As incorporated 8th Jan. 1818" (Plate V) was included in the sixth edition of Kilbourn's *Ohio Gazetteer*, which appeared in 1819. It was engraved by Hiram Platt, probably an employee of Kilbourn.[37] Kilbourn explained that in 1818, the year that the map was made, approximately 1,500 people lived in Columbus and nearly 200 houses had been constructed there. The town could boast of 5 "English" schools and one for young ladies. Business was good, he stated; the 10 merchantile stores were "vending about 200,000 dollars worth of merchandize annually." Also, Kilbourn continued, Columbus had a chemical laboratory which produced, among other items, agreafortis, oil of vitriol, copperas, rubrum and calomel.

The town lots in Columbus were laid out by Joseph Vance of Franklin County and Joel Wright of Warren County,[38] both employees of the state, and were sold at public auction on June 18, 1812, the day war was declared

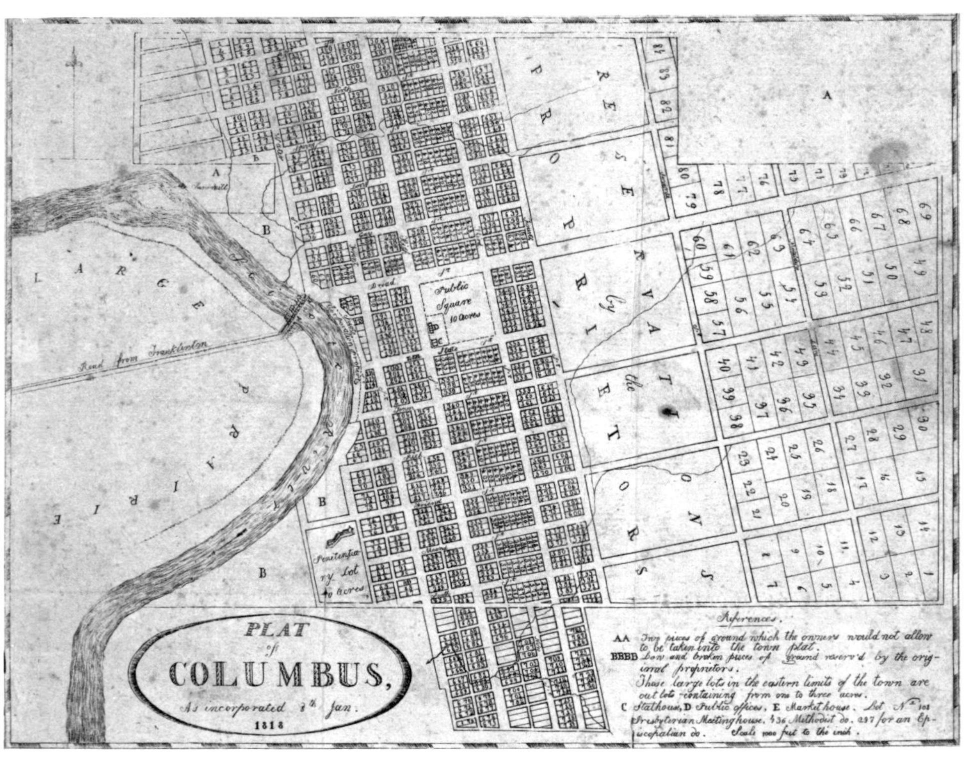

Plate V: Platt, "Plat of Columbus," (1818), in Kilbourn, *Ohio Gazetteer* (1819). Platt's map of Columbus is much improved over the one used by Kilbourn in 1817 and was used to advertise to immigrants the advantage to settling in Ohio.

against Britain by the United States. Some lots, still covered by forest, sold for as high as $1,000 each. On the west side of the public square which stood in the center of the town were constructed the capital building and a state office building. The capital, which Kilbourn showed as "C" in the "References," was built in 1814 as a 2-story, brick federalist structure, 75 feet by 50 feet. The bricks used in its construction were made from a mound that at one time had stood near the intersection of High and Mound streets. A bell tower that rose 106 feet from the ground was located in the center of its hip roof. The state's first capitol was indeed handsome, and the interior was in the style of the day. Unfortunately, the entire building was consumed by fire on February 1, 1852.

Sixty feet from this structure was the State Office Building, referred to in the map as "D". It was constructed in 1815 and was a 2-story building 150 feet by 25. Contained in this building were the offices of the Secretary of State, Treasurer, State Auditor, Governor, Adjutant General, Quarter-Master General and the State Library. It was torn down in 1857 when general improvements were made on the square.

For those who might have considered settling in Columbus, Kilbourn listed the various major congregations located there in 1818. On Lot 103, he indicated, was the "Presbyterian Meeting House." The First Presbyterian Church of Columbus had originated in Franklinton, on the west bank of the Scioto, in 1806. Eight years later, a small log cabin was erected on a lot near the intersection of Third and Spring streets in Columbus, owned by Reverend James Hoge who migrated to Ohio in 1805 and was the first minister of that church. Church services were conducted both in Columbus and in the church in Franklinton. Because the majority of the members lived in Columbus, however, a permanent church was erected at the cost of $1,796.50 near the intersection of Town and Spring streets during the early part of 1818. Another church, the Methodist, was constructed on Lot 436, located on the north side of Town Street, which was donated to that body by the proprietors of Columbus in 1814. By July 1815, a building was planned and a small one-room log cabin was constructed at the cost of $157.53. The building was later improved, and in 1818 a second room was added. The Protestant Episcopal Church, the third church mentioned by Kilbourn was established in Columbus in 1817. Kilbourn was one of the founders of this particular congregation and was elected by its membership as one of its first vestrymen. Although its services were conducted for years in a small church on Third Street, Kilbourn indicated that Lot 297 on the corner of North and High streets belonged to the Episcopal Trinity Church.

The first state penitentiary was built in 1815 on the 10-acre plat of ground along the southwest edge of Columbus reserved for that purpose. William Ludlow, the director of public buildings for the state, supervised the construction of the 3-story brick prison. Thirteen cells were located on the third floor; four were dark, apparently for use as solitary confinement cells, and nine were equipped with windows to the outside. A stone wall 15 to 18 feet high surrounded the structure. In 1818, however, the year the map was made, a new penitentiary was built. The new prison was more elaborate than the first and consisted of an enlarged cell block, prison yard, outbuildings and a double wall to enclose the entire site. When the penitentiary was moved from this location, the 10-acre tract was the subject of a long but interesting court fight to determine its ownership once the state vacated the buildings. In 1858, the land was subdivided and sold at public auction.[39]

The market place indicated on the map as "E" was first located in the middle of High Street near its intersection with Rich. In 1817, however, it was declared a nuisance by the Columbus City Council and removed to a market building on State Street, just west of High. In 1816, Lucas Sullivant received a franchise from the state legislature and built the first bridge, a toll affair, across the Scioto to connect the new state capital with his village of Franklinton. After his death, the enterprise fell to his son, Joseph Sullivant. In 1833, when the National Road passed through Columbus, both the citizens of Columbus and the city itself, no doubt tired of paying for the use of the bridge, paid Sullivant $10,000 for his franchise and the government then built a toll-free bridge in its place.[40]

The map of "Zanesville, Putnam and West Zanesville" (Plate VI) was drawn by William C. Pelham, who migrated to Zanesville from Virginia before 1817. Pelham, who also drew several manuscript maps of the state, lived in Zanesville between 1816 and 1827. Like other cartographers in the west, he was anxious to illustrate for prospective settlers the extent to which manufacturing and commercial activity had developed, as well as to indicate to those interested in settlement the educational and religious opportunities in the town. Although there is no evidence that this map was ever published for general distribution, there is little doubt that Pelham was interested in promoting the opportunities that Zanesville offered. He contributed information about his adopted home of Zanesville to Kilbourn's valuable *Ohio Gazetteer*. Because he made his several maps in Ohio at a time when most cartography in the country was being produced in

Plate VI: Pelham, "Zanesville, Putnam and West Zanesville," (1817), Cincinnati Historical Society, Cincinnati. This map was used to promote land purchases at the famous crossing point of the Muskingum River.

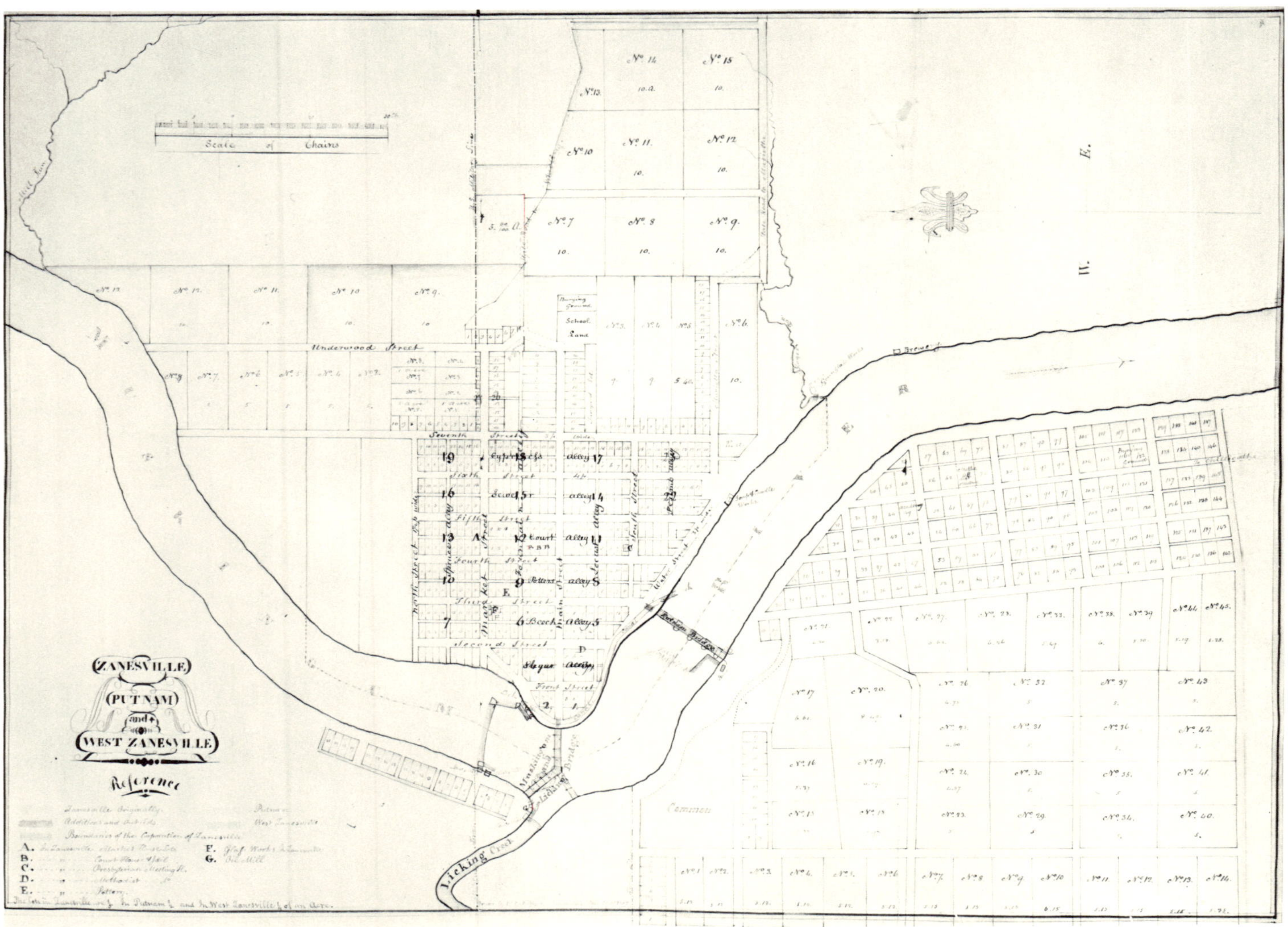

either Philadelphia or Boston, Pelham certainly was one of the state's finest early map makers.

Zanesville, located within a mile-square tract of land granted to Ebenezer Zane by Congress, was laid out in 1799 by Jonathan Zane, a brother of Ebenezer, and John McIntire, a personal friend and business associate of the Zane brothers. The boundaries of the original town plat, which consisted of 19 squares, were Shinnick Street on the north, Seventh Street on the east, South Street on the south and the Muskingum River on the west. The tract lay partially within the United States Military Tract and was one of three grants given by Congress to Ebenezer Zane, who had proposed that in exchange for land at the crossing places of the Muskingum, the Hockhocking and the Scioto rivers, he would open a road between Wheeling, Virginia, through Ohio, to Limestone, Kentucky, maintain a tavern at each tract and operate ferry service for travelers at each crossing place. The village, at first located entirely on the east bank of the Muskingum, was originally named Westbourne, but on January 1, 1801, its name was changed to Zanesville. The lots between Seventh and Underwood streets and those that border Underwood east of the original plat, were part of a larger tract given by the United States government to Robert Underwood, a land speculator, who subdivided the land as an addition to Zanesville. David Harvey purchased Lots 1 and 2 from Underwood; in 1808, Underwood further subdivided his tract and created "Harvey's North Meadow." Squares 26 and 27 on Pelham's map, along both sides of Market Street, were laid out in 1815 by John McIntire and became known as "John McIntire's northeast addition." The 8 lots at the head of Market Street were laid out in February 1817 by David J. Marple. Lot 3 in "Marple's east division" was eventually given up for the extension of Market Street eastward. McIntire also owned and subdivided the large tract of land bordered by Fountain Alley and the State Road on the north, Seventh Street on the south, and Marietta Street and the State Road to Marietta on the south. Lot 6, south of Marietta Street, belonged to McIntire. This subdivision, known as "McIntire's east division," was supposedly not laid out until May 13, 1868,[41] although Pelham's map clearly illustrates the planned plats.

Like so many communities founded along Ohio rivers, it was located on one side only, and soon other surveyors and land speculators established communities either immediately adjacent to the original settlement or directly across the river in order to compete for settlers, business and political favor. Shortly after Zanesville was founded, Dr. Increase Mathews and Levi Whipple, both nephews of Rufus Putnam of Marietta, purchased the land west of the Muskingum and south of the Licking River. There, in 1801, they laid out Springfield, a name which was later changed to Putnam, after their uncle. The original plat consisted of 147 in-lots or town lots and 45 out-lots of nearly 5 acres each. This river town competed directly with Zanesville and was its leading rival for the location of the state's capital in 1810 when the Ohio General Assembly moved the capital from Chillicothe to the east bank of the Muskingum. Another village, West Zanesville, was laid out north of the Licking River in 1809 by John McIntire and was represented by that row of town lots numbered 1 through 20 on Pelham's map.

This map, like others that were drawn to promote local development, indicated what Pelham considered to be the village's most important points. The market place that he showed on Market Street at Court Alley was built in 1814 at the cost of $150 to the town. On Fourth Street, between Main Street and Fountain Alley, was located the State House which served as the second state capitol building between 1810 and 1812. The State House, a 2-story, 50-foot square edifice, was built in 1809 and patterned after Independence Hall in Philadelphia. When the state's capital was removed back to Chillicothe, this building served as the Muskingum County courthouse until 1874, when it was razed.

Two churches were named in his "Reference," yet Pelham indicated three on his map. The first, the Presbyterian Meeting House, located on the corner of Fourth and South streets, was new at the time this map was drawn. Reverend James Culbertson was its first preacher, and on Sunday, August 31, 1817, President James Monroe, who passed through Zanesville on his way to Washington from a visit to Detroit, worshipped with its congregation. The Second Street Methodist Church, the second one Pelham mentioned, was established in 1808 by Reverend James Watts of the Wills Creek Circuit. In Putnam was located the Moxahala Avenue Methodist Church, the third one illustrated, which was built in 1816 on lot 68.

In 1817 industry in Zanesville was in its infancy, yet, in a new land, that fact, along with its potential growth, was the promise that it held for prospective immigrants. The pottery Pelham showed between Third Street and Potters Alley was founded by Samuel Sullivan in 1808 and was the precursor of the pottery industry that subsequently developed and made Zanesville famous worldwide. In 1817, Sullivan was manufacturing what was known as red-ware, a type of pottery so named because of its red appearance, and the factory produced mostly table plates, pitchers, cups and saucers. On May 13, 1815, the Zanesville Glass Manufacturing Company was organized, with its shop at the corner of Market Street and Third. It was popularly known as the White Glass Works because it produced a white glass that was used for medicine bottles and whiskey flasks for commercial use, and sugar bowls and pitchers for the domestic market. The next year, in 1816, two enterprising citizens opened a rival glass company at the mouth of Slago Run. It was known locally as the Green Glass Works because its products were colored green.

Other industries mentioned by Pelham included the brewery built by Joseph Lattimore in 1816 and located along the Muskingum, and the "Soap and Candle Works" founded in 1811 by Daniel Prouty, but owned by N. and C. Wilson at the time this particular map was made. In 1816 Colonel Andrew Jackson and several other businessmen established a grist mill along the west bank of the Muskingum up from the mouth of Licking River. The following year a sawmill and linseed oil mill were added. The Jackson Company Mills, as they were known, apparently received their power from both the Muskingum and Licking.

Two dam sites were built at Zanesville. The first, the Licking Dam, was constructed on the Licking River by Isaac Zane, the son of Jonathan Zane. Tradition has it that the elder Zane promised to his son a half interest in the dam if he would see that it was built. Once the project was completed, however, Father Zane sold the dam from under the not-so-watchful eyes of his son and shared none of the proceeds with him. The second dam was authorized by the state's General Assembly and was built by John McIntire in 1813 up from the mouth of the Licking, at the base of Market Street. Along with this particular dam, the state legislature specified

> John M'Intire and his associates shall construct a good and sufficient lock, calculated to admit and secure the safe and easy passage of all water crafts

> ascending or descending said river [Muskingum], which lock shall be at least twenty feet wide and ninety feet long; and the said John M'Intire and his associates shall at all times keep the said lock in good repair, and shall, upon application made to him or them by any person or persons wishing to pass said lock with any craft as foresaid, without delay open said lock, that such person or persons may pass free of expense. . . . John M'Intire and associates shall construct a good and sufficient lock of stone, not less than twenty feet wide and ninety feet long and of a depth equal to the fall of the river and heighth of the dam. . . . That so soon as the said John M'Intire and his associates shall have completed the canal aforesaid, and the lock at the mouth thereof, they shall be entitled [to] demand and to receive the following rate of toll upon all water craft which pass through said canal: for every loaded boat or water craft of any description, twenty five cents per ton upon her actual load where her load shall exceed one ton, and for every boat or water craft when its actual load is one ton or less, six cents for every hundred weight of such load, and six cents for every ton of her burthen, and for every empty boat or water craft six cents per ton of her burthen, except periogues [*sic*] and canoes, and for every empty periogue [*sic*] or canoe twenty-five cents.[42]

Of course, to help promote commerce, it was necessary to have bridges across rivers rather than rely on the slower ferries. Consequently, on January 21, 1812, the Ohio General Assembly authorized the Muskingum and Licking Bridge Company to build a bridge across the Muskingum.[43] Two years later, in 1814, the first of the famous Y-bridges was built in part by Jacob Houck, a local stone mason, to facilitate travel and commerce along the historic Zane Trace. This bridge fell down several years later and was replaced in 1817 by the Putnam Bridge.

At the corner of Fountain Alley and Underwood Street, David J. Marple and Wyllys Silliman located the first water works in Zanesville in 1817 and attempted to inaugurate perhaps the first piped water system in a municipality in the state. This private utility was authorized by the city authorities and was an attempt to provide a service to the public for a profit. Water from a fresh spring was collected in a reservoir 75 feet long, 25 feet wide and 9 feet deep made from stone and brick. Pipes were made by drilling a 2½ inch hole through selected elm, poplar and oak logs, which were tapered in order to be fitted together. The principal water main from the reservoir through the city ran down Main Street to just below Third. Apparently, hollow logs about 7 feet high fitted vertically into the water main were used as hydrants, strategically placed along the pipe line. To prevent the water from spouting from the top of these wooden hydrants, wooden plugs were driven into the hole at the top. Copper or wooden spigots were inserted into the hydrants so that water could be drawn. The entire utility, however, was more ambitious than successful. Because the reservoir was only a few feet higher than the hydrants, the water pressure was so low that only one hydrant could be used at a time. The water that did flow from the system was so tinctured that it could be used for neither drinking nor cooking.[44] Needless to say, the entire venture was a loss, although the reservoir remained until 1838 when it was disassembled and the stones used in other building projects. Although a failure, this attempt to provide a central water system for Zanesville represented the ambition and imagination held by the city's founding fathers, who not only gambled to make a profit, but, in their serious competition with other western cities, tried to offer something more to those immigrants who desired to settle in the West.

Educational facilities were not wanting in Zanesville. Along with several other private schools, Pelham chose to draw attention to the interesting Stone Academy located on Lot 49 in Putnam. The building itself was a rather large stone structure. It had been constructed as the state capitol by those who had laid out Putnam, hoping that the presence of the building would be enough to have the state's capital located there. Despite these preparations, however, Zanesville received the honor and this structure became Stone's Academy, a private school of some local note. In 1817, David Hall opened the Academy to both boys and girls at the cost of $3 per quarter. It served as a school until 1850 when it was converted to a private residence.[45]

Not every community founded in Ohio was strategically located to take advantage of important physiographical features in order to ensure maximum economic development. Numerous communities were established by religious sects that wished to create a peaceful and isolated society in which their own members would be insulated against the evils and temptations of the surrounding material world, where they could serve God in the manner they believed best. Certainly not least among these was the United Society of Believers in Christ's Second Appearing, or the Shakers. This sect first entered Ohio following the Great Kentucky Revival, which occurred during the last decade of the eighteenth century. Because of the isolation of the frontier settlements, the emotionalism and excitement caused by the Kentucky meetings and serious concern that many had for the salvation of their souls, Shakerism found some fertile ground for its settlements in the Miami Valley in western Ohio. The sect never had much of a following in the state, but its membership was sufficient to establish Union Village in Warren County, Watervliet in Montgomery County and North Union in Cuyahoga County. By the end of the nineteenth century, however, fewer than 90 Shakers remained in the state to witness their villages crumble. The plans of the Shaker villages (Plate VII, VIII and IX) were drawn by an unknown traveler who visited the communities in 1834 and took time to sketch their patterns. These drawings are included as part of a small hand-sewn booklet in the Library of Congress and, because of the colorful history of the Shakers in Ohio and the charm of the illustrations, would have to be included in the history of the mapping of the state.

The map, "Sketches of the Various Situations at Union Village" (Plate VII), was the plan of the community located 3 miles west of Lebanon in Warren County, popularly known as Shakertown. Located on a 4,500-acre tract and first occupied by the Shakers in 1805, this village became the center of the Shaker settlements in the West and members looked to it second only to New Lebanon in New York for leadership and guidance. Union Village was organized primarily under the leadership of Elder David Darrow who, until the time of his death in 1829, was the head of the community. At the time that this map was drawn, Elder Solomon King had been appointed to replace Darrow upon the latter's death.

By 1834 Union Village had reached its zenith. The greatest number of the faithful had already been accumulated within the village and the principal buildings had been constructed. After 1835, shortly after the sketch was made, the settlement began to be plagued with serious difficulties that threatened the existence of the order. The church proper was composed of the families of South House, Center House, North House, and Brick House. West Brick House, West Frame House and East House, the latter not shown on the map, were known as intermediate families, but worshipped with the rest of the community. North Lot and West Lot, both not shown, were what the Shakers called novitiates, or the Gathering Families, where the young were housed and educated before joining one of the principal families. The last family was the Mill Family, which was composed of both the Square House and Grist Mill. The population at Union Village in 1830, 4 years before this map was made, was 502: 238 males, two of whom were black, and 264 females, six of whom were blacks. Each family performed some necessary function for the survival of the community. The surpluses of their labors were sold to the general public, and the merchandise from their blacksmiths, masons, stonecutters, carpenters, tanners, clothiers, spinners, weavers and millers was known far and wide for its high quality.[46]

The second Shaker community sketched was "Watervliet" (Plate VIII), located about 6 miles south of Dayton in Van Buren Township in Montgomery County. Named for a Shaker settlement in New York, this community was founded in 1807 after a series of revival meetings were held along Beaver Creek, the small stream that ran through the village. At the time the map of this community was made, Richard McNemar, one of the most influential members of the faith in the West, was head of the village. It was McNemar who established the printing office just north of the main dwelling house. There in the print shop he began to publish quantities of Shaker literature that was distributed among the nonbelievers, including the famous *Brief Exposition of Shakerism,* which first came off the Shaker presses in Albany, New York, in 1830. Although the exact number of families at Watervliet in 1834 is not clear, there probably were three in the settlement at the time the map was made: West Family, North Family and Center Family. The saw mill along the creek was the first building erected by the Shakers, and between 1812 and 1819 the grist mill, tannery, copper shop, woolen factory, broomshop and blacksmith shop were erected and in operation. In 1814 a wooden ware industry was commenced, and the wagon shop illustrated at the eastern edge of the settlement produced a quality wagon that brought considerable revenue into the society. The large "dwelling house" north of the garden on the main north-south road through Watervliet was probably the large brick structure built in 1820 with the help of Shakers from Union Village. This building housed the Center family. Directly across the street from the dwelling house was a large 2-story meeting house, 40 feet by 50 feet, which was built in 1825. The second floor of this brick structure was never finished. The house on the south side of the road at the western edge of the village was built in 1825 for the Young Believers, or those recently converted to Shakerism. Like the other Shaker settlements in Ohio, it too faced serious problems almost from the outset; by 1900, those Shakers who remained at Waterlviet had sold their property and sadly moved to Union Village.[47]

The third map (Plate IX) is that of the most famous of all the Ohio Shaker settlements, North Union. Today, it is remembered only through the name of the prosperous community of Shaker Heights. This settlement was originally founded in 1822 in Warrensville Township in Cuyahoga County just east of Cleveland on land owned by Ralph Russell. The sketch illustrated most of the Shaker community of North Union was it was in 1834.

That year there were probably only 3 families at North Union, whose presiding elder was Ashbel Kitchell. The first was the Mill Family, known

(Overleaf)

Plate VII: "Union Village," (1834), Library of Congress, Washington, D.C. Drawn by an unknown traveller, Union Village in Warren County, this plan illustrated the headquarters of the Shaker religion in the West at the zenith of its influence.

Plate VIII: "Watervliet," (1834), Library of Congress, Washington, D.C. Once a prosperous Shaker community in Montgomery County, Watervliet's influence decreased as Dayton prospered.

Plate IX: "North Union," (1834), Library of Congress, Washington, D.C. Founded in 1822 in Cuyahoga County, this once influential community is remembered through modern Shaker Heights.

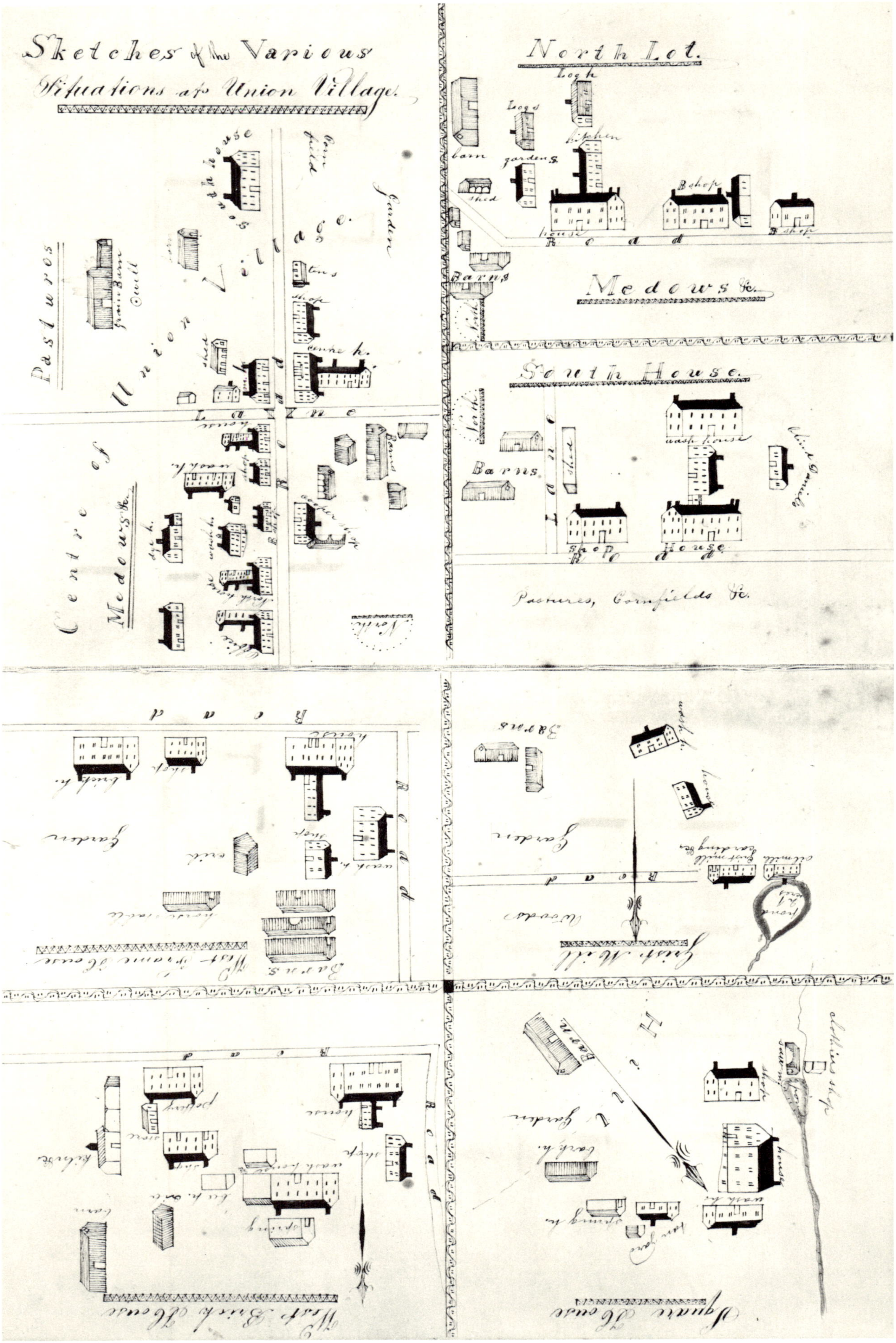
Sketches of the Various Situations at Union Village
North Lot.
Medows &c.
South House.
Pastures, Cornfields &c.
Pastures
Centre of Union Village
Medows &c.
West Frame House
Grist Mill
West Brick House
Square House

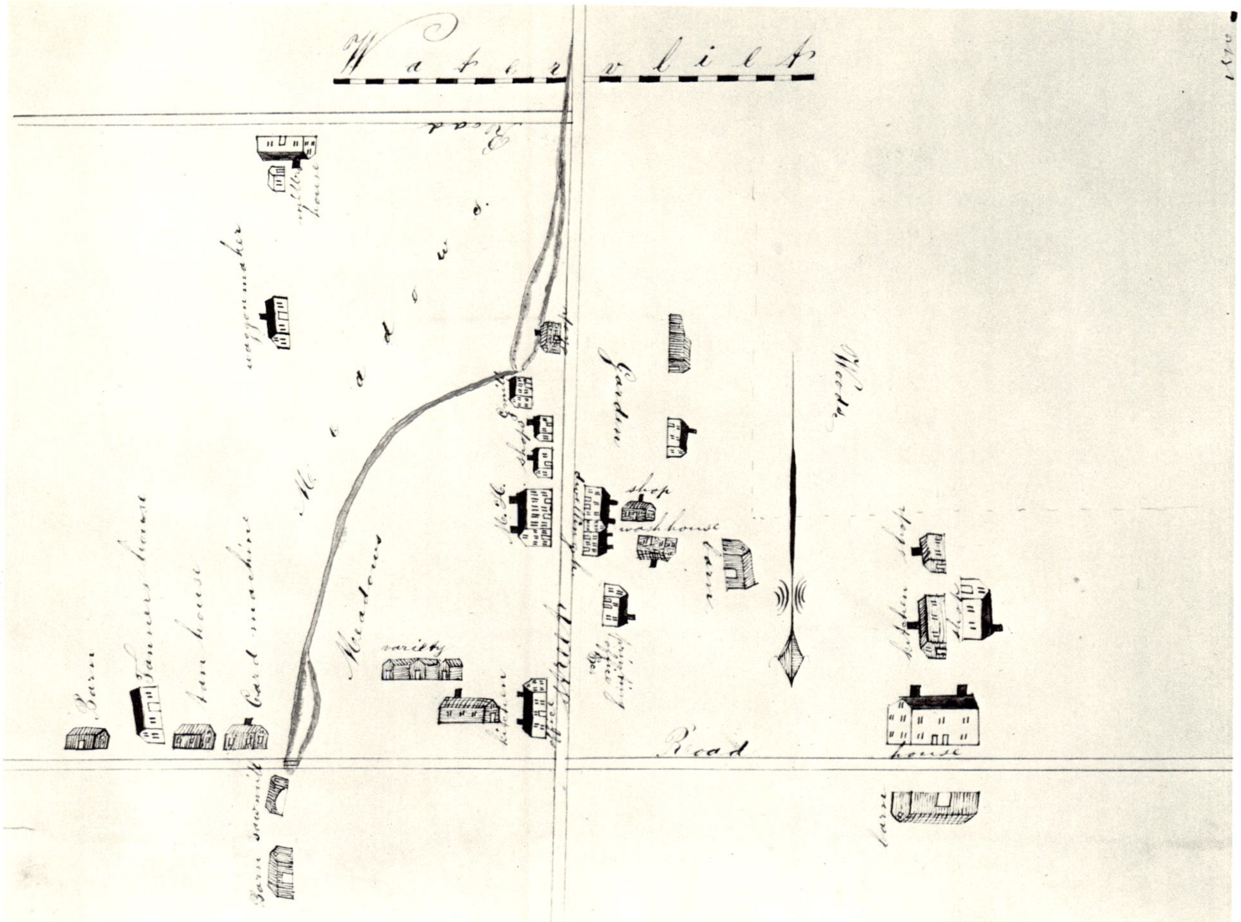
Watervliet
Road
Meadows
Garden
wash house
shop
Road

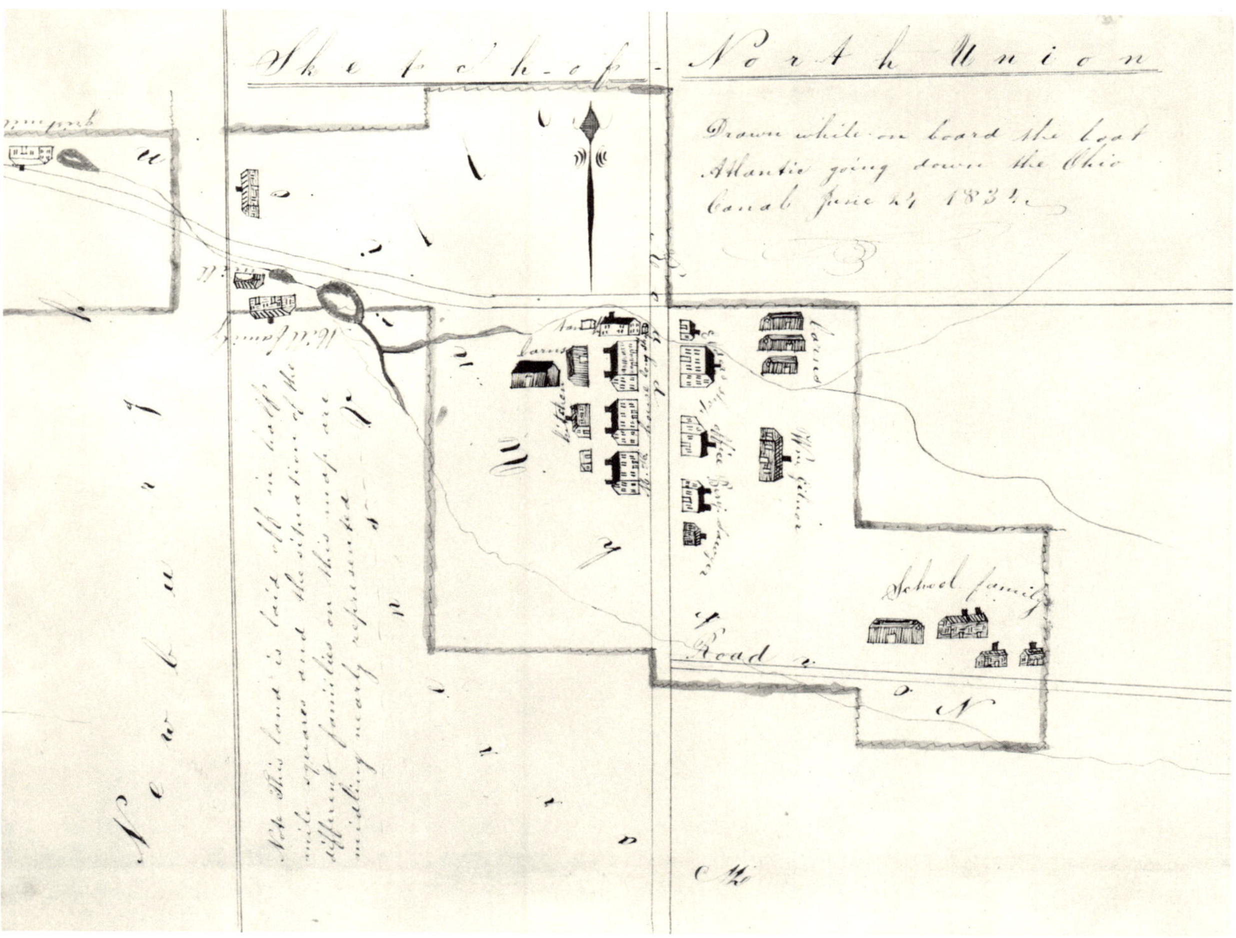
Sketch of North Union
Drawn while on board the boat Atlantic going down the Ohio Canal June 24 1834.
School family
Road

also as North or Second Family, which made its living from the grist mill and sawmill, both located along Doan Creek at the west edge of the village. The lumber and other commodities that came from these mills served as a good source of income for the Shakers and were readily purchased by the farmers in the area and the citizens of Cleveland. The second family, Center Family or Middle Family, was clustered around the crossroads. It made its contribution to the settlement through its agricultural activities and also from the manufacture of brooms, yarn, leather goods, bonnets, gloves and other commodities that were sought by non-Shakers. In September 1832, the Shakers began the construction of the "long shop" directly across from the elders' shop. This structure, completed in 1833, was 30 feet wide and 120 feet long and was used for workshops. Later, it was divided into 3 separate parts. One section was moved from its original site and used as a boys' house. Another was shoved a little to the south and served as a family dwelling, while the third became the broom shop. Situated at the east edge of the village was the School Family or East Family. It was originally used by those who sought to join the Shakers, and it was here that they would receive their instructions and preparations. Because of this, the family was also called the Gathering Order. In 1879, this Family was abandoned and those who remained moved to the Center Family.[48]

Urban growth and development has been steady in Ohio since the state's first permanent settlement was founded along the banks of the Muskingum nearly 200 years ago. Many cities, including Newark, Portsmouth, Circleville and Cincinnati, were laid out on sites which had been first occupied and used by the Indians. Others began as speculative ventures, promoted either by companies or individuals, and at least one city, Columbus, was founded to be a capital. Still others were planned to serve as refuges from the material world where man could seek spiritual peace. Whatever the purposes, the maps that platted the cities also measured the state's urban development, and are records of the urbanization process not only in Ohio, but in the United States as well.

1. Archer Butler Hulbert, ed., *The Records of the Original Proceedings of the Ohio Company,* Vol. I, *Ohio Company Series* (Marietta, Ohio, 1917), 15.

2. *Ibid.,* pp. 19–21.

3. See Charles Whittlesey's map of the Marietta Works, Chapter II, Plate VIII, p. 000.

4. Hulbert, *Ohio Company,* I, 51.

5. *Ibid.,* p. 89.

6. *Ibid.*

7. Hulbert, *Ohio Company,* II, 51–52.

8. *Ibid.*

9. *Ibid.,* p. 79.

10. *Ibid.,* p. 80.

11. *Ibid.*

12. Letter from Josiah Harmar to Thomas Mifflin, November 9, 1789, quoted in Ebenezer Denny, *Military Journal of Major Denny* (Philadelphia, 1859), p. 241.

13. Daniel Drake, *Natural and Statistical View, or Picture of Cincinnati and the Miami Country. Illustrated by Maps. With an Appendix Containing Observations on the Late Earthquakes, The Aurora Borealis, and Southwest Wind* (Cincinnati, 1815). For more information about this remarkable man, see Edward D. Mansfield, *Memoirs of the Life and Services of Daniel Drake, M.D.* (Cincinnati: Applegate & Co., 1860);

Emmet Field Horine, *Daniel Drake (1785–1852), Pioneer Physician of the Midwest* (Philadelphia, 1961); Henry D. Shapiro and Zane I. Miller, eds., *Physician to the West, Selected Writings of Daniel Drake on Science and Society* (Lexington, Kentucky: University Press of Kentucky, 1970).

14. Daniel Drake, *Notices Concerning Cincinnati* (Cincinnati, 1810).

15. Drake, *Natural and Statistical View,* p. ix.

16. *Ibid.*, pp. 134–35.

17. *Ibid.*, pp. 137–38.

18. *Ibid.*, p. 136, 161–63.

19. *Ibid.*, pp. 136, 163.

20. *Ibid.*, pp. 136, 163–164.

21. *Ibid.*, pp. 155–56.

22. *Ibid.*, pp. 136–37.

23. *Ibid.*, p. 150. See Jacob Burnet, *Notes on the Early Settlement of the North-Western-Territory* (New York, 1847), p. 398.

24. Drake, *Natural and Statistical View,* p. 151.

25. *Ibid.*, pp. 140–42.

26. *Ibid.*, pp. 137, 146.

27. *Ibid.*, p. 147.

28. *Ibid.*, p. 146.

29. Letter from Moses Cleaveland to Oliver Phelps, August 5, 1796, quoted in George F. Marshall, "Moses Cleaveland: Some Old and Original Manuscripts," *Annals of the Early Settlers' Association,* III (1882), 74.

30. The original of this map is at the Western Reserve Historical Society. For printed copies of this map, see William Ganson Rose, *Cleveland: The Making of a City* (Cleveland, 1950), p. 28. Charles Whittlesey, *Early History of Cleveland, Ohio* (Cleveland, 1867), p. 239.

31. Whittlesey, *Early History of Cleveland,* 241; S. O. Griswold, "The Corporate Birth and Growth of Cleveland, *Annals of the Early Settler's Association,* V (1884), 30.

32. For a general discussion of the surveys of Cleveland, see Samuel J. Baker, "The Original Surveys of Cleveland," *Association of Engineering Societies,* III (August 1884), 217–31.

33. Gertrude Van Rensselaer Wickham, *The Pioneer Families of Cleveland,* Vol. I (Cleveland, 1914), 39.

34. *Ibid.*, p. 8.

35. *Ibid.*, p. 24. A close study of Wickham's geneological volume can easily identify the ownership of those original town plats.

36. Seth Pease, "Seth Pease's Journals to and from New Connecticut, 1796–1798," *Tract No. 94* of the Western Reserve Historical Society (Cleveland, 1914), p. 67.

37. William T. Martin, *History of Franklin County* (Columbus, 1858), p. 166.

38. Alfred J. Wright, "Joel Wright, City Planner," *The Ohio State Archaeological and Historical Quarterly,* LVI (1947), 287–94.

39. Jacob H. Studer, *Columbus, Ohio: Its History, Resources, and Progress with Numerous Illustrations* (1873), pp. 368–74.

40. *Ibid.*, pp. 38–39.

41. J. F. Everhart, *History of Muskingum County, Ohio with Illustrations and Biographical Sketches of Prominent Men and Pioneers* (Columbus, 1882), pp. 105–6.

42. "An Act to enable John M'Intire and his associates to erect a dam across the Muskingum river," February 21, 1812, *Acts Passed at the First Session of the Tenth General Assembly of the State of Ohio,* Vol. X (Zanesville, 1812), pp. 175–76.

43. "An Act to Authorize Levi Whipple and others to erect a toll-bridge over the Muskingum," January 21, 1812, *ibid.*, pp. 35–40.

44. Everhart, *History of Muskingum County,* p. 216.

45. *Ibid.*, pp. 116–17.

46. J. P. MacLean, *Shakers of Ohio; Fugitive Papers Concerning the Shakers of Ohio, With Unpublished Manuscripts* (Columbus, 1907), pp. 59–111.

47. *Ibid.*, pp. 190–226.

48. *Ibid.*, pp. 112–89.

CHAPTER SEVEN

Internal Improvements

As much as the record of surveying or mapping a state is part of the history of its cartography, so too are the historic illustrations of its internal improvements. It was recognized quite early by those anxious to develop the regions west of the mountains that a transportation system that would provide an avenue over which immigrants and commerce could travel was tantamount to the growth and prosperity of that area and, at the same time, to the strengthening and solidifying of the national union. In Ohio, the first state to be carved from the public domain, such a system was necessary. Immigrants, looking for new opportunities in the lands north of the Ohio River, needed roads into the state's rich interior. These same people, once established as farmers and merchants and manufacturers, demanded easy access to the nation's market places. The founding fathers of the state convinced Congress that it should give 3 percent of the money received from the sale of public land within Ohio to the state for the construction of roads. During the nineteenth century, roads, canals and railroads each went through their building stage. Financed by both private and public monies, they all contributed to the state's growth and development. That development of the state's internal improvements was recorded on plans, plats, stock leaflets and maps, which are part of the history of the mapping of Ohio.

One of the first major internal improvements made in Ohio was the construction of the National Road. By 1818 this famous highway stretched from Baltimore across the mountains to the banks of the Ohio at Wheeling and was planned to be the major route between the settled East and the American West. In 1820, Congress appropriated $10,000 for surveying a route that would extend the National Road from Wheeling to St. Louis. Once these surveys were complete, Congress, on March 3, 1825, set aside $150,000 to construct the road from the Ohio as far as Columbus. For some distance between these two points, the road followed the older Zane's Trace, which cut from Wheeling to Zanesville and then through Lancaster to Aberdeen, across the Ohio River from Maysville, Kentucky. From Zanesville, however, the National Road left the Zane's Trace and was laid out on a most direct, westerly line to Columbus, avoiding Newark to the north. Newark lay on the old road from Zanesville to Columbus, which left Zanesville, followed along the east bank of the Licking River to Newark, passed through Granville and continued southwestward to the state's capital. Because of the loss of this connection to Columbus, public sentiment ran high in Newark, as well as in the capital, where those who had businesses along the original route stood to lose if the new road entered the city directly from the east as planned, rather than from the northeast. Rivalry between towns and individual interests for internal improvements to pass along certain routes was common, and competition for these rights of way was keen.[1] Certainly, the path followed by a particular road would mean fortune for some and want for others.

The "Plat of Columbus As incorporated Jan. 1818"[2] (Plate I), as altered by Horton Howard of Columbus, and his letter to Secretary of War Lewis Cass which accompanied that map best illustrated the sharp competition within the city of Columbus between that city's own citizens for the precise route of the National or Cumberland Road. This particular map first appeared in the 1819 edition of John Kilbourn's *Ohio Gazetteer*. Howard, who in 1828 published the famous map of Ohio that had been compiled by Alfred Kelley, was concerned that the route followed by the National Road westward from Zanesville might possibly be altered to include Newark and thus enter Columbus via Broad Street rather than Friend's Street. In the letter to Cass, who, as Secretary of War in the Jackson administration, had charge of the construction of the road, the Columbus publisher explained the situation as he interpreted it in the state's capital and appealed to the Secretary's common sense that the route should remain as it was first surveyed. In order to illustrate his arguments more clearly, Howard reconstructed the business district of Columbus as it was in 1831 on this map for the Secretary and explained the effect the two routes would have on business. His arguments developed as follows:

Columbus 7th Month 26th 1831

Respected Friend:

It will, no doubt be a source of pleasure to thee to know that thy appointment to the Department over which thou preside, affords very

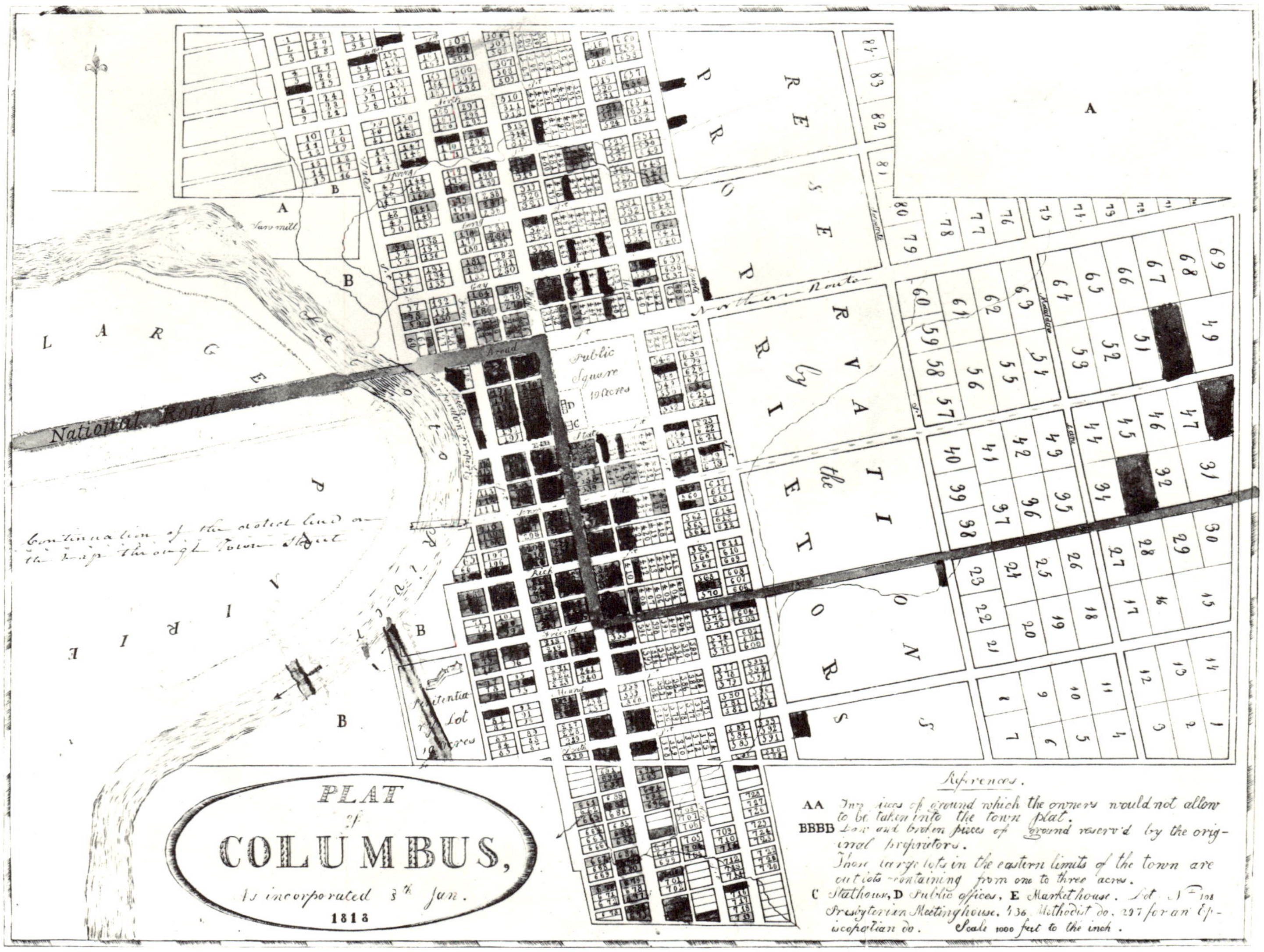

Plate I: Platt, "Plat of Columbus," (1831), National Archives, Washington, D.C. This map, an altered version of the 1813 edition, depicts local boosterism to attract federally funded internal improvements.

general satisfaction to citizens of Ohio. We are pleased because we believe, the general welfare will be thereby promoted. It is peculiarly gratifying to the citizens of Columbus, because we know thou art acquainted with the wants of the western Country and have no doubt thou wilt use thy influence in a proper manner in promoting its best interests.

We are with the exceptions of a few individuals, (without regard to political parties,) extremely disgusted at the treatment we have received in regard to the location and construction of the national road. It was (from the best of my recollection) located by Jonathan Knight[3] from Zanesville through Columbus [in] 1825.

This able and industrious engineer, discovered, that, from the nature of the soil and from the most suitable sites for bridges crossing, Walnut & Allum Creeks it would be proper to bring the road into and with Friend Street (as represented by the town platt which I send thee) to High Street which is the main business street in this town, about 1000 feet south easterly from the centre of the Town platt, and then turning north westerly with this street passed the centre about 1000 feet, and thence turning westerly crossing Scioto on the toll bridge and from there to Indiana. This gave, so far as I have ever known very general satisfaction to the Citizens of Columbus, as it passed the door of every tavern, every store, nearly every mechanic's shop, except a few which lay on the same street still further south east. Also all the public buildings except the jail and penitentiary, and the meeting houses for worship. In fact embracing the whole of what is considered the *business part* of Columbus, and it is highly probable that no attempt would have been made to alter the location, if it had not been for machinations of some of the people of Newark, who if they could succeed in changing the location to suit their views would render it convenient to bring it over David Pughs watched toll bridges crossing Allum & Walnut Creeks, he therefore, inlisted in the strife, and the noise they made, drew an individual or two, in this town who possessed some Property on their contemplated route, to aid in changing the Location as, aforesaid, under the spurious pretense that it ought to pass through Columbus in a straight line, to the frail *toll* bridge across the Scioto west of the town;[4] thus passing through a street, which (even now) has neither a tavern store or public building, and there are now but four houses and they are private residences, which are not passed by the road as it was located by Jonathan Knight.

I suppose that his inducement to carry the road further north than the state house (as he had there passed the business part of the town) was to save the United States, for the present, the expense of building a bridge across Scioto; The principal business is done between State Street and friend Street where the location comes to High Street; and as the Canal is to be opened in a few weeks, from Lake Erie to this town and to the Ohio river, this fall or next summer, the business and improvement of the town must increase southerly and westerly to a great extent from this part of high street and from friend street, which crosses the Canal and to the ford, where almost the whole transportation and travail passes during the whole year, except in time of freshets; when they are forced to cross the toll bridge at a heavy expense. Down the Canal also on both sides the town will no doubt rapidly extend it self, and become a great business place. There will, I presume be no doubt of these facts, in thy mind, when thou takes a view of the Town Platt, river, Canal & from actual survey in this Platt I have carefully marked every lott on which, there is a family residence, store, mechanic's Shop, meeting house or school house, for thy satisfaction and to guard thou against misrepresentations.

Those dissatisfied persons teased and pestered President Adams untill a review was ordered, and the first location confirmed.

It was then rationally considered as finally settled and many other persons, as well as my self, confiding in the known ability, industry and disinterestedness of the road commissioner, and a determination on the part of the Executives and of Congress, not to remove the location, were thus induced to invest Capital to a considerable amount in the purchase of lands or lotts on this location, and particularly in the purchase of lotts and making *substantial improvements* on those parts of the streets in Columbus on which the National road was located, and as it now stands.

The opposition in Congress, to internal improvements by the General government, prevented appropriations for making this part of the road, untill General Jackson became president. Then another attempt was made by some of those troublesome people; but this after stoping the contracts a while and causing some re-examination, also failed, and a new road commissioner, J. Harrison having been appointed, was in like manner teased, until, as I understand, he decided that the location ought to come from Zanesville, into Columbus, as it was made; the ground and sites for bridges being more favorable than any other route, and, that he would sooner continue, it with friend street across the Canal, and cross the river there or turn as it does along high street to the public buildings and then turn with State street west, than to cross our town Platt indirectly on broad street.

This alarmed those in town who had been endeavouring to procure a change, and they joined many of our other Citizens, and the Town Council also, in petitioning to have *no alterations* made in the location of the road. This as we again supposed, finally settled the location.

This security prevented any opposition to a large mail contractor, (who lives in the north part of the town) who last season, aided by some others who wish the road to remain as it is located, prevailed on the war department to admit of the making fifteen miles, extending from the Toll bridge on Scioto which was accordingly done, and notwithstanding there was, as I have understood a considerable of the appropriation remaining unexpended, some *secret influence* prevented any contract from being made for the construction of the road between Hebron and Columbus and, not withstanding the ample additional appropriation made last session to finish the road to Columbus, this secret influence, with others more openly (having secured its coming to the toll bridge as aforesaid) found the means last session of Congress, to introduce into the Senate the subject of changing this location, and although nothing further was done in it, we can have none of this road put under contract.

We only find an *anonimous Publication* in a *Zanesville paper,* (as much as possible to conceal the authors) requesting Contract or to make proposals for constructing the road on J. Knight's location, and also on what they call Pugh's route. By this we understand, another farce is to be acted in order, if possible to practice a fraud on the Department, the bids if made from this call, will, no doubt, be much to high, on the present location, or, they will be too low on the other route, and made by *irresponsible persons* or by those, who if they succeed in changing the location have sense enough to know that an actual relocation must be made, and new proposals invited, or they will *back out.* The Department will therefore do will [*sic*] to guard in time, against their disruptive movements. Thus they are delaying the construction of this road, whilst the ample funds appropriated for it, remain useless and the road already made from Zanesville to within about twenty six miles of Columbus will be of little use untill it is finished to this point. Thus we suffer, to give time for David Pugh or some such person to go again to Washington, with such help as he had before, and if possible to procure a delay for thousands to be spent in debating the subject in Congress and pave the way for much more to be spent by an engineer and his company in reexamination re-locations here and then return, as he did before and either disgust, or delight each of our citizens as will listen to him, whilst he relates that "the secretary of war, took him in his carriage over Washington and then introduced him to the President, who smoked a corn cob pipe with him."

Now thou wilt perceive by the map that if the location were changed it would pass by but four doors in town which it does not now pass, and these four are private residences. It would reduce the 2000 feet lost by the square turn in town, very little, if any, for other crooks must be made to shun unsuitable ground, which so far as I am acquainted, I know to be worse. It would leave the Canal, the Public buildings, stores, taverns the principal part of the mechanics, the most populous parts and the most likely to improve quite at a distance. And for What?

To answer purposes which we believe are no more worthy than private speculations. And Citizens with few exceptions have long been dissatisfied with the unjust unstable policy which has been pursued in regard to this road, and they believe it to have been a part of that system which has caused much dissatisfaction amongst the best friends of General Jackson, and which has sunk the members of his former Cabinet in the public estimation!!! Better things however and a more dignified course is generally expected from those whom the President has recently called to his aid I have therefore with much freedom, from the best of my recollection, and the best data at hand, made as correct a statement of this tedious disagreeable business as I reasonably could, regretting at the same time the heavy tax on thy time which its purusal would require, but which seemed necessary to give this a correct understanding of this subject.

In conclusion, I ask on behalf of my fellow citizens as well as myself, that orders may be given for the immediate construction of this part of the National Road, and remain with much respect and esteem,

Thy assured friend,

Horton Howard

The government persisted and, contrary to the demands of those who wanted the route changed, the National Road entered Columbus as originally surveyed by Jonathan Knight in 1825. During the year following Howard's interesting appeal to Cass, major portions of the National Road were completed between Zanesville and Hebron. Contracts were let to private contractors for the construction of small segments of the road at a time. After clearing and grubbing the land, several metals or strata of stones were laid over the roadway to create a pavement. Labor problems, poor weather, land slips, disease, and defaulted construction contracts plagued the building of the road. By October 1832, however, "A rough grade, sufficient for the passage of carriages between the two places [Hebron and Columbus], will be required to be formed by the 1st of January next, and the full grade completed on the 1st day of June of the coming year."[5]

By the end of 1833, the National Road had reached the eastern bank of the Scioto and major portions of the project already had been pushed to completion further west. The same cholera epidemic that took Howard in 1833 also affected the progress of the road; Henry Brewerton, the superintendent of the Cumberland Road Office in Columbus, reported that "the cholera made its appearance on a portion of this district of the road, which rendered it necessary to suspend operations for about two weeks, at a time most favorable for the advancement of the work."[6] Of major concern to Howard and to the citizens of Columbus was the type of bridge that would cross the Scioto River. The bridge illustrated on the map was owned by Joseph Sullivant who had inherited it from his father Lucas Sullivant, the surveyor. The year that the National Road reached Columbus, the citizens of the city purchased the bridge from its owner and it was replaced by one built by the Federal Government. This new bridge was more substantial than Sullivant's wooden toll bridge and was one of the finest in the West. Brewerton described its construction to Brigadier General C. Grotiot, Chief Engineer of the entire Cumberland Road project:

> The progress made in the masonry of the bridge over the Scioto river at this place, has been as great as could have been expected, considering the very limited time of our operations this season, and the injury by the destruction of the temporary bridge by the freshet [flood] which occurred the day after the recommencement of the work. The pier of this bridge has been raised above the highest known freshet, and the foundation of the west abutment completed and the superstructure commenced. It is hoped the foundation of the one on the east side of the river will be laid this fall, although greater difficulties have been encountered in reaching a good foundation, than was expected, in consequence of wharfing constructed, and made earth, situated immediately in front of the site of this abutment, rendering the usual means resorted to for the exclusion of the water, unavailable. The foundations are all secured upon a grillage of hewed timber and planks 2 ½ feet in thickness. The masonry is built of the best quality of limestone, in masses weighing from 1000 to 8000 pounds each, and in point of workmanship, is intended to be at least equal to any work of similar kind in this country.[7]

The example set by New York State when it successfully completed its Erie Canal encouraged the Ohio General Assembly to make plans of its own to develop some canal system to connect Lake Erie with the Ohio River and, at the same time, to provide those living in the interior of the state with easy and cheap access to exterior markets. On July 4, 1825, in an elaborate ceremony that emphasized the nationalistic sentiments of a young and growing country, New York Governor DeWitt Clinton, who not only was the father of the New York canal system but had encouraged Ohio's Governor Ethan Allen Brown to develop a similar system in his own state, officially turned the first shovelful of dirt on the Licking Summit near Newark, to mark the beginning of the construction of the Ohio-Erie Canal.

When completed some years later, this canal connected the Lake Erie port of Cleveland with Portsmouth on the Ohio River, 308 miles away. Sixteen days later, July 20, 1825, a similar ceremony was conducted in Middletown through which the proposed Miami-Erie Canal was to pass that would eventually connect Toledo with Cincinnati. For almost a quarter of a century, canal construction in the state was energetically pursued so that by 1851, with the addition of a feeder system to the two major routes, a canal network of nearly 1,000 miles stretched across the state.

Even before this ambitious program was completed, however, the canal era in Ohio was at an end. Newly built railroads began to compete with the canals for both their freight and passenger business. Because iron rails could be laid into areas that were ill suited for canals and because railroads provided faster transportation than the slower waterways, the railroads began to amass huge volumes of business. Between 1861, when the state legislature first leased a part of its canal system to private individuals, and 1900, when only the state's two major canals remained in the public's trust, the state had permitted the canals to fall into gross disrepair. Segments of the state's canal network were used for streets or railroad beds and, in Cincinnati, Toledo and Cleveland, the ditches that once transported the state's agrarian and industrial wealth were used to carry sewage.

By 1900, howver, there was a revival of interest in the publicly-owned canals. It was believed that the canal could benefit the state and provide good, cheap transportation for its growing commercial enterprises and, at the same time, provide a sound alternative to the privately controlled railroads which seemed to care little for the public welfare. Leading among those who advocated the renovation of the canals was Charles E. Perkins, Chief Engineer for the Board of Public Works for the State of Ohio. Perkins, born in Akron in 1850 at the time the canals in Ohio reached the peak of their popularity, was the grandson of Simon Perkins, the founder of the "Rubber City" and the great-nephew of David Tod who was one of the state's governors during the Civil War. A graduate of Western Reserve Academy, the Renssalaer Polytechnic Institute in Troy, New York, and the

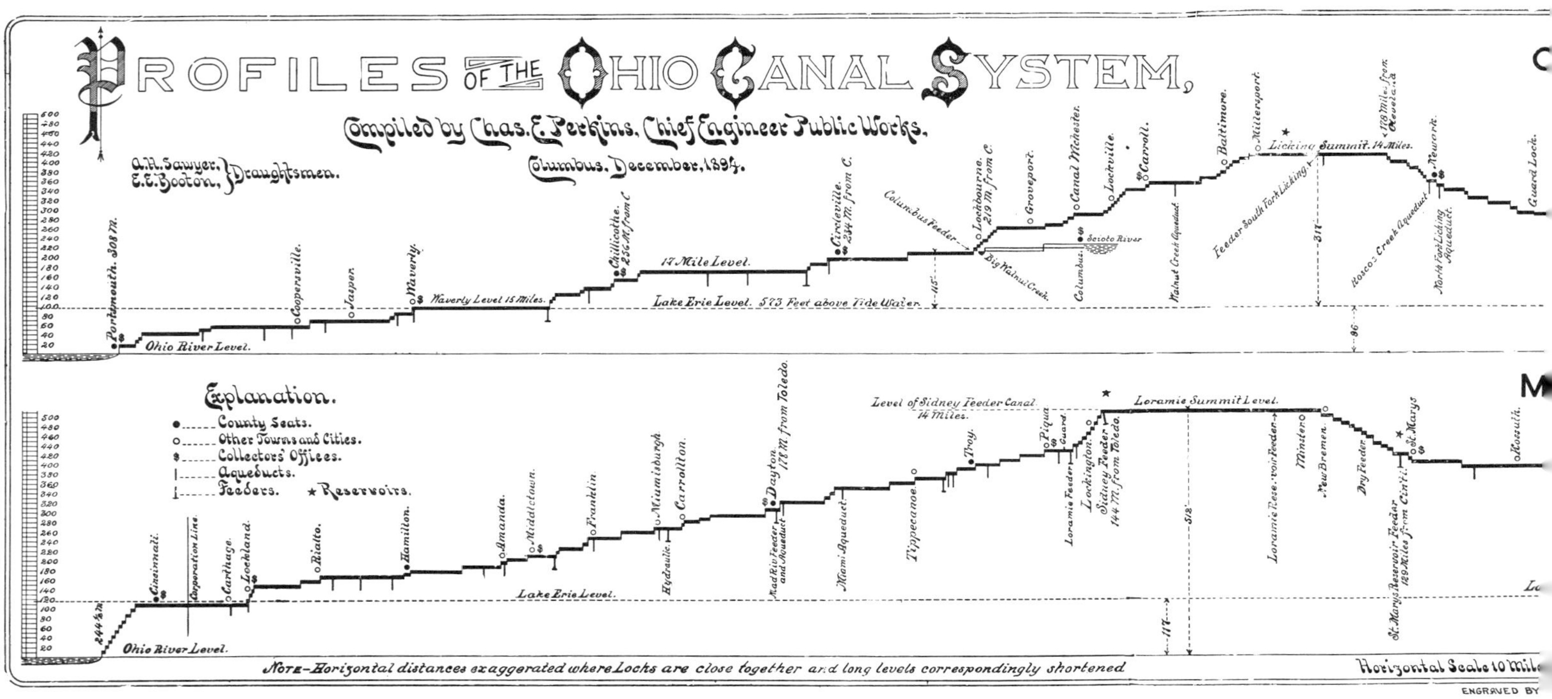

Plate II: Perkins, "Ohio Canal System," (1900), in Perkins, *Report of Chief Engineer* (1900). This map quickly delineates the Ohio canal system which was constructed in the nineteenth century.

Columbia School of Mines, Perkins, after serving as City Engineer for Akron, was rewarded for his loyalty to the Republican Party when he was appointed Chief Engineer to the Board of Public Works in 1892 by Governor William McKinley. Perkins was capable as an engineer and an administrator and was a "zealous believer in inland water ways as a public necessity and advocate of the building, across the Ohio portage, of a modern canal connecting Lake Erie with the southern river."

In his very important *Report of the Chief Engineer of Public Works and General Superintendent of Canals* made in 1900 to the Board of Public Works, Perkins presented poignant arguments for the use of the canals as main arteries for Ohio's inland commerce. He discussed the conditions of the state's canals and described their deplorable state of decay. Included in his report was the "Profiles of the Ohio Canal System" (Plate II) and the "Map of the State of Ohio showing Existing, Abandoned and Proposed Canal Routes and the Lands Granted by the United States to the State of Ohio for Canal Purposes" (Plate III). Both maps had been compiled by Perkins himself earlier, and were included in the report at his own request in order to "again meet the demand for some diagram for ready reference to the canal system of the state, I take the liberty of submitting herewith a (skeleton) map of the state of Ohio, showing existing, abandoned and proposed canal routes with their feeders, as nearly as possible, and the lands granted by the United States to the state of Ohio for canal purposes." The maps were drawn by Amos H. Sawyer who later drew several maps for the Ohio State Archaeological and Historical Society's publication *History of the Ohio Canals* which appeared in 1905.[8] The maps are self-explanatory and represent probably the best capsule summary of the canal system in Ohio.

Another interesting map related to the history of the canals in Ohio was the "Map of North-Western Ohio Showing the Lands granted by the United States to the State of Ohio, to aid in the construction of Ohio Canals, with the original Government Townships and the present County Lines. Made under the direction of the Ohio Canal Commission from

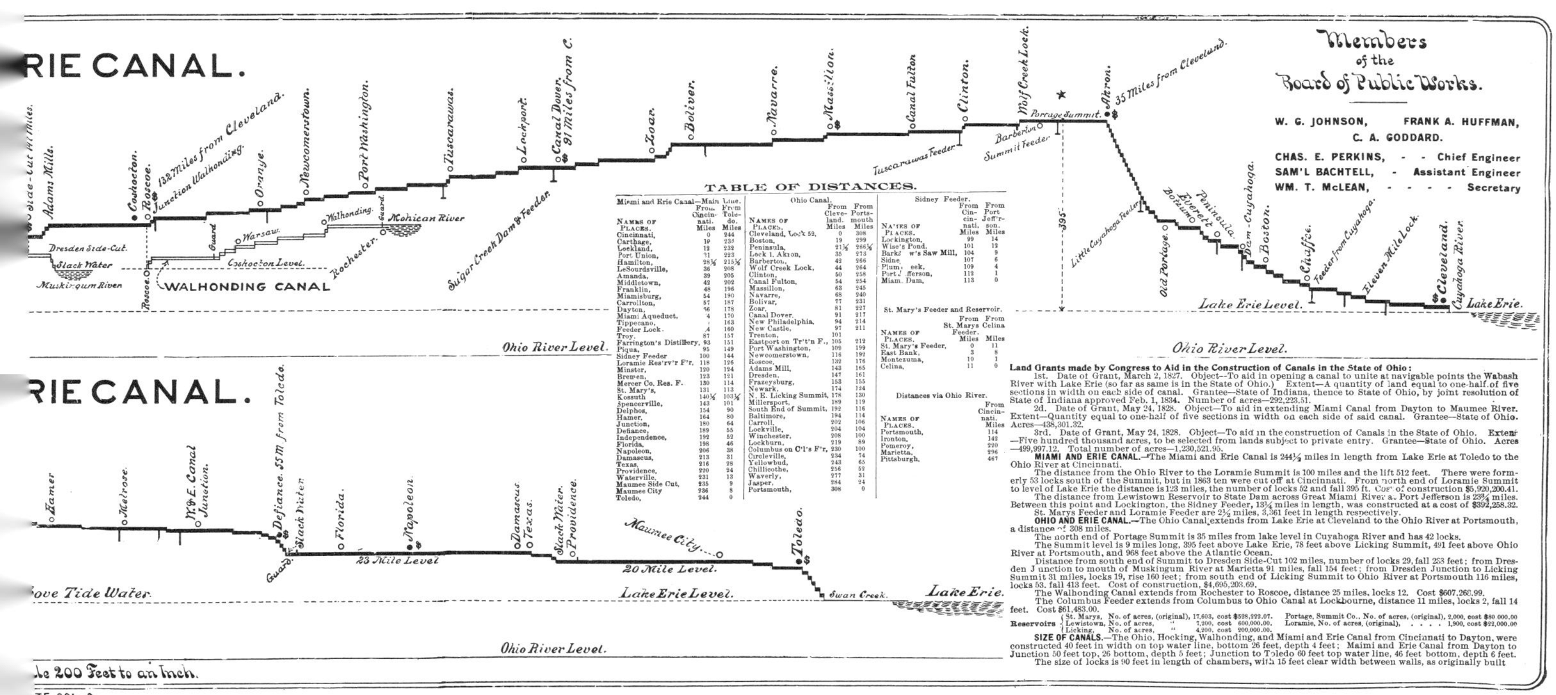

Plate III: Perkins, "Map of the State of Ohio," (1900), in Perkins, *Report of Chief Engineer* (1900). The map clearly illustrates the extent of the canal system in the state.

Official Lists and Maps on file in the office of the Auditor of State in the City of Columbus, Ohio" (Plate IV), which was compiled by Amos H. Sawyer and W. Dildine probably about the same time as Perkins's maps. This map, as the title indicated, illustrated the amount and location of the lands given by the federal government to the state of Ohio in three separate grants to aid in the construction of its canals. At first, the building program was financed by the sale of state bonds to private individuals, but such an internal improvement project was much too great for private financing alone. It was necessary for the federal government to come to

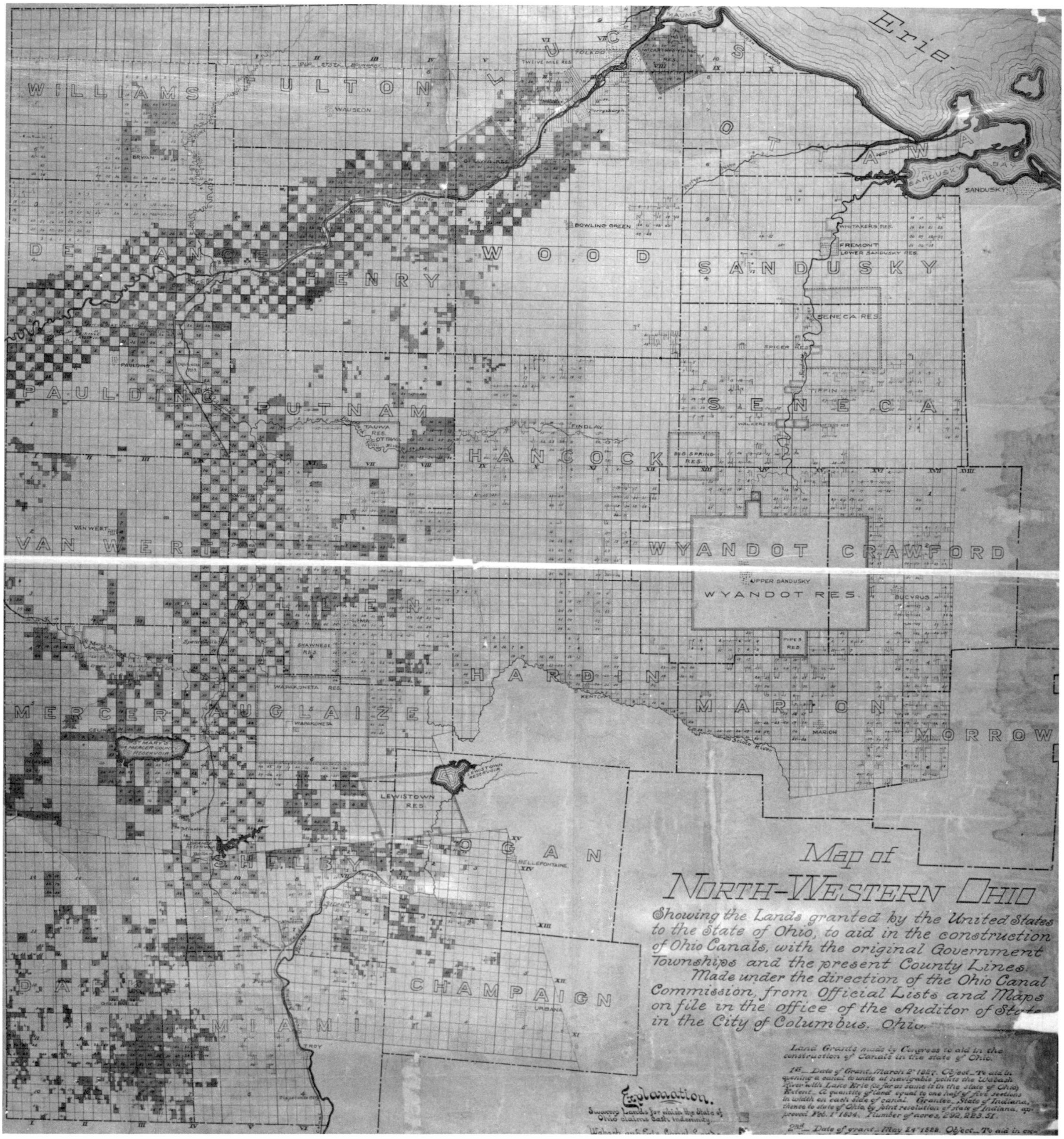

Plate IV: Sayer and Dildine, "Map of North-Western Ohio," Ohio Historical Society, Columbus. After the construction of the Ohio-Erie Canal, financed by state funds, the federal government aided Ohio through land grants to build the Maumee-Erie Canal extension.

the assistance of one of its states. In a report to the House of Representatives on February 11, 1828, the Committee on the Public Lands recommended that the Congress should approve an allotment of land to Ohio. The completion of a canal system in Ohio was both patriotic and in the national interest, the committee stated. "The men, munitions, and provisions, in war, the trade and intercourse in peace, that must pour through this channel of inland commerce," the report argued, "claim for it a very big degree of national advantage."[9] Such a grant of land, "will also aid in the procurement of means to purchase unsold lands; and greatly facilitate the migration not only to such of them as lie in Ohio, but still further westward."[10]

Before Ohio received any lands, however, Congress granted to Indiana, on March 2, 1827, land in both Ohio and Indiana to help finance that state's Wabash-Erie Canal, which was to connect Lake Erie through the Maumee River with the Wabash. This grant was for alternate sections for 5 miles on either side of the canal route. It was understood that Ohio would build that portion of the canal that lay within its own boundaries. After some legislative hassling which involved the question of state rights, Indiana relinquished her rights to those lands located in Ohio. Those Ohio lands amounted to approximately 292,225 acres. The actual construction of the canal began in 1837 and was not completed until 1842. These Wabash-Erie Canal lands are easily identifiable on the map and appear along the Maumee River from the middle of the Twelve Mile Reserve to the Indiana border.[11]

Two separate grants were made by Congress directly to Ohio, even though both were included in the same bill.[12] The first, included in Section 2 of the act, consisted of nearly 440,000 acres and it was made to aid in the financing of the Miami-Erie Canal from Dayton to where it met the Maumee River at Defiance. This grant was "equal to one-half of five sections in width, on each side of the said canal, between Dayton and the Maumee River at the mouth of the Anglaise [*sic*], so far as the same shall be located in public land." Section 5 gave to Ohio an additional 500,000 acres which were to be sold to help with the "payment of debt, or the interest thereon" incurred by the state in the construction of those canals authorized by the General Assembly. These lands were to be located by the Governor of Ohio from the public lands within the state. In all, Congress appropriated some 1,230,521.95 acres to the state to help finance the building of Ohio's ambitious canal system.

As already mentioned, almost as soon as the canal system in Ohio was completed, railroads were built and began to compete with the slower-moving canal barges for both passenger and freight business. Railroad construction had a meager beginning in Ohio during the second quarter of the nineteenth century; but by the end of the third quarter, railroad construction reached an almost fever pitch as the miles of track laid in the state grew from 572 in 1851 to approximately 4,461 in 1875. During this time, most of the major railroad routes in Ohio were built, and through the laying of trunk lines, all parts of the state were linked. In some cases, the construction of these trunk lines encouraged local businessmen to ensure that their own particular section of the state was connected with a major commercial center in order to have easy access to marketplaces outside their locale.

One of these trunk lines was constructed by the Marietta and Cincinnati Railroad Company, which was formed to connect Marietta with the Queen City. This company, chartered on March 8, 1845, as the Belpre and Cincinnati Railroad, had its name change to reflect its eastern terminus in 1850. At that time, William Parker Cutler, the grandson of Manasseh Cutler, who was one of the original proprietors of the Ohio Company, was elected the company's president while he was a member of Ohio's General Assembly. The route planned for the railroad was westward from Harmar, a small community on the west bank of the Muskingum opposite Marietta, through Athens and Chillicothe to just north of Branch Hill in Clermont County where it junctioned with the Little Miami

Railroad. Thus, with the completion of this road by 1857, the coal from Athens and Hocking counties, the iron ore from the historic Hanging Rock Iron District, and the surplus farm produce from the rich soils of the lower Scioto Valley were carried to the markets in Cincinnati and beyond.

The "Map of the Marietta and Cincinnati Railroad and its Connections" (Plate V) accompanied the company's "Synopsis of the Fifth Annual Report" which was released to its stockholders on August 15, 1854.[13] The Marietta and Cincinnati Railroad Company had experienced hard times shortly after it was chartered by the Ohio legislature, but the stock report of 1854 painted an optimistic picture of the company's progress for any prospective investors. The report pointed out that the entire length of the road was to be 258 miles and, once completed, the road would connect the southern part of the state with the "Great South Western System." In 1854, the report continued, the road itself was opened from Loveland on the Little Miami Railroad to Byers, 24 miles east of Chillicothe. It was planned that the road would reach Athens by fall. It was anticipated that the entire line would be finished within the year. The company's stock was valued at $3,555,550, which included some $200,000 invested from Athens County, $300,000 from Ross and $200,000 from investors in Washington. In fact, in 1850, 390 Ohioans had invested $1,261,150 in the Marietta and Cincinnati; the sum, however, was far too small to build a railroad of that size. Of course, a great deal of money for the construction of Ohio's railroads came from eastern sources as well as European. It is of interest to note that this map was printed by Thierry Frères in Paris, France, and probably was used to entice European investors to gamble their money on this venture in southern Ohio. Of particular interest are the "Explanations," which showed that all the other railroads in the region were "Tributary Railways" to the Marietta and Cincinnati. Although this

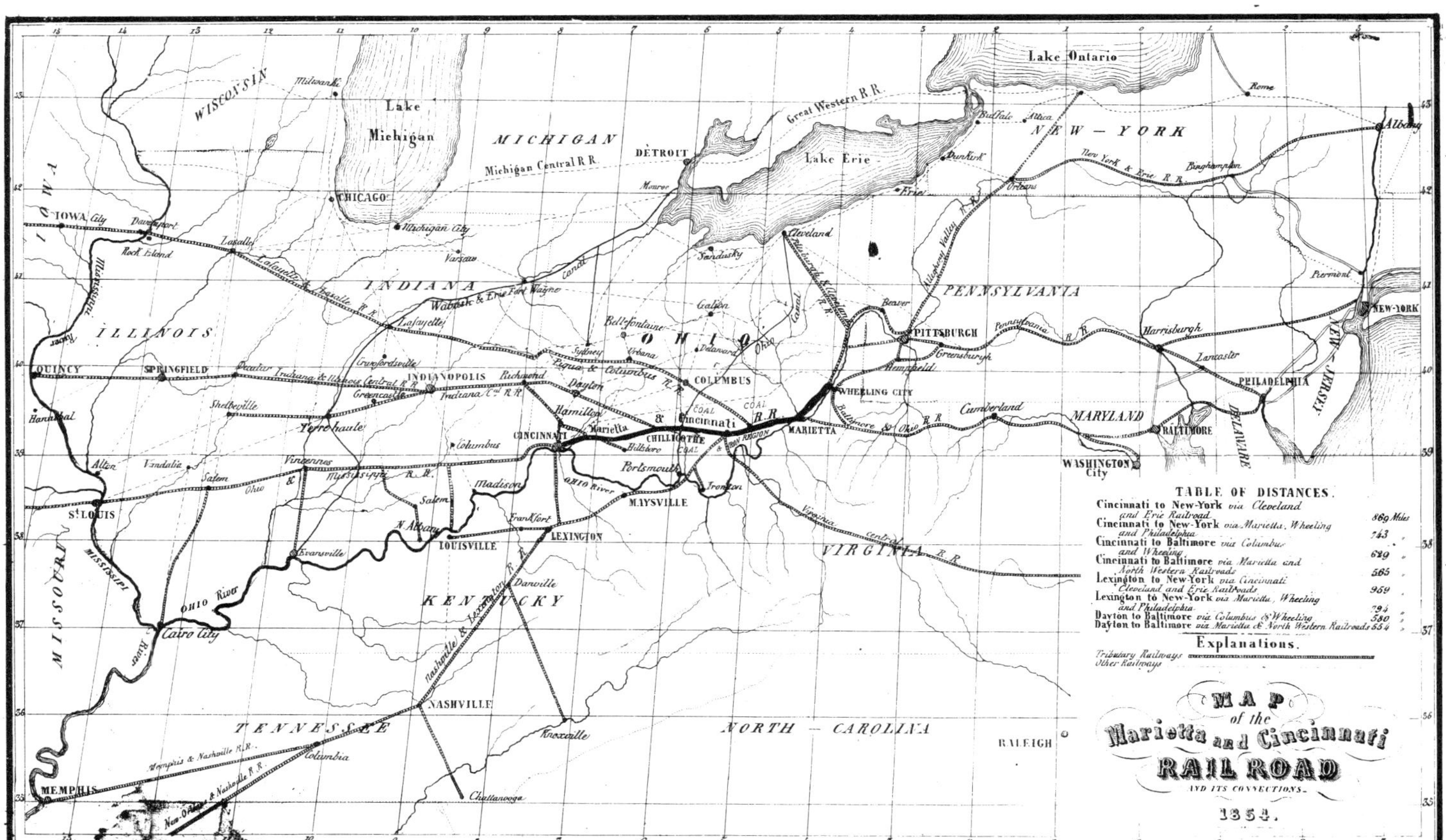

Plate V: "Map of the Marietta and Cincinnati Railroad," (1854), Ohio Historical Society, Columbus. Included with a stock prospectus, this map helped to sell a private internal improvement project during the height of railway construction in Ohio.

map was not one of the official railroad maps of the state for the nineteenth century, it nevertheless represented the aggressive nature of the speculator in attracting capital for railway construction and is part of the story of the development of the internal improvements in Ohio.

The growth of the railroads in Ohio tolled the death of the canals, and during the latter part of the nineteenth century, the iron rails were, in several cases, built upon the very berm of the old waterways. This expansion heralded the state's transition from an economy dependent on agriculture to one dominated by modern industrialism. With industrialization came economic, social and political problems that kept polticians, legislators and reformers busy for decades in search of solutions. The numerous railroad companies were criticized because of their apparent abuses of the public trust and their lack of concern for the general welfare. To help remedy this, the Ohio General Assembly created the office of State Commissioner of Railroads and Telegraph on April 5, 1867. This office had the power to investigate complaints leveled against the railroads; to examine bridges, tracks and sidings, and to require their repair, if necessary; to initiate suits against the companies through local county prosecutors; and to compile an annual report on the condition of railroads in the state that was delivered to the governor. Through effective data gathering and record keeping, it was believed that efficient railroad legislation could be written.

The "Railroad Map of Ohio" of 1875 (Plate VI) accompanied the Railroad Commissioner's Ninth Annual Report to the Governor of Ohio.[14] This particular map was prepared and published for John G. Thompson, the Ohio Railroad Commissioner, by the O. W. Gray and Sons printing house in Philadelphia. The actual cartographer was probably Frank Arnold Gray, the son of Ormando Willis Gray. This 1875 edition of the "Railroad Map of Ohio" was not the first. In 1872, young Gray had published a similar map, "New Rail Road Map of the State of Ohio," in the *New Topographical Atlas in the State of Ohio. With Descriptions historical, scientific, and statistical, together with maps of the United States and territories,* which was compiled by Henry F. Walling[15] and O. W. Gray and published by the Stedman, Brown and Lyon firm in Cincinnati. Later, other railroad maps of Ohio drawn by Frank Arnold Gray appeared as single sheets. Similar maps of Ohio by the younger Gray also appeared in several county atlases, including B. N. Griffing's *Atlas of Defiance County, Ohio* published in 1890 and William J. Kauffman and Orrin E. Kauffman's *Atlas of Stark County, Ohio,* in 1896.[16]

Gray's map illustrated the 56 railroad companies that held track in, or partially within, Ohio as of June 30, 1875. Their cumulative length of track equalled 4,461 miles of main line plus an additional 1,190 miles of siding and other track that was not shown. This trackage represented an average of 1 mile for every 7.07 square miles of area in the state, and 1 mile for every 472 persons. The total amount of money invested in paid-in capital stock was $152,714,519.27 of which $35,536,797.68 was held by 17,569 Ohioans. The railroads' bonded indebtedness equalled $161,247,086.24. Combined, the stock value, plus that of the bonds, represented an aggregate investment of $313,961,605.51. For these monies, the companies paid $6,377,791.09 in dividends to their stockholders, and 37 of the companies paid $12,691,903.72 interest on their bonds.

In 1875, 17,193,948 passengers traveled over Ohio's rails; the average length trip for each passenger was 38.20 miles. Eight riders were killed,

Plate VI: "Railroad Map of Ohio," (1876), in *Report of the Commissioner of Railroads* (1876). By the end of the country's first centennial, railroads were the state's principal means of mass transportation.

and twenty-seven injured. That represented 1 accidental death for every 2,149,244 passengers. Working on the rails, however, was more hazardous. In 1875, 93 employees were killed and 181 injured. Of these, the commissioner reported, 77 deaths occurred because of individual carelessness or misconduct; only 16 were killed "from causes beyond their own control." It was clearly more dangerous, however, to be close to the train rather than a passenger or an employee. Incidents related to trains resulted in death for 115 people while 98 suffered injury of some kind. Of these train-related accidents that resulted in death, 30 were caused by "getting on or off trains which were in motion, at stations, in yards, and stealing rides," while 74 "were upon the track, which, with a reckless prudence, is made use of by them as a common thoroughfare for travel, and a place upon which to lie down for rest, and even sleep."[17]

Aside from the passenger service, the railroads did a large freight business. In 1875, for those companies that reported to the state railroad commissioner, 25,750,501 tons of freight were hauled over the iron rails in Ohio. Of this tonnage, 6,174,775 was coal; 1,004,831 petroleum; 756,962 of ores of various kinds; 648,532 of pig iron; 2,253,888 of timber; 1,726,112 head of livestock; 3,453,689 tons of grain; 996,031 tons of flour; and 871,437 tons of pork, beef and lard.

Most of Ohio's major railroads were built after 1850 with the idea of connecting Ohio's producing areas to the large commercial centers in the East. Railroad construction slowed during the 1860s but revived again during the early 1870s. Most of the construction during the 1860s and 1870s was designed to connect various sections of the state through trunk lines with the major routes and thus tap the interior wealth of the state. This was illustrated on Gray's map in the area east of the Scioto River. The main line through southeastern Ohio was the Columbus and Hocking Valley Railroad, which, when completed in the late 1860s, became the major artery over which coal moved to the factories in the northern part of the state. The major east-west route through that region was the Marietta and Cincinnati Railroad; it tapped the mineral wealth in the Hocking Valley and the famous Hanging Rock Iron District. Shortly after its completion in 1860, it began to acquire small local lines in order to add to its service area. The first trunk line purchased was that which ran from Athens to Belpre through Little Hocking on the Ohio River, in order to link the Marietta and Cincinnati with the Baltimore and Ohio Railroad at Parkersburg, West Virginia. This small line had been built by the Union Railroad Company, and the road itself was known locally as the Union Branch. In January 1864, the Marietta and Cincinnati acquired the tracks between Portsmouth and Perry County from the Cincinnati and Muskingum Valley Railway, and thus connected that important Ohio River city with its main line at Hamden. This section of track, known as the Portsmouth Branch, brought the iron ore to the main east-west road from the Hanging Rock District which, at that time, was considered to be one of the most valuable iron regions in the country. This type of expansion and consolidation took place across the state and became the pattern for the railroad industry during the remainder of the nineteenth century.

Of course, internal improvements within Ohio began when the first improvements were made upon an Indian or animal trail through the wilderness, and have continued to the present day. The Ohio intrastate highway system, which dissects the state, and the local airports, which dot the Ohio landscape, are a continuation of the attempt to reduce travel time between commercial centers and other points in the state. The improvements developed in the nineteenth century, the National Road, the canals, the numerous railroads, including the electric interurban railway system, each had a taste of popularity, and today lay mostly abandoned in proximity to more modern and speedier modes of travel. The surveyors' sketches, the engineers' plans, the illustrations in promotional brochures, and the many commercially made maps that showed internal improvements, while specific in subject, help to explain settlement patterns, commercial and business development, and urban growth in the state of Ohio.

1. See Harry N. Scheiber, "Urban Rivalry and Internal Improvements, 1820–1860," *Ohio History,* LXXI (October 1962), 227–39. For an example of the interest in the competition for that section of the National Road between Springfield, Ohio, and Richmond, Indiana, see "A report of a survey of the National Road between Springfield, in the State of Ohio, and Richmond, in the State of Indiana," House Document #62, January 13, 1836, *House Documents,* 24th Congress, 1st Session, Serial #288 and the accompanying map.

2. This map, along with the Horton Howard letter to Lewis Cass dated July 26, 1831, is located in the Records of the Office of the Chief Engineer, Record Group 77, National Archives, Washington, D.C.

3. Jonathan Knight was a surveyor and one of the road commissioners who determined the line of the National Road between Wheeling and Richmond, Indiana. By May 1828, he no longer was working for the federal government.

4. This was the wood bridge built by Lucas Sullivant in 1815. See "An act to authorise Lucas Sullivant and his associates, to erect a toll bridge across the Scioto river at the town of Columbus," February 3, *Ohio Acts* (1815), pp. 130–32.

5. "Report from the Secretary of War," January 17, 1833, Senate Document #31, *Senate Documents,* 22d Congress, 2d Session, Vol. I, Serial #230, p. 9.

6. "Report of Chief Engineer," November 1, 1834, Senate Document #1, *Senate Documents,* 23d Congress, 2d Session, Vol. I, Serial #266, p. 171.

7. "Report of Chief Engineer," November 1, 1834, p. 169. For an interesting summary of the National Road through Ohio, see Norris F. Schneider, "The National Road: Main Street of America," *Ohio History,* LXXXIII (Spring 1974), 114–46.

8. C. C. Huntington and C. P. McClelland, *History of the Ohio Canals* (Columbus: Ohio State Archaeological and Historical Society, 1905), frontispiece.

9. "Report of Committee on the Public Lands," February 11, 1828, House Document #139, *House Documents,* 20th Congress, 1st Session, Serial #177, p. 1.

10. "Report of Committee on the Public Land," February 11, 1828, p. 3.

11. For the details of this episode during the canal building era in Ohio, see Harry N. Scheiber, *Ohio Canal Era: A Case Study of Government and the Economy, 1820–1861* (Athens, Ohio: 1969), pp. 99–100. See also, Lee Newcomer, "Construction of the Wabash and Erie Canal," *Ohio State Archaeological and Historical Quarterly,* XLVI (1937), 199–208.

12. "An Act to aid the State of Ohio in extending the Miami Canal from Dayton to Lake Erie, and to grant a quantity of land to said State to aid in the construction of the Canals authorized by law; and for making donation lands to certain persons in Arkansas Territory, May 24, 1828, *Laws of the United States of America,* Vol. III (1835), 118–20.

13. This map is in the Charles Dawes Memorial Library Marietta College, Marietta, Ohio.

14. *Ninth Annual Report of the Commissioner of Railroads and Telegraph of Ohio For the Year Ending June 30, 1875* (Columbus, 1876).

15. Henry F. Walling, in 1868, was chairman of the Civil and Topographical Engineering Department, Pardee Scientific School in Lafayette College, Easton, Pennsylvania. While there, he published an *Atlas of the State of Ohio* (New York, 1868). Walling advocated that complete and accurate surveys and maps be made of each county subdivision in the country and was instrumental in the production of several cadastral maps in Ohio. See Walter W. Ristow, "Nineteenth-Century Cadastral Maps in Ohio," *Papers of the Bibliographical Society of America,* LIX (3rd Quarter, 1965), 306–15.

16. Frank Arnold Gray's maps of Ohio also appeared in J. M. Lathrop and H. C. Penny, *Atlas of Highland County, Ohio* (Philadelphia, 1887), p. 72; B. N. Griffing, *Atlas of Darke County, Ohio* (Philadelphia, 1888), pp. 77–78; B. N. Griffing, *Atlas of Fulton County, Ohio* (Philadelphia, 1888), and B. N. Griffing *Atlas of Mercer County, Ohio* (Philadelphia, 1888), pp. 50–51.

17. *Ninth Annual Report of the Commissioner of Railroads and Telegraphs of Ohio,* pp. 15–17. All data concerning this map was taken from the published letter report from John G. Thompson to Governor William Allen, dated December 6, 1875, printed in this report, pp. 3–49.

Appendix: A Chronological Bibliography of Ohio County Atlases

Adams

Arms, W. F., *et al. Caldwell's Illustrated Historical Atlas of Adams County, Ohio, 1797–1880.* Newark, Ohio: J. A. Caldwell, (1880).

Allen

Hardesty, Hiram H. *Historical Atlas of the World (Allen County Edition).* Chicago: H. H. Hardesty & Company, 1875.

Harrison, Robert H. *Atlas of Allen Company, Ohio.* Philadelphia: R. H. Harrison, 1880.

Ashland

Caldwell, Joseph A., *et al. Atlas of Ashland County, Ohio.* Condit, Ohio: J. A. Caldwell, 1874.

Ashtabula

Lake, D. J., *et al. Atlas of Ashtabula County, Ohio.* Philadelphia: Titus, Simmons, & Titus, 1874.

Sill, Tucker and Company. *Ashtabula County Atlas.* Ashtabula, Ohio: Sill, Tucker and Company, 1905.

Athens

Lake, D. J. *Atlas of Athens County, Ohio. From actual surveys . . . to which is added a township map of the State of Ohio, also an Outline and Railroad Map of the United States.* Philadelphia: Titus Simmons & Titus, 1875.

(Bush, Fred W.) *The Centennial Atlas of Athens County, Ohio.* Athens, Ohio: The Centennial Atlas Association, 1905.

Auglaize

Howland, H. G., *et al. Atlas of Auglaize County, Ohio.* Philadelphia: Robert Sutton, 1880.

Meyer, J. Henry. *Atlas and History of Auglaize County with Biographical Sketches.* Wapakoneta: The Magee Brothers, 1917.

— — — . *Atlas and History of Auglaize County.* Piqua, Ohio: Magee Brothers Company, 1917.

Walsh, John B. *Atlas of Auglaize Company with Historical & Bibliographical Sketches.* Wapakoneta, Ohio: Atlas Publishing Company, 1898.

Belmont

Lathrop, J. M., *et al. Atlas of Belmont County, Ohio.* Philadelphia: H. C. Mead & Company, 1888.

Brown

Lake, D. J. and Griffing, B. N. *Atlas of Brown County, Ohio. From Actual Surveys. To which is added a map of the state of Ohio, also an outline and Railroad Map of the United States and the Hemisphere.* Philadelphia: Lake, Griffing & Stevenson, 1876.

Butler

Everts, L. H. *Combination Atlas Map of Butler County, Ohio.* Philadelphia: L. H. Everts, 1875.

Rerick Brothers. *The County of Butler, Ohio, An Imperial Atlas and Art Folio, Including Chronological Charts, Statistical Tables, and Description of Surveys.* Richmond, Indiana: Rerick Brothers, 1895.

Republican Publishing Company. *Butler County Atlas and Pictorial Review, including Soil Surveys, Geological Reports, Township, Village and City Maps.* Hamilton, Ohio: The Republican Publishing Company, 1914.

Carroll

Eberhart, G. A. *Illustrated Historical Atlas of Carroll County, Ohio, From Recent and Actual Surveys, and Records.* Chicago: H. H. Hardesty, 1874.

Lee, R. H. *Lee's Farm Atlas of Carroll County, Ohio.* Ann Arbor, Michigan: 1915.

Champaign

Starr, J. W. and Headington, J. H. *Atlas of Champaign County.* Urbana, Ohio: Starr & Headington, 1874.

Clark

Lake, D. J., *et al. Atlas of Clarke* [*sic*] *County, Ohio.* Philadelphia: C. O. Titus, 1870.

Everts, L. H. *Combination Atlas Map of Clarke* [sic] *County, Ohio.* Philadelphia: L. H. Everts & Company, 1874.

Rerick Brothers. *The County of Clark, Ohio, an Imperial Atlas and Art Folio, Including Chronological Charts, Statistical Tables, and Descriptions of Surveys.* Richmond, Indiana: Rerick Brothers, 1894.

Clermont

Lake, D. J. *Atlas of Clermont County, Ohio.* Philadelphia: C. O. Titus, 1870.

Lake, D. J., *et al. Atlas of Clermont County, Ohio.* Philadelphia: Lake & Gordon, 1891.

Clinton

Lake, Griffing & Stevenson. *An Illustrated Historical Atlas of Clinton County, Ohio.* Philadelphia: Lake, Griffing & Stevenson, 1876.

Mitchell, Thirey and Hahn. *Historical-Directory and Atlas Clinton County, Ohio, giving a list of the House-holders of the County. . . . A Map of the County and each of its Thirteen Townships.* Cincinnati: Mitchell, Thirey & Hahn, 1903.

Columbiana

Lake, D. J. *Atlas of Columbiana County, Ohio.* Philadelphia: C. O. Titus, 1870.

Columbiana County Map and Atlas Company. *Atlas of Surveys of Columbiana County, Ohio.* Lisbon, Ohio: Columbiana County Map and Atlas Co., 1902.

Coshocton

Lake, D. J., *et al. Atlas of Coshocton County, Ohio.* Philadelphia: C. O. Titus, 1872.

Bock, George J. Jr. *Atlas of Coshocton County, Ohio. From Deed Records and Actual Surveys.* Coshocton, Ohio: George J. Bock, Jr., 1910.

Crawford

Gould, H. T., Starr, C. E., and Starr, J. W. *Illustrated Atlas of Crawford County.* Bucyrus, Ohio: Gould & Starr, 1873.

Weber, H. L. and Swingley, C. D. *The 1894 Atlas of Crawford County, Ohio.* Bucyrus, Ohio: H. L. Weber & C. D. Swingley, 1894.

Hopley Printing Co. *Atlas (of) Crawford County, (Ohio).* Bucyrus, Ohio: Hopley Printing Co., 1912.

Cuyahoga

Hopkins, G. M. *Map of Cuyahoga County, Ohio.* Philadelphia: S. H. Matthews, 1858.

Lake, D. J. *Atlas of Cuyahoga County, Ohio.* Philadelphia: Titus, Simmons & Titus, 1874.

Cram, G. F., *et al. Atlas of Cuyahoga County and the City of Cleveland, Ohio.* Chicago: G. F. Cram & Company, 1892.

Atlas of Cuyahoga County Outside Cleveland. Cleveland: Basalt & Company, 1903.

Stranahan, H. B. and Company. *Maps of Cuyahoga County Outside of Cleveland.* Cleveland: H. B. Stranahan & Company, 1903.

Darke

Lake, D. J. *Atlas of Darke County, Ohio. From Actual Surveys. . . . To which is added a map of the State of Ohio, also, an Outline and Railroad map of the United States.* Philadelphia: Lake, Griffing & Stevenson, 1875.

Griffing, B. N. *Atlas of Darke County, Ohio.* Philadelphia: Griffing, Gordon and Company, 1888.

Defiance

Hardesty, Hiram H. *Atlas of Fulton and Defiance Counties, Ohio.* Chicago: H. H. Hardesty and Company, 1876.

— — —. *Historical Atlas of the World Illustrated (Defiance County).* Chicago: H. H. Hardesty & Company, 1876.

Griffing, B. N. *Atlas of Defiance County, Ohio.* Philadelphia: Griffing, Gordon, and Company, 1890.

Delaware

Beers, F. W., Leavenworth, A. and Warner, G. E. *Atlas of Delaware County, Ohio.* New York: Beers, Ellis and Soule, 1866.

Everts, L. H. *Illustrated Historical Atlas of Delaware County, Ohio, with an Atlas of Ohio and general Maps of the United States and Grand Divisions.* Philadelphia: L. H. Everts & Company, 1875.

Modie, F. Burr. *Modie's Centennial Atlas and History of Delaware County, Ohio.* Columbus: F. Burr Modie, 1908.

Erie

Stewart and Page. *Combination Atlas Map of Erie County, Ohio.* Philadelphia: Stewart & Page, 1874.

The Atlas Publishing Company. *Illustrated Atlas and Directory of Erie County, Ohio. Including a Directory of the Freeholders of the County, Compiled and Published from Official Records and Personal Examination.* Battle Creek, Michigan: The Atlas Publishing Company, 1896.

Fairfield

Hannum, E. S. *Hannum's Atlas of Fairfield County, Ohio.* Lancaster, Ohio: E. S. Hannum, 1866.

Everts, L. H. *Combination Atlas Map of Fairfield County, Ohio.* Philadelphia: L. H. Everts, 1875.

Fayette

Everts, L. H. *Illustrated Historical Atlas of Fayette County, Ohio, with an Atlas of Ohio, and General Maps of the United States and Grand Divisions.* Philadelphia: L. H. Everts & Company, 1875.

Franklin

Graham, John, *Map of Franklin County, Ohio.* 1853.

Caldwell, J. A., Gould, H. T., *et al. Caldwell's Atlas of Franklin County and of the City of Columbus, Ohio.* Columbus, Ohio: J. A. Caldwell & H. T. Gould, 1872.

Modie & Kilmer. *Modie and Kilmer's Folio Atlas of Franklin County.* Columbus, Ohio, 1910.

Fulton

Jones, Charles H. *Historical Atlas of the World Illustrated (Fulton County).* Chicago: H. H. Hardesty and Company, 1875.

Hardesty, Hiram H. *Atlas of Fulton and Defiance Counties, Ohio.* Chicago: H. H. Hardesty and Company, 1876.

Griffing, B. N. *Atlas of Fulton County, Ohio.* Philadelphia: Griffing, Gordon, & Company, 1888.

Gallia

Griffith, William, Jr. *Illustrated Atlas of Gallia County, Ohio.* Cincinnati: The County Commissioners, 1874.

Hardesty, Hiram H. *Historical Hand Atlas Accompanied by Histories of Lawrence and Gallia Counties, Ohio.* Chicago & Toledo: H. H. Hardesty and Company, 1882.

Geauga

Lake, D. J. *Atlas of Lake and Geauga Counties, Ohio.* Philadelphia: Titus, Simmons, & Titus, 1874.

Stranahan, H. B. *Atlas of Geauga County, Ohio.* Cleveland: H. B. Stranahan and Company, 1900.

Zethmayr, Frank R. *Geauga County.* Chardon, Ohio: F. R. Zethmayr, 1928.

Greene

Everts, L. H. *Combination Atlas Map of Greene County, Ohio.* Chicago: L. H. Everts, 1874.

Riddell, Levi and Riddell, W. D. *Riddell's Atlas of Greene County including United States Map.* Xenia, Ohio, 1896.

Guernsey

Lake, D. J. *Atlas of Guernsey County, Ohio.* Philadelphia: C. O. Titus, 1870.

Atlas Publishing Company. *Illustrated Atlas of Guernsey County, Ohio.* Wooster, Ohio: Atlas Publishing Company, 1902.

Luse, W. J. *Farm Atlas of Guernsey County, Ohio.* Columbus: The W. J. Luse Company, 1916.

Hamilton

Harrison, R. H., *et al. Titus' Atlas of Hamilton County, Ohio.* Philadelphia: C. O. Titus, 1869.

Moessinger, George, & Birtsch, Fred. *Map of Hamilton County, Ohio.* Cincinnati: 1884.

Hancock

Eberhart, G. A. *Illustrated Historical Atlas of Hancock County, Ohio.* Chicago: H. H. Hardesty and Company, 1875.

Hancock County, Ohio. *Atlas of Hancock County, Ohio.* Findlay, Ohio, 1945.

Hardin

Howland, H. G. *Atlas of Hardin County, Ohio.* Philadelphia: R. Sutton and Company, 1879.

Harrison

Caldwell, J. A., *et al. Caldwell's Atlas of Harrison County, Ohio.* Condit, Ohio: J. A. Caldwell, 1875.

Henry

Hardesty, Hiram H. *Historical Atlas of the World Illustrated (Henry County).* Chicago: H. H. Hardesty and Company, 1875.

Henry County, Ohio. *Henry County, Ohio.* Napoleon, Ohio: 1947.

Highland

Lake, D. J., *et al. Atlas of Highland County, Ohio.* Philadelphia: Titus, Simmons & Titus, 1871.

Lathrop, J. M. and Penny, H. C. *Atlas of Highland County, Ohio.* Philadelphia: R. C. Mead and Company, 1887.

Hunter, H. W. *New Atlas of Highland County, Ohio.* Hillsboro, Ohio: N. W. Hunter, 1916.

Hocking

Lake, D. J. *Atlas of Hocking County, Ohio. From Actual Surveys . . . to which is added a Township Map of the State of Ohio, also an Outline and Railroad Map of the United States.* Philadelphia: Titus, Simmons & Titus, 1876.

Holmes

Caldwell, Joseph A. *Caldwell's Atlas of Holmes County, Ohio.* Condit, Ohio: J. A. Caldwell Pub., 1875.

Stiffler, Andrew Jackson. *The Standard Atlas of Holmes County, Ohio. To which is added an Authentic History of the County, Historical Sketches of each Township and Village, Biographies and Portraits of Representative Men and Valuable Statistics, etc., together with a Map of Ohio and a Map of the United States.* Cincinnati: The Standard Atlas Publishing Company, 1907.

Huron

Lake, D. J. *Atlas of Huron County, Ohio.* Philadelphia: Titus, Simmons & Titus, 1873.

Mesnard, H. W. and Perrin, William N. *Atlas of Huron County, Ohio.* Cleveland: L. B. Mesnard, Son & Company, 1891.

Jackson

Lake, D. J. *Atlas of Jackson County, Ohio. From Actual Surveys . . ., to which is added a township map of the State of Ohio, also an Outline and Railroad map of the United States.* Philadelphia: Titus, Simmons & Titus, 1875.

Jefferson

Beers, Frederick W., Warner, George E., *et al. Atlas of Jefferson County, Ohio.* New York: F. W. Beers & Company, 1871.

Knox

Caldwell, Joseph A., Starr, J. W., *et al. Atlas of Knox County, Ohio.* Granville, Ohio: J. A. Caldwell & J. W. Starr, 1871.

Caldwell, Joseph A. *Atlas of Knox County, Ohio.* Philadelphia: E. R. Caldwell, 1896.

Lake

Lake, D. J. *Atlas of Lake and Geauga Counties, Ohio.* Philadelphia: Titus, Simmons, and Titus, 1874.

Stranahan, H. B. *Atlas of Lake County, Ohio.* Cleveland: H. B. Stranahan & Company 1898.

Lawrence

Hardesty, Hiram H. *Historical Hand Atlas Accompanied by Histories of Lawrence and Gallia Counties, Ohio.* Chicago & Toledo: H. H. Hardesty and Company, 1882.

Lake, D. J. *Atlas of Lawrence County, Ohio.* Philadelphia: D. J. Lake & Company, 1887.

Licking

Beers, F. W., Nichols, B., *et al. Atlas of Licking County, Ohio.* New York: Beers, Soule & Company, 1866.

Everts, L. H. *Combination Atlas Map of Licking County, Ohio.* Philadelphia: L. H. Everts, 1875.

Lake, D. J. *Atlas of Licking County, Ohio.* Philadelphia: Titus, Simmons & Titus, 1876.

Logan

Stewart, D. J. *Combination Atlas Map of Logan County, Ohio.* Philadelphia: D. J. Stewart, 1875.

Lathrop, J. M. *Atlas of Logan County, Ohio.* Philadelphia: D. J. Lake & Company, 1890.

Lorain

Lake, D. J. *Atlas of Lorain County, Ohio.* Philadelphia: Titus, Simmons & Titus, 1874.

American Atlas Company. *Atlas and Directory of Lorain County, Ohio.* Cleveland: American Atlas Company, 1896.

Lawrence, C. H. and Griffin, L. W. *Atlas of Lorain County, Ohio.* Elyria, Ohio: Century Printing Co., 1912.

Lucas

Andreas and Baskin. (Durant, S. W. and Durant, P. A.) *An Illustrated Historical Atlas of Lucas and part of Wood Counties, Ohio.* Chicago: 1875.

Hardesty, Hiram H. *Historical Hand-Atlas, Illustrated. (Lucas County)* Chicago: H. H. Hardesty and Company, 1882.

Uhl, John B. *The Official Atlas of Lucas County, Ohio.* Toledo, Ohio: Uhl Bros. & Company, 1900.

Madison

Caldwell, Joseph A., *et al. Caldwell's Atlas of Madison County, Ohio.* Condit, Ohio: J. A. Caldwell, 1875.

Mahoning

Lake, D. J. *Atlas of Mahoning County, Ohio.* Philadelphia: Titus, Simmons & Titus, 1874.

Mueller, A. H., Co. *Atlas of Surveys of Mahoning County, Ohio.* Philadelphia: A. H. Mueller Company, 1899–1900.

Gutknecht, William J., *et al. Atlas of Mahoning County, Ohio and Part of Trumbull County, Ohio.* Youngstown, Ohio: The Youngstown Arc Engraving Company, 1915.

Sidwell Studio. *Atlas of Mahoning County, Ohio.* Lombard, Illinois: Map Department of Mahoning County by Sidwell Studio, 1938.

Marion

Harrison, Sutton & Hare. *Atlas of Marion County, Ohio, from Records and Original Surveys, to which is added a map of the United States and a Railroad map of Ohio.* Philadelphia: Harrison, Sutton & Hare, 1878.

Medina

Everts, L. H. *Combination Atlas Map of Medina County, Ohio.* Chicago: L. H. Everts, 1874.

American Atlas Company. *Atlas and Directory of Medina County, Ohio.* Cleveland: American Atlas Co., 1897.

Meigs

Hardesty, Hiram H. *Hardesty's Historical and Geographical Encyclopedia* (Meigs County). Chicago: H. H. Hardesty and Company, 1883.

Mercer

Griffing, B. N. *Atlas of Mercer County, Ohio.* Philadelphia: Griffing, Gordon & Company, 1888.

Lutz, Martin. *Atlas of Mercer County, Ohio.* Celina, Ohio: J. E. Hamburger, 1900.

Miami

Lake, D. J., *et al. Atlas of Miami County, Ohio.* Philadelphia: C. O. Titus, 1871.

Everts, L. H. *Illustrated Historical Atlas of Miami County, Ohio, with an Atlas of Ohio, and general maps of the United States and Grand Divisions.* Philadelphia: L. H. Everts, 1875.

(Miami County). *Maps of its Different Townships and School Districts to Accompany Combined Atlas.* Directory of Miami County, 1883.

Furnas, B. E. *A Combined Atlas-Directory of Miami County, Ohio.* Springfield, Ohio:

Republic Printing Company, 1883.
Rerick Brothers. *The County of Miami, Ohio, an Imperial Atlas and Art Folio, Including Chronological Chart, Statistical Tables and Description of Surveys.* Richmond, Indiana: Rerick Brothers, 1894.

Monroe

Hardesty, Hiram H. *Historical Hand-Atlas Accompanied by History of Monroe County, Ohio.* Chicago & Toledo: H. H. Hardesty and Company, 1882.
Caldwell, Joseph A. *Caldwell's Atlas of Monroe County, Ohio.* Mount Vernon, Ohio: Atlas Publishing Company, 1898.

Montgomery

Everts, L. H. *Combination Atlas Map of Montgomery County, Ohio.* Philadelphia: Hunter Printers, 1875.

Morgan

Lake, D. J., *et al.* Atlas of Morgan County, Ohio, from actual surveys . . . to which is added a township map of the State of Ohio; also an Outline and Railroad Map of the United States. Philadelphia: Titus, Simmons & Titus, 1875.
Murray, P. *Atlas of Morgan County, Ohio.* New Philadelphia, Ohio: P. R. Murray, 1902.

Morrow

Lake, D. J., *et al. Atlas of Morrow County, Ohio.* Philadelphia: C. O. Titus, 1871.
Buck, Thaddeus E. *Atlas of Morrow County, Ohio.* Mt. Gilead, Ohio: The Berlin Printing Co., 1901.

Muskingum

Beers, Frederick W., Nichols, B., *et al. Atlas of Muskingum County, Ohio.* New York: Beers, Soule & Company, 1866.
Everts, L. H. *Combination Atlas Map of Muskingum County, Ohio.* Philadelphia: L. H. Everts, 1875.
(Geographic Institute). *Atlas of Muskingum County, Ohio.* Indianapolis: The Geographic Institute, 1916.

Noble

Wall, Mann & Hall. *Illustrated Atlas of Noble County, Ohio.* Philadelphia: Wall, Mann & Hall, 1876.

Ottawa

Hardesty, Hiram H. *Illustrated Historical Atlas of Ottawa County.* Chicago: H. H. Hardesty & Company, 1874. (Also a German edition, 1874.)
Jaeger, Godfrey. *Historical Hand Atlas Illustrated* (Ottawa County). Chicago: Godfrey Jaeger & Co., 1881.
Goodman, H. J. *Illustrated Historical Atlas of Ottawa County, Ohio.* Port Clinton, Ohio: H. J. Goodman, 1900.

Paulding

Morrow, Oliver and Bashore, F. W. *Historical Atlas of Paulding County, Ohio, Containing maps of Paulding County, townships, towns and villages . . . also maps of the United States and the State of Ohio.* Madison, Wisconsin: Western Publishing Co., 1892.
Morrow, Oliver. *Atlas and Directory of Paulding County, Ohio.* Paulding, Ohio: Miller & Shaffer, 1905.
———. *Atlas and Directory of Paulding County, Ohio.* Paulding, Ohio: F. Miller & W. N. Shaffer, 1906.
Ogle, George, *et al. Standard Atlas of Paulding County, Ohio.* Chicago: George A. Ogle & Co. Pub., 1917.

Perry

Lake, D. J., *et al. Atlas of Perry County, Ohio.* Philadelphia: Titus, Simmons & Titus, 1875.

Pickaway

Lake, D. J., *et al. Atlas of Pickaway County, Ohio.* Philadelphia: C. O. Titus, 1871.

Portage

Everts, L. H. *Combination Atlas Map of Portage County, Ohio.* Chicago: L. H. Everts, 1874.

Stranahan, H. B., *et al. Atlas of Portage County, Ohio.* Cleveland: H. B. Stranahan & Company, 1900.

Preble

Lake, D. J., *et al. Atlas of Preble County, Ohio.* Philadelphia: C. O. Titus, 1871.

Putnam

Hardesty, Hiram H. *Historical Hand-Atlas, Illustrated (Putnam County).* Chicago: H. H. Hardesty & Company, 1880.

Seitz, Daniel W. and Talbot, O. C. *The Putnam County Atlas . . . containing Sectional Outline Maps of the County . . . Farm Maps of the Fifteen Townships . . . Together with Maps of all the Towns reduced to single page scale; Half Tone views of County Buildings, Churches, School Buildings, Business Houses, Private Dwellings, etc.; Half Tone Portraits with Brief Biographies of Putnam County People. . . .* Leipsic, Ohio: Free Press Publishing House, 1895.

Richland

Andreas, Alfred T. *Atlas Map of Richland County, Ohio.* Chicago: A. T. Andreas, 1873.

Rerick Brothers. *The County of Richland, Ohio. An Imperial Atlas and Art Folio, Including Chronological Charts, Statistical Tables and Descriptive Surveys.* Richmond, Ind.: Rerick Brothers, 1896.

Ross

Gould, Hueston T., *et al. Illustrated Atlas of Ross County and Chillicothe, Ohio.* Columbus, Ohio: H. T. Gould & Co., 1875.

Sandusky

Everts, Stewart & Co. *Combination Atlas Map of Sandusky County, Ohio.* Chicago: Everts, Stewart & Company, 1874.

Hopple, William H. *The County of Sandusky Ohio. An Imperial Atlas and Art Folio.* Fremont, Ohio: William H. Hopple, 1898.

Seneca

Stewart, David I. *Combination Atlas Map of Seneca County, Ohio.* Philadelphia: D. J. Stewart, 1874.

Rerick Brothers. *The County of Seneca, Ohio. An Imperial Atlas and Art Folio, Including Chronological Chart, Statistical Tables, and Description of Surveys.* Richmond, Indiana: Rerick Brothers, 1896.

Shelby

Page & Smith. *Combination Atlas Map of Shelby County, Ohio.* Philadelphia: Hunter, 1875.

American Atlas Company. *Atlas and Directory of Shelby County, Ohio.* Cleveland: American Atlas Co., 1900.

Stark

Beers, F. W., Sanford, G. P., *et al. Atlas of Stark County, Ohio.* New York: F. W. Beers & Company, 1870.

Everts, L. H., *et al. Combined Atlas Map of Stark County, Ohio.* Philadelphia: L. H. Everts & Company, 1875. (Also a German edition, 1875).

Kauffman, William J. and Kauffman, Orrin F. *Atlas of Stark County, Ohio.* Canton, Ohio: The Ohio Map & Atlas Company, 1896.

Summit

Tackabury, Mead & Moffett. *Combination Atlas Map of Summit County, Ohio.* Philadelphia: Tackabury, Mead & Moffett, 1874.

Akron Map and Atlas Company. *Illustrated Summit County, Ohio, Representing Her Manufacturing Interests, Commercial Houses, Public Institutions, Farms, Homes and People, with History, Statistics and General Information, Maps of United States, Ohio, Summit County Townships, Towns, Villages and City of Akron.* Akron, Ohio: Akron Map and Atlas Company, 1891.

Rectigraph Abstract and Title Company. *Atlas and Industrial Geography of Summit County, Ohio.* Akron, Ohio: The Rectigraph Abstract & Title Company, 1910.

Trumbull

Everts, L. H., *et al. Combination Atlas Map of Trumbull County, Ohio.* Chicago: L. H. Everts, 1874.

American Atlas Company. *Atlas and Directory of Trumbull County, Ohio.* Cleveland: American Atlas Co., 1899.

Gutknecht, William J. *Atlas of Mahoning County, Ohio and Part of Trumbull County, Ohio.* Youngstown, Ohio: The Youngstown Arc Engraving Company, 1915.

Tuscarawas

Everts, L. H., *et al. Combination Atlas Map of Tuscarawas County, Ohio.* Philadelphia: L. H. Everts & Company, 1875.

Rhodes, Edwin S. *First Centennial History and Atlas of Tuscarawas County, Ohio.* New Philadelphia, Ohio: E. S. Rhodes, 1908.

Union

Mowry, A. S., *et al. Atlas of Union County, Ohio.* Philadelphia: Harrison, Sutton, & Hare, 1877.

Vinton

Lake, D. J., *et al. Atlas of Vinton County, Ohio. From Actual Surveys. . . . To which is added a Township Map of the State of Ohio, also an Outline and Railroad Map of the United States.* Philadelphia: Titus, Simmons, & Titus, 1876.

Warren

Everts, L. H., *et al. Combination Atlas Map of Warren County, Ohio.* Philadelphia: L. H. Everts, 1875.

Bone, Frank A. *Complete Atlas of Warren County, Ohio, Containing Maps of the County and of its Townships and Villages.* Lebanon, Ohio: 1891.

McKay, William S. *The Centennial Atlas of Warren County, Ohio.* Lebanon, Ohio: The Centennial Atlas Assoc., 1903.

Washington

Lake, D. J., *et al. Atlas of Washington County, Ohio, From Actual Surveys. . . . To which is added a Township Map of the State of Ohio, also an Outline and Railroad Map of the United States.* Philadelphia: Titus, Simmons & Titus, 1875.

Roe, Frederick B. *Atlas of the City of Marietta, Washington County, Ohio, and Vicinity Including Williamstown, Wood County, W. Va.* Chicago: George F. Cram & Company, 1902.

Wayne

Caldwell, Joseph A. *Caldwell's Atlas of Wayne County and of the City of Wooster, Ohio.* Sunbury, Ohio: J. A. Caldwell, 1873.

— — — . *Caldwell's Atlas of Wayne County, Ohio.* Mount Vernon, Ohio: Atlas Publishing Company, 1897.

— — — . *Atlas of Wayne County, Ohio.* Mount Vernon, Ohio: Atlas Publishing Company, 1908.

Williams

Andreas and Baskin. *An Illustrated Historical Atlas of Williams County, Ohio.* Chicago: Lakeside Press, 1874.

Lathrop, James M. and Grim, John C. *Atlas of Williams County, Ohio.* Cleveland: Gordon, Lathrop and Company, 1894.

Democratic Publishing Co. *Plat Book of Williams County, Ohio.* Bryan, Ohio: 1904.

Ogle, George A., *et al. Standard Atlas of Williams County, Ohio.* Chicago: G. A. Ogle and Company, 1918.

Wood

Durant, S. W. and Durant, P. A. *Illustrated Historical Atlas of Lucas and Part of Wood County, Ohio.* Chicago: Andreas & Baskin, 1875.

Griffing B. *An Atlas of Wood County, Ohio.* Philadelphia: Griffing, Gordon and Co., 1886.

Wyandot

Hare, A. J. *Atlas of Wyandot County, Ohio.* Philadelphia: Harrison & Hare, 1879.

A Selected Bibliography

GENERAL REFERENCE

Baldwin, Charles C. "Early Maps of Ohio and the West," *Tract No. 25* of The Western Reserve and Northern Ohio Historical Society (Cleveland), April 1875.

Beers, Henry Putney. *The French and British In the Old Northwest: A Bibliographical Guide to Archives and Manuscript Sources.* Detroit: Wayne State University Press, 1964.

Brun, Christian. *Guide to the Manuscript Maps in the William L. Clements Library.* Ann Arbor: The University of Michigan Press, 1959.

Claussen, Martin P., and Friss, Herman R. *Descriptive Catalog of Maps Published by Congress, 1817–1843.* Washington, D.C., 1941.

Douthit, Ruth Long. *Ohio Resources for Genealogists with some References for Genealogical Searching in Ohio.* Detroit: The Detroit Society for Genealogical Research, 1966.

Ladd, Richard S. *Maps Showing Explorers' Routes, Trails and Early Roads in the United States.* Washington, D.C.: The Library of Congress, 1962.

LeGear, Clara Egli. *A List of Geographical Atlases in the Library of Congress.* Vol. 5. Washington, D.C.: The Library of Congress, 1958.

Miller, Hunter. *Treaties and other International Acts of the United States of America.* Washington, D.C., 1933.

National Archives. *Civil War Maps in the National Archives.* Washington, D.C., 1964.

Ohio Maps in the National Archives. Typewritten list at the Ohio Historical Society, Columbus, Ohio.

Ohio, The Sesquicentennial of Statehood, 1803–1953. An Exhibition in the Library of Congress. Washington, D.C.: The Library of Congress, 1953.

Parkman, Francis. Appendix 1, "Early Unpublished Maps of the Mississippi and the Great Lakes," in *LaSalle and the Discovery of the Great West.* Boston: Little, Brown and Company, 1893.

Phillips, Philip Lee. *A List of Geographical Atlases in the Library of Congress.* 4 vols. Washington, D.C., 1909–.

———. *A List of Maps of America in the Library of Congress.* Washington, D.C., 1901.

Raisz, Erwin. "Outline of the History of American Cartography," *Isis,* XXVI (March 1937), 378–90.

Smith, Clara A. *List of Manuscript Maps in the Edward E. Ayer Collection.* Chicago: Newberry Library, 1927.

Special Libraries Association. *Map Collections in the United States and Canada.* New York: Special Collections Association, 1954.

Stephenson, Richard. *Land Ownership Maps.* Washington, D.C., 1967.

Winsor, Justin. *French Exploration and Settlement in North America and Those of the Portugese, Dutch, and Swedes, 1500–1700.* Vol. IV of *Narrative and Critical History of North America.* Boston: Houghton, Mifflin and Company, 1884, pp. 163–246.

PRIMARY SOURCES

American State Papers. Indian Affairs, Vols. I-IV. Washington, D.C., 1832.

"A Plan of an Old Fort and Intrenchment in the Shawanese country, taken on horse back, by computation only," *Royal American Magazine for January 1775* (Boston, 1775), pp. 29–30.

Atwater, Caleb. "Description of the Antiquities Discovered in the State of Ohio and other Western States," in American Antiquarian Society, *Archaeologia Americana. Transactions and Collections,* I (1820).

———. *Writings of Caleb Atwater.* Columbus: Scott and Wright for the author, 1833.

Baldwin, Charles C., ed. "Map and Description of Northeastern Ohio," *Magazine of Western History,* I (December 1884), 109–14.

Barbiere, Joe. *Scraps from the Prison Table.* Doylestown, Pennsylvania: W. W. H. Davis, printer, 1868.

Beatty, Erkuries, "The Diary of Major Erkuries Beatty," *The Magazine of American History,* I (1877), 175–79; 235–43; 309–15; 432–38.

Bierce, Lucius V. *Historical Reminiscenses of Summit County.* Akron: T. & H. G. Canfield, 1854.

Blazier, George Jordan, ed. *Joseph Barker: Recollections of the First Settlement of Ohio.* Marietta: Marietta College, 1958.

Bond, Beverley W., Jr., ed. *The Courses of the Ohio River.* Cincinnati: Historical and Philosophical Society of Ohio, 1942.

Bonnécamps, Pierre Jean [Joseph Pierre?] de. *Relations du voyage de la Belle rivière fait in 1749, sous les ordres de M. de Celoron. Pierre Jean de Bonnecamps: a Quebec, October 17, 1750.* Vol. LXIX of *The Jesuit Relations and Allied Documents. Travels and Explorations of the Jesuit Missionaries in New France,* ed. Reuben Gold Thwaites. Cleveland: Burrow Brothers, 1896–1901.

Boyer, John. "General Wayne's Daily Encampment," *American Pioneer,* II(July 1843), 291–92.

Bradley, Glenn D. "Fort Meigs in the War of 1812," *Quarterly Bulletin* of the Historical Society of Northwestern Ohio, (January 1930), n.p.

Burnet, Jacob. *Notes on the Early Settlement of the North-western Territory.* New York: D. Appleton & Company, 1847.

Butterfield, Consul Willshire. *Journal of Capt. Jonathan Heart.* Albany, New York: Joel Munell's Sons, 1885.

Carey, Mathew. *Carey's American Pocket Atlas.* Philadelphia: Mathew Carey, 1805.

Charlevoix, Pierre Francois Xavier de. *Histoire et Description Generale de la Nouvelle*

France, avec le Journal historique d'un Voyage fait par ordre du Roi dans l'Amérique Septentrionnale. 3 vols. Paris, 1744.

The Cincinnati Directory Advertiser for the Year 1836–37. Cincinnati: J. H. Woodruff, 1836.

Croghan, George. "George Croghan's Journal, February 28, 1765–October 7, 1765," in Clarence Walworth Alvord and Clarence Edwin Carter, eds. *The New Regime: 1765–1767,* Vol. II, *British Series. Collections* of the Illinois State Historical Society Library, Vol. XI. Springfield: Illinois State Historical Library, 1916, pp. 1–64.

———. "Croghan's Journal," *Aspinwall Papers* (Part I), Vol. IX, *Collections* of the Massachusetts Historical Society (Fourth Series). Boston: Massachusetts Historical Society, 1871, pp. 362–78.

Cuming, Fortescue. *Sketches of a Tour to the Western Country Through the States of Ohio and Kentucky; a Voyage down the Ohio and Mississippi Rivers and a Trip through the Mississippi Territory and part of West Florida. Commenced at Philadelphia in the Winter of 1807, and Concluded in 1809.* Vol. IV of *Early Western Travels, 1748–1846,* ed. Reuben Gold Thwaites. Cleveland: The Arthur H. Clark Company, 1904.

Cuppy, John. "Recollections of John Cuppy," in Louise Phelps Kellogg, ed., *Frontier Advance on the Upper Ohio: 1778–1779.* Vol. XXIII of *Publications of the State Historical Society of Wisconsin,* ed. Milo M. Quaife. Madison: State Historical Society of Wisconsin, 1916, pp. 158–61.

Cutler, Manasseh. *An Explanation of the Map of Federal Lands.* (March of America Facsimile Series, Number 51. University Microfilms, Inc.: Ann Arbor, Michigan.)

Darlington, William M. *Christopher Gist's Journals.* Cleveland: The Arthur H. Clark Company, 1893.

"Defence of Fort Sandusky," *The Port folio,* Third Series V(March 1815), 209–20.

Denny, Ebenezer. *Military Journal of Major Ebenezer Denny, an officer—the Revolutionary and Indian Wars with an introductory memoir by W. H. Denny.* Philadelphia, Historical Society of Pennsylvania, 1859.

"A Description of Fort Harmar," *Magazine of Western History,* I(November 1884), 26–31.

Drake, Daniel. *Natural and Statistical View, or Picture of Cincinnati and the Miami Country. Illustrated by Maps. With an Appendix Containing Observations on the Late Earthquakes, the Aurora Borealis, and Southwest Wind.* Cincinnati: Printed by Looker & Wallace, 1815.

———. *Notices Concerning Cincinnati.* Cincinnati: Printed for the author, at the press of John W. Browne & Company, 1810.

Evans, Lewis. *Geographical, Historical, Political, Philosophical and Mechanical Essays: The First Containing an Analysis of a General Map of the Middle British Colonies in America; and of the Country of the Confederate Indians; A Description of the Face of the Country; the Boundaries of the Confederates; and the Maritime and Inland Navigations of the Several Rivers and Lakes Contained therein.* Philadelphia: Printed by B. Franklin and D. Hall, 1755.

Foster, Robert H., "Depreciation Lands," *Pennsylvania Archives,* Third Series, III(1894), 759–71.

Galbreath, Charles Burleigh. *Expedition of Celoron to the Ohio Country in 1749.* Columbus: The F. J. Heer Printing Company, 1921.

Gipson, Lawrence Henry. *Lewis Evans.* Philadelphia: The Historical Society of Pennsylvania, 1939.

Gist, Charles. *Sketches and Statistics of Cincinnati, in 1851.* Cincinnati: Wm. H. Moore & Company, 1851.

Harmar, Josiah. Papers, Draper Mss., Vols. 1–2 (microfilm), Ohio Historical Society: Columbus, Ohio.

Harris, Thaddeus M. *The Journal of a Tour into the Territory Northwest of the Alleghany Mountains; Made in the Spring of the Year 1803. With a Geographical and Historical Account of the State of Ohio. Illustrated with Original Maps and Views.* Boston: Printed by Manning and Loring, 1805.

Heckewelder, John. "Map and Description of Northeastern Ohio," *Tract No. 64* of the Western Reserve Historical Society (Cleveland), November 1884.

Heitman, F. B. *Historical Register of the United States Army.* Washington, D.C., 1890.

Hicks, Frederick Charles, ed. *Thomas Hutchins. A Topographical Description of Virginia, Pennsylvania, Maryland, and North Carolina.* Cleveland: The Arthur H. Clark Company, 1904.

Hildreth, Samuel Prescott. "Fort Harmar," *The American Pioneer,* I(February 1842), 25–28.

— — — . "A Description of Campus Martius," *The American Pioneer,* I(March 1842), 83–89.

— — — . *Pioneer History Being An Account of the First Examinations of the Ohio Valley and the Early Settlement of the Northwest Territory.* Cincinnati: H. W. Derby & Company, 1848.

Hulbert, Archer Butler, ed. *Ohio Company Series.* Vols. I and II: *The Records of the Original Proceedings of the Ohio Company.* Vol. III: *Ohio In the Time of the Confederation. Marietta College Historical Collections.* Marietta, Ohio: Marietta Historical Commission, 1917–1918.

Hutchins, Thomas. "A Description of part of the Country Westward of the River Ohio, with the Distances Computed from Fort Pitt to the several Indian Towns by Land & Water," in Vol. I, Thomas Hutchins Mss. Pennsylvania Historical Society, Philadelphia, Pennsylvania.

— — — . "Journal from Fort Pitt to the Mouth of the Ohio in the Year 1768," *Indiana Historical Society Publications,* II(1895), 418–21.

— — — . "The Rout from Fort Pitt to Sandusky, and thence to Detroit," Thomas Hutchins Mss., Pennsylvania Historical Society, Philadelphia, Pennsylvania.

Jenkins, Warren. *The Ohio Gazetteer and Traveler's Guide with an appendix, or General Register.* Columbus: I. N. Whiting, 1837.

— — — . *The Ohio Gazetteer and Traveler's Guide with an appendix, or General Register.* Columbus: I. N. Whiting, 1839.

— — — . *The Ohio Gazetteer, and Traveler's Guide, Revised edition, with a second appendix, Containing the Census of the state for 1840, as taken by Order of Congress.* Columbus: I. N. Whiting, 1841.

Jones, David. *A Journal of Two Visits Made to Some Nations of Indians on the West Side of the River Ohio, in the Years 1772 and 1773 . . . With a Notice of the Author, By Horatio Gates Jones.* New York: Reprinted for Joseph Sabin, 1865.

Jordan, John W. "Notes of Travel of William Henry, John Heckewelder, John Rothrock and Christian Clewell, to Gnadenhutten on the Muskingum, in the Early Summer of 1797," *The Pennsylvania Magazine of History and Biography,* X(1882), 125–57.

Kilbourn, John. *Columbian Geography or A Description of the United States of America to which is Prefixed a Concise Introduction to the Science of Geography in General.* Chillicothe: Nashee & Denny, 1815.

— — — . *Compendious System of Universal Geography.* Worthington, Ohio: Putnam & Israel, 1813.

— — — . *The Ohio Gazetteer; or topographical dictionary, containing a description of the several counties, towns, villages, settlements, roads, rivers, lakes, springs, mines &c. in the state of Ohio; alphabetically arranged.* Columbus: P. H. Olmsted & Company, (July) 1816.

— — — . *The Ohio Gazetteer; or topographical dictionary, containing a description of the several counties, towns, villages, settlements, roads, rivers, lakes, springs, mines &c. in the state of Ohio; alphabetically arranged.* Columbus: Printed for J. Kilbourn by Smith & Griswold (November), 1816.

— — — . *The Ohio Gazetteer; or topographical dictionary, containing a description of the several counties, towns, villages, settlements, roads, rivers, lakes, springs, mines &c. in the state of Ohio; alphabetically arranged.* Albany, New York: Joshua Fish, 1817.

— — — . *The Ohio Gazetteer; or topographical dictionary, containing a description of the several counties, towns, villages, settlements, roads, rivers, lakes, springs, mines &c. in the state of Ohio; alphabetically arranged.* Columbus: P. H. Olmsted, Office of Intelligencer and Gazette, 1817.

— — — . *The Ohio Gazetteer; or topographical dictionary, containing a description of the several counties, towns, villages, settlements, roads, rivers, lakes, springs, mines &c. in the state of Ohio; alphabetically arranged.* Columbus: J. Kilbourn, 1818.

— — — . *The Ohio Gazetteer; or topographical dictionary, containing a description of the several counties, towns, villages, settlements, roads, rivers, lakes, springs, mines &c. in the state of Ohio; alphabetically arranged.* Columbus: Published and sold by J. Kilbourn, 1819.

— — — . *The Ohio Gazetteer; or topographical dictionary, containing a description of the several counties, towns, villages, settlements, roads, rivers, lakes, springs, mines &c., in the state of Ohio; alphabetically arranged.* Columbus: J. Kilbourn, 1821.

———. *The Ohio Gazetteer; or Topographical Dictionary.* Columbus: Published by J. Kilbourn, 1829.

———. *The Ohio Gazetteer; or Topographical Dictionary.* Columbus: Published by John Kilbourn, 1831.

———. *The Ohio Gazetteer; or Topographical Dictionary; being a continuation of the work originally compiled by the late John Kilbourn. 11th edition and enlarged by a citizen of Columbus.* Columbus: Scott and Wright, 1833.

King, John H. *Three Hundred Days in a Yankee Prison: Reminiscenses of War Life Captivity Imprisonment at Camp Chase Ohio.* Kennesaw, Georgia: Continental Book Company, 1959.

Knopf, Richard C., ed. *Anthony Wayne: A Name in Arms, The Wayne-Knox-Pickering-McHenry Correspondence.* Pittsburgh: The University of Pittsburgh Press, 1960.

Marshall, George F. "Moses Cleaveland: Some Old and Original Manuscripts," *Annals of the Early Settlers' Association,* III(1882), 68–75.

Mather, W. W. *Second Annual Report on the Geological Survey of the State of Ohio.* Columbus: Samuel Medary, printer to the State, 1838.

Mathew, Alfred. "A Description of Campus Martius," *Magazine of Western History,* I(December 1884), 115–18.

McAfee, Robert B. *History of the Late War in the Western Country.* Bowling Green, Ohio: Historical Publications Company, 1919.

McBride, James. *Pioneer Biography: Sketches of the Lives of some of the Early Settlers of Butler County, Ohio,* Vol. I. Cincinnati: Robert Clarke & Company, 1869.

McClure, David. *Diary of David McClure, Doctor of Divinity, 1748–1820.* New York: The Knickerbocker Press, 1899.

McGrane, R. C., ed. "William Clark's Journal of General Wayne's Campaign," *Mississippi Valley Historical Review,* I(December 1914), 418–44.

Melish, John. *A Geographical Description of the United States.* Philadelphia: T. H. Palmer, publisher, 1816.

———. *Travels in the United States of America in the Years 1806 & 1807, and 1809, 1810, & 1811.* 2 vols. Philadelphia, 1818.

Miller, Andrew. *New States and Territories, or the Ohio, Indiana, Illinois, Michigan, North-Western Missouri, Louisiana, Mississippi, and Alabama in their Real Characters, in 1818; Showing, in a new and shortway, the situation, size, number of Inhabitants, Whites and Indians—the number of Counties, Villages, Printing Offices, Banks, Factories, Furnaces, Forges, Mills, &c. of each; and the name, situation, extent, and number of Inhabitants of each County, with its County-Town, & number of Houses, Stores, Banks &c., in each, by a Map Table. Also, a description of the Rivers, Roads, Settlements, qualities and prices of Lands; the Timber, Water, Climate, Diseases, prices of Produce, Stock and Goods—and the advantages and disadvantages of each, and of their particular parts; and of the new parts of York State, Pennsylvania, Virginia and Kentucky; with a few words concerning the impositions and difficulties experienced in moving, settling, &c.* (Printed for the Benefit of Emigrants, and others, Intending to Visit the Western Country, 1819).

Morse, Jedidiah. *The American Gazetteer.* Charlestown: Thomas & Andrews, 1804.

———. *The American Universal Geography, or, a View of the Present State of all the Empires, Kingdoms, States and Republics in the Known World and of the United States in Particular.* Boston: Isaiah Thomas & Ebenezer T. Andrews (June), 1802.

Mulkearn, Lois, ed. *A Topographical Description of the Dominions of The United States of America [Being a Revised and Enlarged Edition of] A Topographical Description of Such Parts of North America as are Contained in The (Annexed) Map of the Middle British Colonies, &c. in North America by T. Pownall, M. P.* Pittsburgh: University of Pittsburgh Press, 1949.

Ninth Annual Report of the Commissioners of Railroads and Telegraphs of Ohio. Columbus: Nevins & Myers, 1876.

Orton, Edward. *First Annual Report of the Geological Survey of Ohio* (Third Organization). Columbus: The Westbote Company, State Printers, 1890.

Pease, Seth. "Seth Pease's Journal to and from New Connecticut, 1796–1798," *Tract No. 94.* of the Western Reserve Historical Society (Cleveland), 1914.

Perkins, Charles E. *Report of the Chief Engineer of Public Works of Ohio and General Superintendent of Canals.* Columbus, 1900.

"Plan of an Ancient Fortification on the East Bank of the Little Miami River, about four miles above the mouth of Todd's Fork, and 30 miles North East from Cincinnati, in the county of Warren, State of Ohio," *The Portfolio,* Third Series, III(1810), 485–86.

Pownall, Thomas. *A Topographical Description of Such Parts of North America as are Contained in the (Annexed) Map of the Middle Colonies, &c., in North America.* London: J. Almon, 1776.

"Report from the Secretary of War," January 17, 1833, Senate Document #31. *Senate Documents,* 22d Congress, 2d Session, 1833.

"A Report of a Survey of the National Road between Springfield, in the State of Ohio, and Richmond, in the State of Indiana," January 13, 1836, House Document #62. *House Documents,* 24th Congress, 1st Session, 1836.

"Report of Chief Engineer," November 1, 1834, Senate Document #1. *Senate Documents,* 23d Congress, 2d Session, 1834.

"Report of Committee on the Public Lands," February 11, 1828, House Document # *House Documents,* 20th Congress, 1st Session, 1828.

Robinson & Jones' Cincinnati Directory for 1846. Cincinnati: Robinson and Jones, 1846.

Salsich, Neil E., ed. "The Siege of Fort Meigs, Year 1813 by Alexander Bourne," *Northwest Ohio Quarterly,* XVII(October 1945), 139–54.

Sargent, Winthrop, "Plan of an Ancient Fortification at Marietta, Ohio," *Memoirs* of the American Academy of Arts and Science, V(1853–1855), 25–28.

de Schweinitz, Edmund. *The Life and Times of David Zeisberger, The Western Pioneer and Apostle of the Indians.* Philadelphia: J. B. Lippincott and Company, 1870.

Scott, Joseph. *The United States Gazetteer.* Philadelphia: F. and R. Bailey, 1795.

Shepard, Claude L., ed. "The Connecticut Land Company and Accompanying Papers." *Tract No. 96* of the Western Reserve Historical Society (Cleveland), 1916.

[Smith, William]. *Historical Account of Bouquet's Expedition Against the Ohio Indians, In 1764.* Cincinnati: Robert Clark & Company, 1868.

Smith, William Henry, ed. *The St. Clair Papers. The Life and Public Services of Arthur St. Clair, Soldier of the Revolutionary War; President of the Continental Congress; and Governor of the North-western Territory; With His Correspondence and Other Papers, Arranged and Annotated by William Henry Smith.* 2 vols. Cincinnati: Robert Clarke & Company, 1882.

Squier, Ephraim George and Davis, Edwin Hamilton. *Ancient Monuments of the Mississippi Valley: Comprising the Results of Extensive Original Surveys and Explorations* in Vol. I of *Smithsonian Contributions to Knowledge.* Washington, D. C.: Smithsonian Institution, 1848.

Stevens, Henry N. *Lewis Evans, His Map of the Middle British Colonies in America: A Comparative Account of Ten Different Editions Published Between 1755 and 1807.* London: Henry Stevens, Son, and Stiles, 1905.

Tanner, Henry S. *The American Traveller; or Guide Through the United States.* Philadelphia: H. S. Tanner, 1836.

— — —. *Description of the Canals and Railroads of the United States Comprehending Notices of all the Works of Internal Improvements throughout the Several States.* New York: T. R. Tanner & J. Disturnell, 1840.

— — —. *Memoir on the Recent Surveys, Observations and Internal Improvements in the United States.* Philadelphia: The author, 1829.

Walmsley, Mildred M. *The Heckewelder Map: 1796.* Cleveland: Western Reserve Historical Society, 1968.

Whittlesey, Charles. "Descriptions of Ancient Works in Ohio," in *Smithsonian Contributions to Knowledge,* III, Washington, D. C. Smithsonian Institution, 1851.

Williams, C. S. *Cincinnati Directory and Business Advertiser,* 1851–52. Cincinnati: C. S. Williams, 1851.

"Mr. Will's Letter," *The American Pioneer,* II(July 1843), 291–92.

Wilson, Frazer E. *Journal of Capt. Daniel Bradley.* Greenville, Ohio, 1935.

Van Cleve, Benjamin, "Extracts from B. Van Cleve's Memoranda," *The American Pioneer,* II(July 1843), 293–96.

Zeisberger, David. "David Zeisberger's History of North American Indians," ed. Archer Butler Hulbert and William Nathaniel Schwarze, *Archaeological and Historical Publications,* XIX (1910), 1–189.

SECONDARY SOURCES

Albach, James R. *Annals of the West: Embracing A Concise Account of Principal Events which have occurred in the Western States and Territories from the Discovery of the Mississippi Valley to the Year Eighteen Hundred and Fifty-six.* Pittsburgh: W. S. Haven, Book & Job Printer, 1857.

Anderson, Russell H. "The Pease Map of the Connecticut Western Reserve," *Ohio State Archaeological and Historical Quarterly,* LXIII (July 1954), 270–78.

Baker, Samuel J. "The Original Surveys of Cleveland," *Journal of the Association of Engineering Societies,* III(August 1884), 217–31.

Bald, F. Clever. "Fort Miami," *Quarterly Bulletin* of the Historical Society of Northwestern Ohio, XV(July 1943), 127–38.

Bond, Beverley, ed. "Dr. Daniel Drake's Memoir of the Miami Country: 1779–1794," *Quarterly Publication of the Historical and Philosophical Society of Ohio,* XVIII(April–September 1923).

Brown, Lloyd Arnold. *Early Maps of the Ohio Valley.* Pittsburgh: University of Pittsburgh Press, 1959.

A History and Biographical Cyclopedia of Butler County, Ohio, with Illustrations and Sketches of its Representative Men and Pioneers. Cincinnati: Western Biographical Publishing Company, 1882.

Butterfield, Consul Willshire. *An Historical Account of the Expedition against Sandusky under Col. William Crawford in 1782.* Cincinnati: Robert Clarke & Company, 1873.

Cook, S. N. "Johnson's Island in War Days," *The Ohio Magazine,* I (September 1906), 226.

Cutler, William Parker and Cutler, Julia Perkins. *Life, Journal and Correspondence of Rev. Manasseh Cutler, LL. D.* 2 vols. Cincinnati: R. Clarke & Co., 1888.

Downes, Randolph C. "Evolution of Ohio County Boundaries," *Ohio Archaeological and Historical Publications,* XXXVI (1927), 340–77.

Ehler, Ralph B. "The Story of Fort St. Clair," *Ohio Archaeological and Historical Society Publications,* XXXII(1923), 515–19.

Everhart, J. F. *History of Muskingum County, Ohio with Illustrations and Biographical Sketches of Prominent Men and Pioneers.* Columbus: J. F. Everhart & Company, 1882.

Ford, Henry A. and Ford, Kate B. *History of Hamilton County, Ohio.* Cleveland: L. A. Williams & Company, 1881.

Fouré-Selter, Hélène. *Gallipolis Ohio, Histoire de l'éstablissement de cinq cents Français dans la vallée de l'Ohio à la fin du XVIIIe siècle.* Paris: Jouve & Cie, 1939.

Griswold, S. O. "The Corporate Birth and Growth of Cleveland," *Annals of the Early Settler's Association,* V(1884), 30–63.

Hanna, Charles A. *The Wilderness Trail or the Ventures and Adventures of the Pennsylvania Traders on the Allegheny Paths.* 2 vols. New York: G. P. Putnam's Sons, 1911.

Hatcher, Harlan. *The Western Reserve.* Indianapolis: The Bobbs-Merrill Company, 1949.

Hibbert, Wilfrid. "The Recently Discovered Pictorial Map of Fort Meigs and Environs," *Bulletin* of the Historical Society of Northwestern Ohio, VI(October 1934), n.p.

Hunt, William E. "The Forks of the Muskingum," *Magazine of Western History,* I(February 1885), 278–88.

Jenks, William L. "The 'Hutchins' Map of Michigan," *Michigan History Magazine,* X(1926), 358–73.

Joerg, W. L. G. "Henry S. Tanner of Philadelphia: His Place in American Cartography, 1815–1850," *Annals of the Association of American Geographers,* XXV(1935), 46.

Jones, Robert Ralston. *Fort Washington at Cincinnati, Ohio.* Published by the Society of Colonial Wars in the State of Ohio, 1902.

Keeler, Lucy Elliot. "Old Fort Sandusky and the DeLery Portage," *Ohio Archaeological and Historical Publications,* XXI(1912), 345–78.

— — —. "Old Fort Sanduski of 1745 and the 'Sandusky Country,'" *Ohio Archaeological and Historical Publications,* XVII(1908), 357–430.

Kennedy, James Harrison. *A History of The City of Cleveland.* Cleveland: The Imperial Press, 1896.

King, Arthur G. "The Earliest Map of Cincinnati," *Bulletin* of the Historical and Philosophical Society of Ohio, XV(October 1957), 302–10.

———. "The Exact Site of Fort Washington and Daniel Drake's Error," *Bulletin* of the Historical and Philosophical Society of Ohio, XI(April 1953), 128–46.

Knapp, H. S. *History of the Maumee Valley Commencing with its Occupation by the French in 1680.* Toledo: Slade Mammoth Printing and Publishing, 1872.

Knauss, William H. *The Story of Camp Chase.* Nashville, Tennessee and Dallas Texas: Publishing House of the Methodist Episcopal Church, South, 1906.

———. "The Story of Camp Chase," *The Ohio Magazine,* I(September 1906), 233–40.

Knopf, Richard C. *The Fort Adams Incident.* Columbus: The Anthony Wayne Parkway Board, State Museum, 1957.

———. "Fort Miami: The International Background," *Ohio Archaeological and Historical Quarterly,* LXI(April 1952), 146–66.

Knopf, Richard C., Baby, Raymond S., Smith, Dwight L. "The Re-Discovery of Fort Washington," *Bulletin* of the Historical and Philosophical Society of Ohio, XI(January 1953), 3–12.

Laning, J. F. "The Evolution of Ohio Counties," *Ohio Archaeological and Historical Publications,* V(1897), 326–50.

Lee, Alfred E. *History of the City of Columbus.* New York: Munsell & Company, 1892.

Lossing, Benson J. *The Pictorial Field-Book of the War of 1812.* New York: Harper & Brothers, Publishers, 1868.

MacLean, John Patterson. "Ancient Work at Marietta, Ohio," *Ohio Archaeological and Historical Publications,* XII(1903), 37–66.

———. *Shakers of Ohio, Fugitive Papers Concerning the Shakers of Ohio, With Unpublished Manuscripts.* Columbus: The F. J. Heering Printing Company, 1907.

———. *The Mound Builders.* Cincinnati: Robert Clarke & Company, 1879.

McPherson, Harry Raymond. "An Alphabetical List of Early Forts, Stockades and Blockhouses in Ohio During the Historic Period." Typed list prepared January 1, 1931. Ohio Historical Society, Columbus, Ohio.

Martin, William T. *History of Franklin County.* Columbus: Follett, Foster & Company, 1858.

Mathews, Alfred. "A Description of Campus Martius," *Magazine of Western History,* I(December 1884), 115–18.

Meek, Basil. *Twentieth Century History of Sandusky County, Ohio and Representative Citizens.* Chicago: Richmond-Arnold Publishing Company, 1909.

Miller, W. C. "History of Fort Hamilton," *Ohio Archaeological and Historical Publications,* XIII(1904), 67–111.

Mills, William C. *Archaeological Atlas of Ohio.* Columbus: Fred J. Heer, 1914.

"Baum Prehistoric Village," *Ohio Archaeological and Historical Publications,* XV (1906), 45–136.

———. "The Exploration of the Edwin Harness Mound," *Ohio Archaeological and Historical Publications,* XVI(1907), 113–93.

———. "Exploration of the Gartner Mound and Village Site," *Ohio Archaeological and Historical Publications,* XIII(1904), 129–89.

———. "Exploration of the Mound City Group," *Ohio Archaeological and Historical Publications,* XXXI(1922), 423–584.

———. "Exploration of the Mound City Group, Ross County, Ohio," *The American Anthropologist,* XXIV(October, November 1922), 397–431.

Mitchener, Charles H. *Ohio Annals. Historic Events in the Tuscarawas and Muskingum Valleys and in the other portions of the State of Ohio.* Dayton, Ohio: Thomas W. O'Dell, 1876.

Moorehead, Warren King. "Section 7: An Archaeological Map of Ohio," in "Report of Field Work," *Ohio Archaeological and Historical Publications,* V(1897), 256–74.

Newcomer, Lee. "Construction of the Wabash and the Erie Canal," *Ohio State Archaeological and Historical Quarterly,* XLVI(1937), 199–208.

"Other Maps," *The Firelands' Pioneer,* XIX(October 1, 1915), 1971–75.

Paul, Hosea. "Maps and Atlases of the Firelands' Counties and Cities," *The Firelands' Pioneer,* XX(December 25, 1918), 2059–62.

Peet, Stephen D. "The Discovery of the Ohio," *The American Antiquarian,* I(April 1878), 21–35.

———. "The Study of the Maps," *The American Antiquarian,* XVII (July 1895), 219–36.

Peters, William E. *Ohio Lands and Their History.* Athens, Ohio: The Lawhead Press, 1930.

Phillips, Josephine E. "The Naming of Marietta," *Ohio State Archaeological and Historical Quarterly,* LV(January-March 1946), 106–37.

Phillips, Philip Lee. *The First Map and Description of Ohio, 1787, by Manasseh Cutler: A Bibliographical Account.* Washington, D. C.: W. H. Lowdermilk & Company, 1918.

———. *The Rare Map of the Northwest, 1785 by John Fitch.* Washington, D. C.: W. H. Lowdermilk & Company, 1916.

Pickham, Howard. "Fort Miami and the Maumee Communication," *Northwestern Ohio Quarterly Bulletin,* XIV(April 1942), 30–41.

Potter, John E. "The Pennsylvania and Virginia Boundary Controversy," *Pennsylvania Magazine of History and Biography,* XXXVIII (1914), 407–26.

Reed, Susan M. "British Cartography of the Mississippi Valley in the Eighteenth Century," *Mississippi Valley Historical Review,* II(September 1915), 213–24.

Reid, Whitelaw. *Ohio in the War.* 2 vols. New York: Moore, Wilstach & Baldwin, 1868.

Renick, Felix. "Mr. Renick's Letter," *The American Pioneer,* II(January 1843), 38.

Ristow, Walter W. "John Melish and His Map of the United States," *The Library of Congress Quarterly Journal of Current Acquisition,* XIX(September 1962), 159–78.

———. "Nineteenth-Century Cadastral Maps in Ohio," *Papers of the Bibliographical Society of America,* LIX(1965), 306–15.

Rose, William Ganson. *Cleveland: The Making of a City.* Cleveland: The World Publishing Company, 1950.

Scheiber, Harry N. *Ohio Canal Era: A Case Study of Government and the Economy, 1820–1861.* Athens, Ohio: The Ohio University Press, 1969.

———. "Urban Rivalry and Internal Improvements, 1820–1860," *Ohio History,* LXXI(October 1962), 227–39.

Shepherd, Lee. "Fort Finney," *Bulletin* of the Historical and Philosophical Society of Ohio, IX(December 1946), 18–22.

Sherman, Charles E. "Appendix A: Maps of Ohio," *Miscellaneous Data,* Vol. IV of *Final Report: Ohio Cooperative Topographic Survey.* Columbus: The Ohio State University Press, 1933, 153–66.

———. *The Ohio-Michigan Boundary,* Vol. I of *Final Report of the Ohio Cooperative Topographic Survey.* Columbus: The Ohio State Reformatory, 1916.

Shetrone, Henry Clyde. "Explorations of the Hopewell Group of Prehistoric Earthworks," *Ohio Archaeological and Historical Publications,* XXXV(1926), 5–227.

———. *The Mound-Builders.* New York: D. Appleton and Company, 1931.

Slocum, Charles E. "Fort Miami and Fort Industry," *Ohio Archaeological and Historical Quarterly,* XII(1904), 120–27.

———. "Some Errors Corrected," *Ohio Archaeological and Historical Publications,* X(1902), 481–83.

Stauffer, David McNeely. *American Engravers upon Copper and Steel.* 2 vols. New York: The Grolier Club of the City of New York, 1907.

Studer, Jacob H. *Columbus, Ohio: Its History, Resources, and Progress with Numerous Illustrations,* n.p., 1873.

Thrower, Norman J. W. "The County Atlas of the United States," *Surveying and Mapping,* XXI(September 1961), 365–78.

Volwiler, Albert T. *George Croghan and the Westward Movement.* Cleveland: Arthur H. Clark Company, 1926.

Weisenburger, Francis P. *Columbus During the Civil War.* (Ohio Civil War Centennial Commission Series, No. 12). Columbus: The Ohio Historical Society, 1963.

Whittlesey, Charles. *Early History of Cleveland, Ohio.* Cleveland: Fairbanks, Benedict & Company, 1867.

Wickham, Gertrude Van Rensselaer. *The Pioneer Families of Cleveland.* 2 vols. Cleveland: Evangelical Publishing House, 1914.

Wilgus, James Alva. "Evolution of Township Government in Ohio," *Annual Report, American Historical Association,* 1894, pp. 403–12.

Williams, Charles Richards. "George Croghan," *Ohio Archaeological and Historical Publications,* XVI(1907), 1–112.

Wilson, Frazer Ells. *Fort Jefferson, The Frontier Post of the Upper Miami Valley.* Lancaster, Pennsylvania: The Intelligencer Printing Company, 1950.

———. "St. Clair's Defeat," *Ohio Archaeological and Historical Publications,* XI (1903), 30–43.

Winsor, Justin. *The English and French in North America: 1689–1763.* Vol. V of *Narrative and Critical History of America.* Boston: Houghton, Mifflin and Company, 1887.

———. *The Mississippi Basin: The Struggle in America between England and France, 1697–1763.* Boston: Houghton, Mifflin and Company, 1895.

Wood, Richard G. "Research Materials in the National Archives Pertaining to Pennsylvania," *Pennsylvania Magazine of History and Biography,* LXIX(1945), 89–102.

Wright, Alfred J. "Joel Wright, City Planner," *The Ohio State Archaeological and Historical Quarterly,* LVI(1947), 287–94.

Yeager, Henry J. "Nouvelles du Scioto: The Story of a Fraud," *Ohio History,* LXXVIII(Autumn 1969), 261–72.

Index